STEEL SELECTION

A Guide for
Improving Performance and Profits

STEEL SELECTION

A Guide for
Improving Performance and Profits

ROY F. KERN
Kern Engineering Company
Peoria, Illinois

MANFRED E. SUESS
Technimet Corporation
Milwaukee, Wisconsin

A WILEY-INTERSCIENCE PUBLICATION

JOHN WILEY & SONS
New York • Chichester • Brisbane • Toronto • Singapore

Library of Congress Cataloging in Publication Data:

Kern, Roy F
  Steel selection.

  "A Wiley-Interscience publication."
  Includes Index.

  1. Steel. I. Suess, Manfred E., joint author.
II. Title.

TA472.K47   620.1'7   78-13610

ISBN 0-471-04287-0

Printed in the United States of America

10 9 8 7 6 5 4 3 2

To All Metallurgists and Materials Engineers

PREFACE

IN HIS BOOK *Managing for Results** Peter F. Drucker states: "Results are obtained by exploiting opportunities, not by solving problems." The proper design of parts for heat treatment and the correct selection of steel for these parts offer practical opportunities for an organization to make itself look good "on the bottom line."

Mr. Drucker goes on to say: "Resources to produce results must be allocated to these opportunities, not to solving problems." In other words, managements of successful firms will allocate adequate resources to an understanding and application of these basic fundamentals rather than to solving individual problems on an emergency basis. Knowledge concerning materials engineering and metallurgical processing is a valuable resource, yet most firms are severely understaffed in this kind of talent and their annual statements reflect the fact.

It has been said that the most important engineering materials in our present economy are SAE 1010 steel and gray iron castings. Although this statement is no doubt an oversimplification, the point is that our most indispensable engineering materials are the constructional grades of steel. To build the machinery needed at home, in the field, and in the factory at the lowest cost the engineering demands on these rather unsophisticated materials are being crowded higher and higher, and to be successful a high degree of materials engineering expertise is required. Inevitably, problems occur in processing and sometimes in operation, and when they do a few individuals have to make a large number of important decisions without adequate time for full investigation. The dedicated group of materials engineers who must make sure that these steels perform properly must often do so without the benefit of extensive metallurgical staff. Nor do they themselves have the time to investigate in depth much of the published research that would be relevant. This often results in a never-ending stream of shop and

*Harper & Row Publishers, New York and Evanston, Illinois, 1964.

field problems, not because of incompetence but simply because of an excessive workload on the qualified few.

It is in this environment that we hope this book will be useful and effective in producing "bottom line" results. Much of the content is based on practical experience, but a good share has been culled from relevant research publications, condensed and presented in the language of the practical materials engineer who must select steels for specific purposes.

We wish to acknowledge the assistance of many friends and associates in the preparation of this book. Perhaps most important is that of our colleagues at Caterpillar Tractor Company and the Allis-Chalmers Manufacturing Company. With apologies to the many whose names are given in the text but not mentioned here, we wish to thank the following:

Jack LaBelle, Allison-Detroit Diesel Division, GMC

F. Thomas, American Steel Foundries, Hammond, Indiana

C. Larson, Associated Spring Corporation, Bristol, Connecticut

George Melloy, Bethlehem Steel Corporation, Bethlehem, Pennsylvania

W. Lehman, Chrysler Corporation, Detroit, Michigan

D. Doane, Climax Molybdenum Company of Michigan, Ann Arbor, Michigan

A. Rausch, Deere & Company, Moline, Illinois

R. Adair, Gleason Works, Rochester, New York

R. Saxton, Gogan Machine Company, Cleveland, Ohio

W. Koppi, International Nickel Company, Oak Brook, Illinois

E. F. Frederick and J. H. Berlien, Keystone Steel & Wire Company, Peoria, Illinois

N. Kates, Lindberg Corporation, Chicago, Illinois

Dr. Morris Cohen, Massachusetts Institute of Technology, Cambridge, Massachusetts

G. Nachman, Metal Improvement Company, Teaneck, New Jersey

E. Peters and F. Barber, Stanadyne Division, Bellwood, Illinois

J. Lincoln Surface Combustion Division, Midland-Ross Corporation, Toledo, Ohio

Dr. H. B. Osborn, formerly with Tocco Division, Park-Ohio Industries, Cleveland, Ohio

Tom Cooper and H. Zoeller, U.S. Air Force Materials Laboratory, WPAFB, Ohio

R. Wallace and Dan Hayes, U.S. Steel Corporation, Pittsburgh, Pennsylvania

Dr. Herman Greenberg, Westinghouse Electric Corporation, Pittsburgh, Pennsylvania

We also wish to thank Sharon Garretts, Jackie Nunes, and Helen Gorski for a superb job in the preparation of the original manuscript.

The encouragement and support of our wives, Pat and Gail, are also gratefully acknowledged.

ROY F. KERN

MANFRED E. SUESS

Peoria, Illinois
Milwaukee, Wisconsin
August 1978

CONTENTS

STEEL SELECTION

A Guide for
Improving Performance and Profits

CHAPTER ONE

THE USEFUL ENGINEERING
CHARACTERISTICS OF STEEL

DURING 1976 WORLD PRODUC-
tion of steel, by far our most widely used metal for engineering purposes, was
approximately 760 million net tons.* Steel's popularity is due principally to the
following, either in part on in combination:

1 Low cost.
2 Wide range of mechanical properties attainable.
3 High modulus of elasticity.

The constructional grades of steel, even in the currently inflated economy,
range in price from less than 15¢ to 40¢/lb, depending on grade, quality, size,
and order quantity; structural steel castings, however, cost 60¢ to $1/lb. Nodular
gray iron castings cost nearly as much as steel castings and regular gray iron or
malleable, not a great deal less. Forgings are priced at approximately 40¢/lb,
plus the basic cost of the steel, which makes them a bargain in today's market.
Aluminum, which has enjoyed an ever expanding share of the market, is a substantially more expensive material, as listed for a 3-in. round bar, 12 ft long.

It must be remembered, of course, that in addition to its lower density, aluminum has many desirable qualities that often make it the preferred choice; for
example, hydraulic cylinder pistons can be made of aluminum to advantage
because aluminum is compatible with the steel barrel and can be machined at
a rate approximately nine times as fast as steel.

*See Appendix II for conversion to SI units.

	Raw Material	If Part Results in 50% Chips
6061–T6 aluminum	$114.80	$109.10
SAE 1045 steel	$ 86.52	$ 85.08

The range of mechanical properties attainable in steel is truly remarkable: as-rolled, high-strength, low-alloy plate is manufactured in yield-strength levels of 25,000 to 80,000 psi and low-cost heat-treated C-Mn-B steel is being used in dynamically loaded fasteners at a strength of 200,000 psi minimum yield point. Improvements in steelmaking practice that produce cleaner steels have resulted in contact stresses of carburized gears and rolling bearings in excess of 350,000 psi.

In contrast to other alloys the constructional grades of steel have the capability of being heat-treated by a wide assortment of methods to impart unique engineering characteristics and control distortion.

1 Quench and temper.
2 Carburize.
3 Carbonitride.
4 Nitride.
5 Cyanide.
6 Nicarb.
7 Tuftride.
8 Martemper.
9 Austemper.
10 Ausform.

In addition, steel can be coated readily with various other metals and chemical compounds to increase corrosion, wear, and scoring resistance. Numerous processes are available to join components into an assembly.

Engineering characteristics of the constructional grades of steel of major significance to designers follow.

HARDNESS

Besides its weight (e.g., to enhance the traction of a vehicle) perhaps the most important engineering property of steel is its hardness. The maximum hardness attainable with constructional steels is a function of the carbon content (see Fig. 1.1). The pioneering effort to demonstrate this function was made by Burns, Moore, and Archer [1], who austenitized thin strips of steel in lead and then quenched them in brine. Quenching was followed by a low-temperature treatment that transformed the soft retained austenite into very hard martensite.

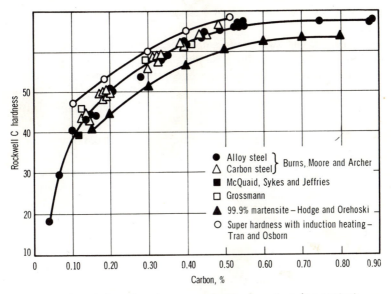

Fig. 1.1 Relation of maximum quenched hardness to carbon content.

Hardnesses of this level would be difficult to achieve in production quenching from a furnace, mainly because of the "mass effect." Tran and Osborn [2], however, reported that even higher hardness levels could be consistently attained by induction heating. Dr. Osborn has advanced a theory that these higher hardnesses may be due to several factors, among which are the extremely fine martensite produced and the presence of high compressive stresses. Hodge and Orehoski [3] reported a hardness of 99.9% martensite with various carbon contents, also shown in Fig. 1.1. Using regression-analysis hardenability data from 338 heats of C-Mn-B steels containing 0.28 to 0.46% carbon, we found that the hardness at (J 1)* on the end-quench specimen was

$$Rc = (204) \times (\%C) + (4.3) \times (\%Si) + (8.32) \times (\%Cu) \\ - (241) \times (\%C^2) + 11.00 \qquad (1.1)$$

For carbon contents between 0.25 and 0.35% Breen, Walter, and Sponzilli [4] established the following relationship:

$$Rc = (37.5) \times (\%C) + 39.5 \qquad (1.2)$$

For production furnace heat treatment the J 1 hardness* is usually the maximum that can be consistently attained even with drastic quenches. Suggested

*Hardness at 1/16 in. from quenched end of end-quench specimen.

hardness ranges for various types of heat treatment are covered elsewhere in this book.

The engineering significance of hardness per se is its relation to abrasive wear resistance and scoring. This is discussed in detail in Chapter 14. Typical specified hardness ranges for most heat treatments of constructional steels are listed.

Test	Unit of Measure	Typical Specified Range
Brinell	Diameter of impression	0.3 to 0.4 mm
(3000 kg)	in millimeters (Bmm)	(or minimum only)
	Brinell hardness number	40 to 60 Bhn
Rockwell	C scale	5 to 6 points
		(or minimum only)
Rockwell	B scale	8 to 10 points
Rockwell	A scale	3 to 5 points
Rockwell	15–N	4 points
Superficial		(or minimum only)
Tukon	Knoop	200 to 225 points
		(or minimum only)
Vickers (500 g)	DPH	120 to 150 points
		(or minimum only)

The Brinell test is ordinarily used for parts like forgings or stock such as plate which is heat-treated before machining. The use of Bmm (diameter of Brinell impression in millimeters) instead of Bhn (Brinell hardness number) in specifications or on part drawings eliminates the need for a conversion chart and the time required to convert to Bhn. The Rockwell test on the C scale is generally used on finished-machined, direct-hardened, carburized, or induction- (or flame-) hardened parts. When the indiscrimate location of the Rockwell impression is likely to interfere with the intended use of a part, the test location is specified on the drawing. Rockwell A and 15–N are most commonly used on finish-machined parts case-hardened by carburizing or carbonitriding. The Tukon and Vickers tests are often applied to very thinly cased parts typical of nitriding. Conversion tables for several methods of hardness testing appear in Appendix I.

All of these tests are based on the penetration of an indentor with a given load and the results are in terms of depth of penetration or size of impression. When the impression is surrounded by a raised periphery (such as that of a moon crater), a soft surface over a harder layer usually exists. This is typical of decarburized surfaces or those on which the surface microstructure contains excessive amounts of austenite (which might occur on a carburized gear). A common method of verification of the soft surface is testing with Rockwell 15–N or A and comparing the Rockwell C conversion with the results of an actual Rockwell C test.

Crude as it may seem, the file hardness test is most useful, especially for carburized or carbonitrided parts. For specification purposes test files of a stated hardness must be used. For fast production testing, however, to check for hardness, partial decarburization, or upper transformation products in parts such as a carburized gear, a high-quality mill bastard file is suitable. The file, along with the Rockwell test, will also detect excessive austenite in carburized parts. A surface that is file hard but Rockwell soft (52 to 55 Rc) usually contains more than 40% austenite.

TENSILE STRENGTH

The purpose of quenching and tempering is to change a steel's microstructure, almost always to bring about an increase in hardness and strength. When austenitization is complete and the combination of quench vigor and hardenability is such that it will provide at least 80% martensite, the tensile strength will almost always be approximately 500 times the Brinell hardness number. Typical tensile properties of as-rolled or cold-drawn carbon steels are described in Appendix III.

The tensile strength of steel is of minor engineering significance except as it is related to the endurance-limit fatigue strength (more than 200,000 cycles). Even this relationship applies only when there is little or no surface residual stress. The work of Garwood et al. [5], shown in Fig. 1.2, illustrates the effect

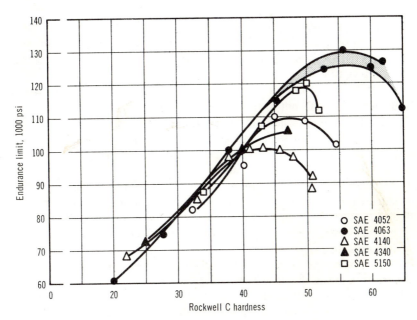

Fig. 1.2 Fatigue strength of several alloy steels (after Garwood, Zurburg, and Erickson [5].

of hardness (or tensile strength) and carbon content on the endurance-limit fatigue strength of several high-hardenability alloy steels. (This work explains the success of nominal 0.60% carbon steels in heat-treated springs and torsion bars.) The endurance-limit fatigue strength of reasonably clean steel free of surface defects is approximately 45% of the tensile strength. This assumes that no surface compressive stress remains from heat treatment or has been induced by shot peening or rolling, which usually increase fatigue strength dramatically. The presence of seams or large inclusions in or just beneath the surface can seriously reduce the 45% figure.

Tensile strength is sometimes a significant engineering characteristic for non-reusable fasteners in which torquing exceeds the yield point of the part. Maximum clamping stress can be obtained by torquing so that the stress is close to the tensile strength of the fastener. Note that this type of tightening should be done only when the stretch in the bolt can be measured accurately, when the steel used contains less than 0.40% carbon, and when the microstructure is more than 95% tempered martensite throughout. For further details see Chapter 13.

Because the tensile strength is the stress at which a part will break, design stresses should never equal this value.

PROPORTIONAL LIMIT (TENSION)

Of perhaps equal importance to the hardness of steel is the proportional limit in tension. The reason is that a part stressed to this point will deflect proportionately with load, and on release of the load will return to its original dimensions. This characteristic is of vital importance in reusable fasteners to provide the necessary clamping load without yielding. Yielding in many parts, even to the amount of 0.2%, will not keep a joint tight and can result in failure. Proportional limit is also an important engineering characteristic when surface compressive stress is present to increase endurance-limit fatigue strength. Yielding to the point of only 0.2% will remove these desirable residual stresses and result in an accompanying loss of fatigue strength.

Most designers, however, are content to work to a percentage of the yield strength in types of loading such as bending and torsion, assuming that their safety factor will keep them within the elastic range of the steel. This works well for low-strength steels but is dangerous for steels with more than 40,000 psi yield strength.

Proportional limit in particular, and yield point as well, is highly sensitive to microstructure, as shown in Table 1.1.

When high proportional limit or yield strength is a major engineering priority, the materials engineer should select steel based on its "microstructural

Table 1.1   Typical Effect of Microstructure on Strength of Steel[a] (SAE 8630 — 1-in. rounds — 400 Bhn)

Microstructure	Tempered Martensite and Bainite	Tempered Martensite	Ferrite and Martensite
Austenitizing temperature	1600 F	1600 F	1425 F
Vigorous quench in	Oil	Water	Water
Tempering temperature	400 F	725 F	None
Proportional limit (psi)	162,500	175,600	127,500
Yield point [0.2% offset (psi)]	178,000	187,200	134,500
Tensile strength (psi)	202,000	200,500	198,500
Yield/tensile ratio	0.88	0.93	0.67

[a]Courtesy of U.S. Steel Corporation.

capability" (ability to attain a desired microstructure by heat treatment). Unfortunately, we do not know this capability precisely and can only estimate it from hardness. This method is described in the following example:

Assume that we have a 2.00-in.-diameter stud that can be water-quenched and that a yield strength of 160,000 psi must be maintained. We have shown that a hardness of 40 Rc with a microstructure of 95% minimum tempered martensite must be provided. With mildly agitated water quenching, the cooling rate in the center of a 2-in. round is approximately 7/16 in. (J 7). Therefore we should select a steel that, even with carbon on the high side and hardenability on the low, still shows a minimum of 95% martensite at J 7. If we think in terms of nominal 0.30% steel, the end-quench hardenability band should show a minimum of 48 Rc at J 7 (48 Rc = 95% martensite for 0.33% carbon). However,

1   SAE 1330 has only 28 Rc at J 7.
2   SAE 4130 has only 29 Rc at J 7.
3   SAE 5130 has only 30 Rc at J 7.
4   SAE 8630 has only 39 Rc at J 7.
5   SAE 94B30H has only 44 Rc at J 7.

In other words, there is no standard "H" steel of 0.30% nominal carbon content that will meet this requirement, but two nonstandard grades, 4330 and 86B30, can achieve it.

PROPORTIONAL LIMIT (COMPRESSION)

The proportional limit, or yield point, in compression is an engineering quality of steel often overlooked, many times with disastrous consequences. The reason for its being overlooked is that it is assumed to be the same as the proportional limit in tension. Leslie and Sober [6] have found that it is not the same and that it is affected by carbon content (Fig. 1.3). (The difference between the proportional limit in compression and that in tension is known as the S-D effect and is 190 × % C in ksi.) This is an important consideration in applications such as rolling bearings and gears in which the contact stress is usually more than 150,000 psi. The lack of compressive yield strength that might occur with a 0.40% carbon steel even at 55 Rc would result in cold working of the surface with subsequence pitting or, at a minimum, undesirable size changes due to the residual stresses developed.

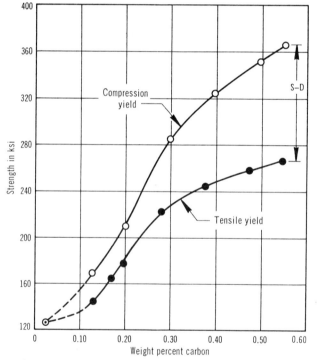

Fig. 1.3 Typical differences in tensile and compression yield strength [6].

TOUGHNESS

The toughness of heat-treated steel is usually thought of in terms of its ability to yield plastically under a high localized stress such as that which might occur at the root of a notch, in a sharp fillet, or at a surface seam. These notches may also be the accidental result of tool marks, voids in welds, or, the most damaging of all, cracks developing in quenching and the beginning of fatigue failure.

Most testing for toughness has been done by using Izod or Charpy specimens in a pendulum-type impact machine. Correlation between test results and field performance has been quite poor, however, especially in quenched and tempered fine-grained steels. Evidence indicates that the notches in the Charpy and Izod specimens are far too blunt and that the loading rate is excessively high. Substantial progress has been made in this area in the last 20 years and today's fracture mechanics techniques can accurately predict materials behavior in the presence of a sharp notch. Properties such as nil ductility temperature and plane strain fracture toughness (KI_c) are rapidly replacing traditional impact energy in characterizing the toughness of heat-treated steels. Fracture toughness is a measure of the quality of microstructure, and, as shown in Chapter 15, the boron-treated steels are excellent in this regard. Besides the effect of boron in providing optimum microstructural quality, only the addition of nickel to steel will increase its toughness; all other alloying elements embrittle. When maximum toughness is required in heat-treated steel, sulfur and phosphorus should be specified at under 0.010% even as low as 0.005% max.

When heat treating steel for maximum toughness, the 600 F tempering embrittlement effect should be avoided. This effect can be seen in Fig. 1.4. The

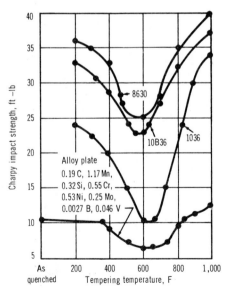

Fig. 1.4 Relation of Charpy impact strength to the tempering temperature of several hardened steels (*Selecting Steels and Designing Parts for Heat Treatment,* American Society for Metals, 1969).

temperature range of 475 to 700 F should not be used when tempering steels with carbon contents up to 0.40%.

A common result in heat treating of parts for maximum toughness is accidental carburization in an atmosphere furnace. It is usually better to austenitize in vacuum or in an air atmosphere than to risk poor atmosphere control. Scaling of finish machined parts, however, is a problem in air atmosphere.

Finally, whether or not equipment has been abused, the ever increasing expense of legal defense in product liability suits means that it behooves engineers to design parts that do not fail in a brittle manner. A part that bends in an accident, for example, or fails in a ductile manner is a strong factor in defending these litigations.

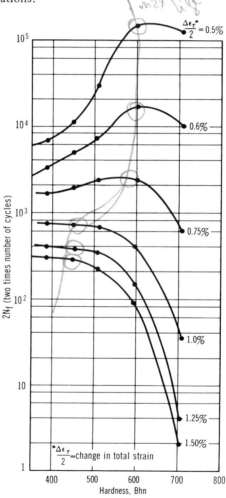

Fig. 1.5 Effects of hardness and strain on short-cycle fatigue life of quenched and tempered 1045 steel (from Morrow and Landgraf [7]).

SHORT-CYCLE FATIGUE STRENGTH

It has been believed until quite recently that failure in cyclic loading was a phenomenon known as fatigue and that material, heat treatment, and subsequent processing such as shot peening could be used to maximize part life under the entire range of operating stresses. Unfortunately, many parts failed to respond as predicted, and as a last resort they were simply made heavier to obtain the required life; that is, to reduce the stress. Morrow and Landgraf [7] found that a different set of rules applied to ensure part life up to 100,000 cycles than when strain was such that it provided life beyond this amount. The Morrow and Landgraf work is shown in Fig. 1.5., in which it can be seen that as the percentage of strain is increased the point of maximum life shifts to the left (lower hardness). In other words, the mechanical property most important to short-cycle fatigue strength is ductility (toughness) whereas tensile strength, surface compressive stress, and (as shown later) clean steel greatly affect long-cycle fatigue strength.

We found the following effects of composition on short-cycle fatigue strength of two parts:

1 41B30 – Q&T to 45–52 Rc   2000 cycles minimum
   86B30 – Q&T to 45–52 Rc   5000 cycles minimum
2 10B16 carburized – 60 Rc   160,000 to 300,000 cycles
   10B16 + 1.5% Ni carburized – 60 Rc 600,000 cycles minimum

ENDURANCE-LIMIT FATIGUE STRENGTH

To a designer perhaps the most significant characteristic of heat-treated steel, next to hardness and yield strength, is its endurance-limit fatigue strength. This is a situation in which stresses ordinarily can be held low enough to expect a life of 200,000 cycles minimum up to 10 million or more. In this range many factors are important, such as steel composition and hardness (Fig. 1.2), steel cleanliness (Fig. 1.6), mean inclusion size (Fig. 1.7), surface finish, and even relative humidity. Morgan [8], for example, showed that fatigue life could be related to "ultrasonic noise" (steel cleanliness).

These factors are all less important than residual stress in and immediately beneath the surface, as shown in work done by Liss, Massieon, and McKloskey [9]. The development of high compressive stresses in hardening the SAE 1045 steel shown in Fig. 1.8 is due to two reactions. First is the microstructural change in a layer approximately 0.125 in. deep to martensite with its accompanying expansion. Second, and of approximately equal effect, according to these authors, is the thermal contraction of the unhardened core inward from the hardened layer. It was once believed that the core of a shell-hardened shaft made in this manner was useful only to hold the surface in place. This is definitely not

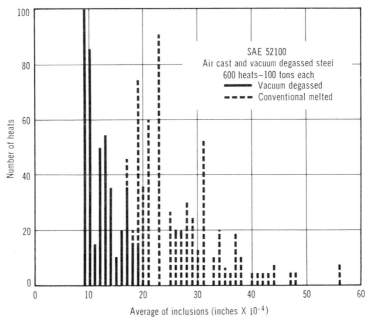

Fig. 1.6 Results of survey showing average inclusion size of 52100 steel, melted conventionally and vacuum degassed (from Church, Krebs, and Rowe [10]).

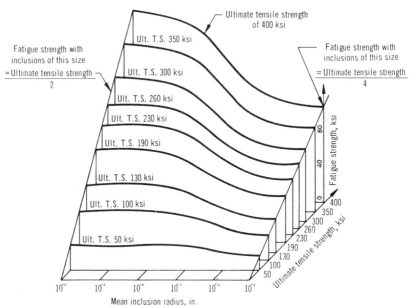

Fig. 1.7 Effect of mean inclusion size on bending fatigue strength (from Morrow and Landgraf [7]).

12

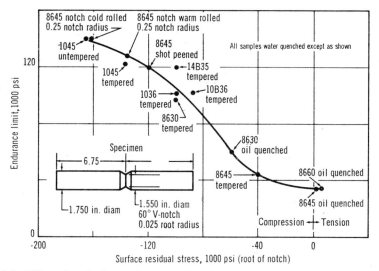

Fig. 1.8 Effect of grade, heat treatment and surface processing on the endurance limit of various steels (from *Selecting Steels and Designing Parts for Heat Treatment,* American Society for Metals, 1969).

so. An axial 0.375–in. hole in a 2-in.-diameter shaft, for example, will reduce the compressive stress on the surface when shell-hardened to less than 50% that of a solid shaft and, even worse, will expose the surface of the hole, which is often residually stressed in tension. The coefficient of thermal expansion, therefore, can be made to contribute to engineering usefulness; however, this physical property can be very troublesome, as described in Chapter 2 on designing for heat treatment.

Next to strength and residual stress, the most important factor in the life of a heat-treated constructional steel is its cleanliness. Particularly damaging are the large, hard alumina inclusions that form from aluminum additions to the molten steel and which might wash off the refractory of the furnaces, ladle linings, ladle nozzles, and so on. When these inclusions are exposed or nearly exposed in a highly stressed area of a part, fatigue failure is almost certain to occur. When maximum endurance limit fatigue strength is required, steels should be as clean as possible and parts should be inspected by magnetic-particle or similar methods to make sure that there are no nonmetallic stringers in areas of maximum stress.

MODULUS OF ELASTICITY

In addition to the broad range of mechanical properties available in the constructional grades of steel, an outstanding engineering characteristic is their high modulus of elasticity. For those who may never have been exposed to this term, it is their "stiffness," or a measure of the elastic deflection under a given load. Unfortunately, heat treating of steel does not increase its stiffness, but only its strength. Typical values of modulus of elasticity for steel and other metals are the following:

	Modulus of Elasticity (psi)
Sintered carbide	75 to 105 $\times 10^6$
Steel	30 $\times 10^6$
Stainless steel	28 to 30 $\times 10^6$
Malleable iron	24 to 28 $\times 10^6$
Gray iron	10 to 23 $\times 10^6$
Nodular iron	21 to 25 $\times 10^6$
Invar	21 to 22 $\times 10^6$
Aluminum	10.5 $\times 10^6$

REFERENCES

1 J. L. Burns, T. L. Moore, and R. S. Archer, Quantitative Hardenability, *Trans ASM,* **26,** 1 (1938)

2 M. A. Tran and H. B. Osborn, Inherent Characteristics of Induction Hardening, Preprint No. 49, American Society for Metals, 1940.

3 J. M. Hodge and M. A. Orehoski, Relationship Between Hardenability and Percentage of Martensite in Some Low Alloy Steels, *Trans. American Institute of Mining, Metallurgical, and Petroleum Engineers,* **167,** 627 (1946).

4 D. H. Breen, G. H. Walter, and J. T. Sponzilli, Computer-Based System Selects Optimum Cost Steels — V., *Metal Progr.,* **104,** No. 6, 43 (1973).

5 M. F. Garwood, H. H. Zurburg, and M. A. Erickson, Interpretation of Tests and Correlation with Service, American Society for Metals (1951).

6 W. C. Leslie and R. J. Sober, The Strength of Ferrite and of Martensite as Functions of Composition, Temperature, and Strain Rate, *Trans ASM,* **60,** 459 (1967).

7 J. Morrow and R. W. Landgraf, Cyclic Deformation and Fatigue Behavior of Hardened Steels, University of Illinois, Department of Theoretical and Applied Mechanics, Report 320, November 1968, an SAE Technical Report.

8 J. B. Morgan, Ultrasonics Can Rate Cleanliness, Fatigue Life of Steel, *Metals Eng. Quart.* **13,** No. 1, 1 (1973).

9 R. Liss, C. Massieon and A. McKloskey, The Development of Heat Treat Stresses and Their Affect on the Fatigue Strength of Hardened Steel, SAE Mid-Year Meeting, Chicago, May 1965.

10 C. P. Church, T. M. Krebs, and J. P. Rowe, The Application of Vacuum Degassing to Bearing Steels, *J. Metals,* American Institute of Mining, Metallurgical, and Petroleum Engineers, **1,** 1966, 62.

CHAPTER TWO

GENERAL ASPECTS OF PROPER
DESIGN FOR HEAT TREATMENT:
AUSTENITIZATION

THE OBJECTIVE OF PROPER DESIGN
for heat treatment is to provide the minimum engineering requirements at the
lowest total cost and, in particular, to minimize the expense of scrap or rework
on parts that have distorted excessively or cracked.

Perhaps the most important physical (not mechanical) property of steel to
be considered in design is its coefficient of thermal expansion. Most heat-treating problems could be solved if this coefficient could be controlled. Because it
cannot, we must learn to design with it, and it is quite substantial (approximately 0.125 in./ft from room temperature to 1550 F).

A simple example of how this physical property affects heat treating is given
in Fig. 2.1. As the shaft is quenched, the corner cools first, and as it shrinks
it mechanically upsets the hot steel beneath it. As the quench progresses, the
entire shaft cools down; but now, because the end is hot-upset, the diameter is
too small to accommodate the circumference. As a result, the end is (usually)
in a high state of residual tensile stress, and if the steel is brittle, quench cracks
may develop.

The coefficient of expansion is a factor that requires serious design consideration because it affects the part during austenitization. With furnace heating a part comes to the austenitization temperature mainly by radiation (80 to
98%) and partly by convection (2 to 20%). By radiation heating particular
portions of a part with thin sections heat fastest, especially those that expose a
large surface area such as a spline or gear. A typical example is the gear on the

15

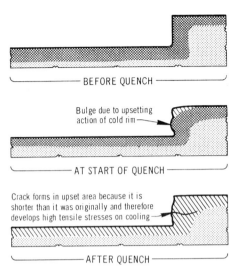

Fig. 2.1 Effect of coefficient of thermal expansion in heat treating a shaft (from *Selecting Steels and Designing Parts for Heat Treatment,* American Society for Metals, 1969).

left in Fig. 2.2. Because its thin sections heated so rapidly, this part could not be made to the required tolerances. The redesign shown on the right was better, but the large access holes required were still troublesome.

Fig. 2.2 Two gear designs showing effect of coefficient of thermal expansion. At left is a widely used design, very troublesome to heat-treat. A preferred design is shown at right (from *Selecting Steels and Designing Parts for Heat Treatment,* American Society for Metals, 1969).

Several other factors at the design stage can contribute to problems trace-able to austenitization:

1 Excessive combinations of components that result in grossly nonuniform sections.
2 Designs requiring contact with furnace hearths or placement near walls.
3 Designs requiring processing that leaves parts in a state of high residual stress before austenitization.
4 Parts very thin and long or large in area that are difficult to heat-treat because of distortion during austenitization.
5 Designs unsuitable for the type of furnace equipment available.

PART COMBINATIONS

Sometimes designers are prone to make designs in which combinations of parts intended to reduce costs can actually increase them due to problems in the austenitization step. The classic example is the gear-and-hub combination shown in Fig. 2.3a. As this part is subjected to the heat of an austenitizing furnace the thin extremities at the top of the hub heat faster than the sections near the gear. Accordingly, this area tries to increase in size but is restrained by the colder metal nearer the gear; therefore it upsets itself (yields in compression). Finally the entire part comes to the prescribed temperature. On cooling, however, and even without quenching, the top end of the hub pinches in because it has upset itself. This upsetting can result in a serious taper condition in the bore. (If the bore is broached before heat treatment, the extremities of the hub will stretch and then close in, thus causing additional taper.) The top portion of the hub of the pinion in Fig. 2.3a is used only as a spacer and need not be heat-treated. A much shorter hub with a steel tubing spacer (Fig.2.3b) would solve the problem in austenitization, make the forging easier, and in most cases reduce the total cost.

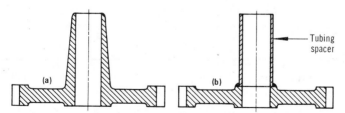

Fig. 2.3 Two designs for gear and hub combinations. (a) Difficult to heat treat without excessive taper in the bore. (b) A preferred design (from *Selecting Steels and Designing Parts for Heat Treatment,* American Society for Metals, 1969).

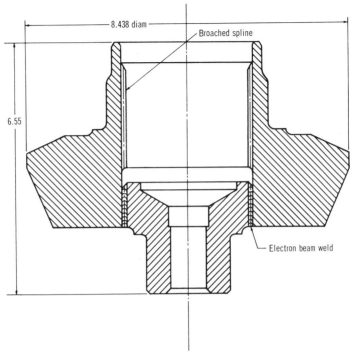

Fig. 2.4. Redesign of a bevel pinion utilizing electron beam welding that was impossible to heat-treat in one piece (from *Selecting Steels and Designing Parts for Heat Treatment,* American Society for Metals, 1969).

The bevel pinion shown in Fig. 2.4 presents a similar problem, although the hub extension is necessary. Here a steel or, preferably, a heat-resistant alloy cap that will create mass can be put over the thin hub before austenitization to retard its heating rate in carburizing.

CONTACT WITH FURNACE HEARTH OR PLACEMENT NEAR WALLS

Designs are sometimes made without regard for the furnace equipment available. One of the results is that parts must be charged into a furnace in such a manner that cold steel is placed on a hearth at a temperature of 1500 to 1700 F. A typical example is the bulldozer cutting edge shown in Fig. 2.5. As the edge touches the hot surface, the face in contact is rapidly heated by conduction. It tries to expand but is restrained by the cold steel behind it and eventually some hot upsetting takes place. Even with a uniform quench the parts were bowed as much

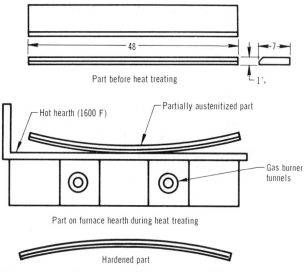

Fig. 2.5 Effect of nonuniform austenitizing on a bulldozer cutting edge with resultant distortion.

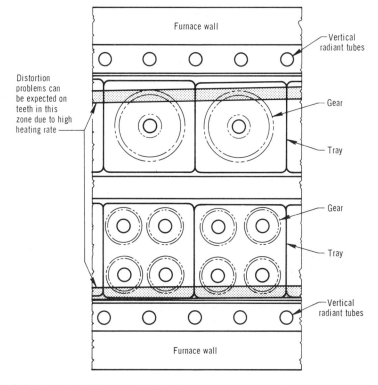

Fig. 2.6 Gears loaded in a carburizing furnace where hot walls result in nonuniform austenitization.

19

as 0.5 in., the concave side being that which rested on the hearth. Ideally, these edges should be induction-hardened or hung in a vertical position in a pit or salt-bath furnace.

A similar condition develops when portions of the outside diameters of gears loaded on trays are close to vertical radiant tubes or to the electric heating elements in the furnace wall; the balance of the gear or gears is exposed only to radiation from the roof. (See Fig. 2.6.) This type of nonuniform heating can create numerous dimensional control problems, particularly on large gears.

These problems are not, of course, due to design per se but are mentioned to show that proper furance equipment must be used to obtain consistent quality, as prescribed, at minimum cost.

DESIGNS IN WHICH RESIDUAL STRESS CAUSES PROBLEMS IN THE AUSTENITIZATION PROCESS

Residual stress from such operations as casting, forging, coining, extruding, welding, and machining can produce distortion during austenitization because the temperatures required cause stress relief. The magnitude of residual stress in a part before hardening is generally low because the yield point of the steel in the unhardened condition is low. This level of stress can still make trouble when the distortion permissible in heat treatment is small, as in a carburized gear.

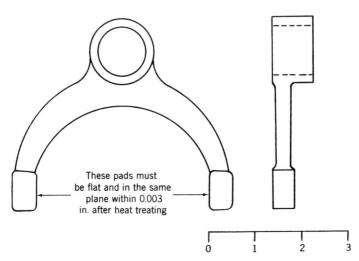

These pads must be flat and in the same plane within 0.003 in. after heat treating

Fig. 2.7 Forged and coined shifter fork.

A typical shifter fork with coined pads is shown in Fig. 2.7. A similar fork forged of 1018 steel was initially processed as follows:

1 Drill and ream shaft bore.
2 Mill pads.
3 Carbonitride 0.020 to 0.030 deep.
4 Grind pads.

Milling the pads was an expensive operation, and hard grinding often removed most of the carbonitride case. It was decided to coin the pads to size, only drill and ream the shaft bore, and then carbonitride. The cold-coining operation introduced stresses in the part of such magnitude that their relief in austenitizing was the major reason for nearly 100% rejection due to excessive distortion. Hot coining reduced the rejection rate to 8%; however, a light grind was necessary to eliminate scrap.

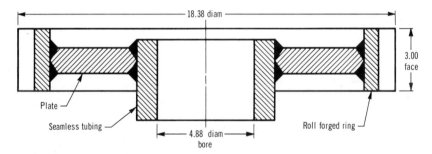

Fig. 2.8 Weld fabricated gear where residual stresses due to welding caused excessive heat treat distortion.

Welded gears must almost always be stress-relieved before machining to hold the tolerances required after carburizing. Figure 2.8 shows such a gear. Its rim is usually 8620 or 4320 steel, its web, 1020, and its hub, 5120. Typical distortion in heat treating both with and without stress relief before machining is as follows:

	Pitch Line Runout		Tooth Taper	
	$\bar{X}$	Standard Deviation	$\bar{X}$	Standard Deviation
Not stress-relieved	0.011	0.0016	0.004	0.007
Stress-relieved	0.007	0.0007	0.0015	0.0003

Although it is not a clear function of design, the manufacture of close tolerance gears requires the relief of machining stresses between processing operations. This treatment is required to a significant extent because the steels often specified, for example, 9310, are more difficult to machine than, say, 8620.

A typical precision spiral bevel gear is shown in Fig. 2.9. Extensive amounts of metal are removed in cutting the teeth of a spiral bevel gear in which a practical indication of both degree and scatter of distortion is the lapping time required after carburizing to obtain the required contact pattern. Statistical analysis of results on 2350 gears (not the gear illustrated in Fig. 2.9) from 24 different gears of 8620H gave the following results:

	Lapping Time $\bar{X}$ (Minutes)	Standard Deviation
Cut and carburized	11.46	0.72
Rough-cut, stress-relieved at 900 F, finish cut, carburized	10.98	0.66

Fig. 2.9   A typical hypoid gear (courtesy of Gleason Works).

LONG PARTS OF SMALL CROSS SECTION
AND LARGE THIN PLATES

Designs that require long small-diameter bars or thin flat bars are difficult to austenitize in a furnace without excessive distortion unless they can be suspended vertically from a fixture. Even with the best fixtures horizontal loading usually results in sagging, especially at carburizing temperatures. At 1700 F practical experience has shown that with horizontal loading the maximum unsupported lengths of bars cannot exceed $6t$, where t = thickness or diameter, with one-half the length between two supports and one-fourth over each end. One solution to this problem considers bars heat-treated at the mill, where machine-straightening equipment is used after hardening. The usual maximum hardness in cold-finished bars is 269 to 321 Bhn, although somewhat greater hardness can be achieved. In hot-rolled carbon steel bars the usual maximum is 25 to 32 Rc, and for alloy, 38 to 45 Rc. Another solution, when toughness of quenched and tempered steel is not required, is to call for material cold-

Fig. 2.10 Gogan quench machine (courtesy of Gogan Machine Corp.).

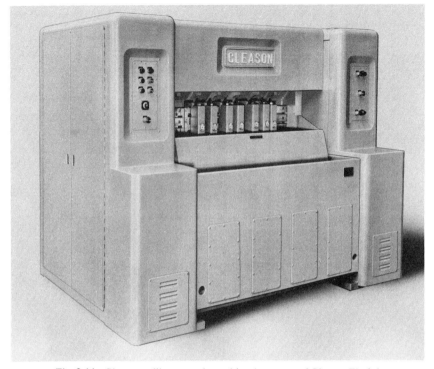

Fig. 2.11   Gleason rolling quench machine (courtesy of Gleason Works).

drawn to mechanical properties. This grade of steel is discussed in greater detail in Chapters 4 and 11.

When designs require a long thin configuration and no furnace equipment is available with sufficient working height in which to suspend the parts, a roller-type quench machine may be used. A Gogan machine hot-straightens the part by rolling it between several sets of rollers before quenching (see Fig. 2.10). A similar machine is manufactured by Gleason Works (Fig. 2.11). The maximum capacity of the Gogan machine is 6 in. in diameter by 120 in. in length. The Gleason unit will effectively quench shafts up to 4 in. in diameter and 40 in. long. Because the part is straight before the quench and because it is also rolled while being quenched, the following straightness tolerances for symmetrical parts can, according to Gleason Works, be held in production:

$$\text{TIR} = K \ \frac{l}{d} \qquad (2.1)$$

where TIR = total indicator reading,
$\quad l$ = length in inches,
$\quad d$ = diameter in inches,
$\quad K = 10^{-4}$ inches

This type of machine is important in metal working for the added reasons given in Chapter 11.

Induction heat treating, particularly by scanning, is a preferred method for small-production quantities of long parts of small cross section if suitable equipment is available. A typical installation is shown in Fig. 2.12. Straightness capability is approximately the same as that for roller die quenched parts. When quantities are very large, such as in the production of automobile rear axle shafts, induction equipment can be made to heat the entire surface of the part in a matter of seconds. With this equipment it is possible, for example, to hold a shaft of 1.25-in. nominal diameter by 30-in. length to a straightness of 0.035 TIR. If no in-house induction-hardening equipment is available, a number of commercial heat-treating firms have induction scanning units that can handle up to 5 in. or more in diameter by 7 ft or more in length. Austenitization by

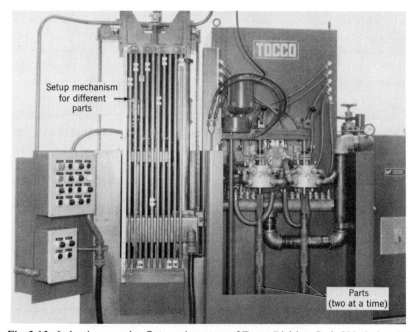

Fig. 2.12 Induction scanning fixtures (courtesy of Tocco Division, Park-Ohio Industries).

induction heating holds distortion to a minimum because as a rule only a small percentage of the total mass is being heated at one time. Steel can also be "through" heat-treated in bar lengths to 20 ft or more with induction-heating equipment.

A difficult heat-treating problem occurs in wear strips, for example, 0.25 in. thick by 2 in. wide by 30 in. long. Several equally spaced holes are drilled through the strips and hardness of more than 55 Rc and straightness of 0.010 in. max camber and bow are desired. Distortion in furnace austenitization and the danger in attempting to straighten parts like these make die quenching the only answer. If the wear involves battering, such as on the bottom of a drag-line bucket, austenitic manganese steel may be used, although welded construction is difficult even with stainless-steel welding rod. Often the solution is to use steels such as 1085 austempered (or oil-quenched and tempered) to a reduced hardness level such as 40 to 48 Rc and to substitute bolting or riveting for welding.

Thin plates such as clutch and harrow discs are very difficult for the average heat-treating department to handle because of distortion in austenitization. Obviously, there is little that can be done in designing to alleviate the problem. A quench press must be used to hot-straighten the part before quenching (and also to provide a uniform quench) in order to prevent excessive distortion. Accordingly, it is often best to buy these parts in finished form from firms specializing in their manufacture. Smaller parts such as rotary meat knives and saw blades may be austempered and finish-ground after heat treatment.

A method of correcting distortion in thin flat plates, whether from austenitizing or quenching, is to stack-temper (for stress relief). This is accomplished by clamping a stack 24 to 36 in. (or more) high in a fixture and heating the assembly to 400 F or higher.

For heavy parts, such as the 30 by 48 by 1 in. plate in Fig. 2.13, a common solution to the problem of distortion in austenitization is to burn them from plate that has been heat-treated by the supplier. The burned edges can create difficulties in slab milling, but these parts are not always machined after burning. Readily weldable C-Mn-B and alloy steels are available in hardness ranges of 321 to 383 and 360 to 400 Bhn, respectively; high carbon grades are available with even greater hardnesses. A problem with burning parts from heat-treated plate is distortion during cutting, although it can be largely eliminated by flooding the plate surface with water during oxyacetylene cutting. Heat-treated plates also present some problems not found with as-rolled steel:

1   The strength of the plates requires that machines such as shears, press brakes, and punches have adequate capacity to handle the steel.

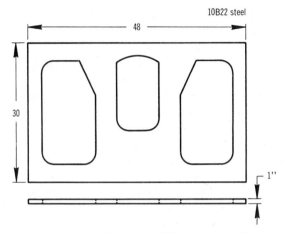

Fig. 2.13 A relatively thin part of large area difficult to austenitize because of excessive distortion (10B22 steel) (from *Selecting Steels and Designing Parts for Heat Treatment*, American Society for Metals, 1969).

2 Permissible radii must be increased as follows:

	Inside Radius of Bend (recommended minima)
SAE 1010	$1T$ (1)
SAE 1020	$3T$
High-strength low alloy	$3T$ to $0.25T$, to $4T$ over 0.25 thick
Q & T 0.20% max C	$3T$ to $0.25T$, to $6T$ over 0.25 thick (2)
	$3T$ to $0.25T$, to $4T$ over 0.25 thick (3)

(1) T = material thickness in inches
(2) Bend with major rolling direction
(3) Bend across major rolling direction

3 Heat-treated plates sometimes have adherent heat-treating scale that may have to be removed by shot blasting or pickling before painting for the sake of appearance.

4 The greater waviness of heat-treated steel plates sometimes results in large variations in the distance between the steel surface and the burner tip, making it difficult to maintain tolerances.

DESIGN EFFECTS ON AUSTENITIZING
FURNACE SUITABILITY

All too often a design is released that cannot be properly accommodated for austenitization in the equipment available and management is not willing to spend money for new furances. A commercial heat-treating firm may have the proper equipment but it may also be many miles away. This is just one of the reasons for having personnel from all departments concerned participate in major engineering design releases.

Most commonly the part is too high or too wide to pass through the door of the furnace or it may be too long to be correctly charged into a salt bath. As a result, the part has to be austenitized in a furnace without a protective atmosphere or, even worse, with one that is improperly controlled. This can lead to a number of undesirable surface effects, such as decarburization or carburization and scaling, in addition to excessive distortion, all of which can adversely affect engineering performance of a specific design.

Sometimes the furnace capable of heating the part, such as a batch-type carbonitriding unit, is not suitable for individual quenching on a plug or in a die.

DESIGN EFFECTS IN AUSTENITIZATION BY
INDUCTION HEATING

Because of the extremely rapid heating rate, austenitization with high-frequency electric current can be likened to reverse rapid quenching. Accordingly, design is of utmost importance. As a general rule, steel exposed to the flux of the inductor will heat fastest on corners (as on the shaft in Fig. 2.1), around holes (as shown in Fig. 2.14), and through thin sections. The bottoms of keyways and the roots of gear teeth and splines are austenitized last, often mainly by conduction from adjacent areas.

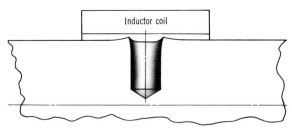

Fig. 2.14 Section through a hole in a part following rapid heating in an induction coil, showing distortion which leads to cracking (from *Selecting Steels and Designing Parts for Heat Treatment,* American Society for Metals, 1969).

As in quenching, however, induction tooling can be designed to concentrate flux by using appropriate coil configuration and laminated core material. The different frequencies available provide not only for various depths of heating but also for the sharpness of the heating effect. Because the induction-heated layer is often much thinner than the hardened depth required, conduction is relied on to provide the required depth of austenitized steel. The extent of conduction is a function of the differential between the surface temperature and that of the core. Thus preheating, either in the induction coil with a suitable delay or in a furnace, can be employed to reduce heat transfer inward. The heat generated by induction will then tend to stay in the surface and fill out irregularities, as shown in Fig. 2.15. The parts shown are a face-type seal ring and a crawler-tractor track roller. In order to obtain the required depths on these parts with the flux concentrating in certain areas, it was necessary to heat so deeply that excessive distortion, grain coarsening, and cracking took place. On the seal ring with the regular cycle (without preheat) the result on the left was for 30 sec heat with 9600 cycle power. By preheating for 5 sec at the same power setting, delaying for 10 sec, and then austenitizing for 15 sec the pattern improved, as shown in the right in Fig. 2.15. On the track roller the furnace preheat eliminated scrap. The austenitization of fillets, groover, keyways, and other depressions can often be improved with a preheat. A preheat can also be used to increase the compressive stress in the hardened layer. The stress increase is caused by the shrinkage occurring as the preheated mass contracts inward from the hardened surface.

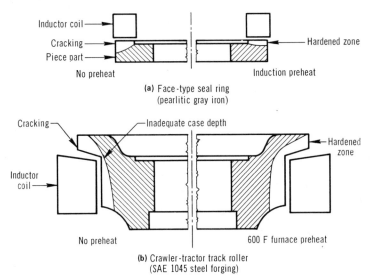

Fig. 2.15 Typical effect of preheating in induction hardening.

Designing a part for most efficient austenitization by induction heating is a complex problem. For general guidance the section on induction and flame hardening, in the *Metals Handbook,** pages 167–202, Vol. 2, eighth edition, is recommended. Beyond this, especially when large capital investments must be made, the expertise of the equipment manufacturer should be relied on.

*American Society for Metals, Metals Park, Ohio, 1964.

CHAPTER THREE

GENERAL ASPECTS OF PROPER
DESIGN FOR HEAT TREATMENT:
QUENCHING EFFECTS

IN THE QUENCHING PROCESS THE importance of design to a successful heat-treating operation is many times greater than it is in austenitization. There are several reasons for this:

1   The quench cooling rate is much faster and usually much less uniform than the austenitization heating rate.
2   Stresses in quenching are a result not only of a nonuniform cooling rate but also of transformation. For example, in a water-quenched 1040 steel shaft 2 in. in diameter an external shell approximately 0.080 in. deep transforms to martensite with an accompanying volume increase, whereas the core is essentially unhardened. The volume increase produces compressive stresses of the order of 150,000 psi in the hardened shell. An area of the surface that is not fully and simultaneously hardened will be residually stressed in tension and this stress may be sufficient to crack the hard shell at a seam or large inclusion.
3   Part drawings are not concerned with distortion in austenitizing but are usually confined to dimensions of the quenched part.
4   Because quenching is often used to develop high surface compressive stresses as well as to harden straightening is highly restricted or not permitted.

5 The part is larger by at least 0.125 in./ft of part size (in all dimensions) before it is quenched than after it is hardened. In other words, the geometry of the part must be maintained while it is brought to room temperature with accompanying shrinkage.

The configuration that a designer can employ is controlled by the austenitizing and quenching equipment at his disposal and by the cost of the steels he can use; a part that is improperly designed for heat treatment, however, will be a problem even with maximum latitude in part cost or available facilities. Non-uniform cooling in the quench (Fig. 2.1) or symmetrical cooling (Fig. 3.1) usually results in hot upsetting or a shift in the final transformation from the center of the section. Both conditions can result in undesirable residual stresses, excessive distortion, or cracking.

It should be kept in mind that a design configuration that takes the quenching operation into consideration permits more vigorous quenching (e.g., in water as opposed to oil), which in turn permits the use of a less expensive steel; for example, 4130 water-quenched can be used instead of 4340 oil-quenched. Water quenching, if uniform, almost always leaves desirable residual compressive stresses in the surface, as shown in Fig. 1.8.

To minimize hot upsetting designs should avoid radical section differences whenever possible. A classic violation is the flanged shaft shown in Fig. 3.2. It is difficult to water-quench a part made of steels such as 1045, 1050, or 10B40

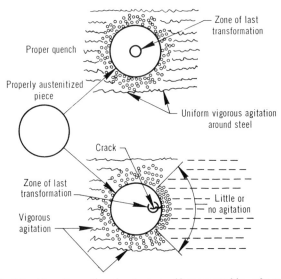

Fig. 3.1 Shift of final transformation due to nonuniform quenching of a round steel bar.

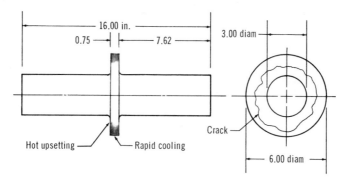

Fig. 3.2 Flanged shaft, showing quench cracking as a result of radical section differences (from *Selecting Steels and Designing Parts for Heat Treatment,* American Society for Metals, 1969).

without cracking the flange. When the outer surface of the thin flange cools, the metal adjacent to the shaft is rapidly upset. As the shaft cools, there is insufficient flange diameter to maintain part geometry and the part cracks. A lower carbon, higher alloy steel might have been satisfactory, but for expediency a high-hardenability alloy steel, oil-quenched, was selected. Another solution to the cracking problem on a part such as this is to shield the flange and vigorously water-quench the body (Fig. 3.3).

Even with perfectly uniform sections, parts can easily crack if made of high-carbon, high-hardenability steels such as 4150. A typical example is the tube in Fig. 3.4. The reasons for this kind of cracking are the following:

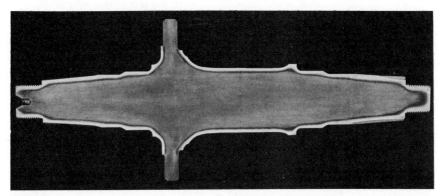

Fig. 3.3 Hardness pattern on shell-hardened bevel gear shaft (weight, 55 lb; nital etch) (from *Selecting Steels and Designing Parts for Heat Treatment,* American Society for Metals, 1969).

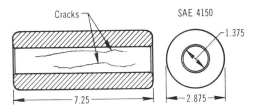

Fig. 3.4 Cracks in a hardened bushing of SAE 4150 steel due to delayed cooling in the bore.

1 The inside diameter transforms last because the high pressure of the quenchant vapors in the bore prevents its cooling at the same rate as the outer surface. This results in the development of high tension stresses in the surface of the bore.

2 If the part is made by boring from bar stock or using seamless tubing, the probability is greater that large nonmetallic inclusions will be exposed at the inner surface than at the outer. These inclusions act as stress concentrations and nucleii for cracks.

3 When the bore-to-wall-thickness ratio is less than 2.25 and the length is more than five times the bore, internal cracking may occur even when oil quenching is used with steels of more than 0.45% nominal carbon content. A solution to this problem, if there is no way to alter the design, is to prequench the bore.

4 Certain alloys such as 4150 are prone to alloy segregation which shows up as "banding." These "bands" may contain several times the nominal alloy content and exhibit extremely high hardenability and depressed transformation temperature compared with the rest of the structure.

Some important questions to consider in proper quenching follow:

1 Does the part really have to be quenched and tempered?
2 Can a lower carbon steel be used?
3 Can the part be designed so that it is symmetrical about all axes?
4 Can holes, grooves, deep splines, and keyways be kept to an absolute minimum?
5 Can sharp corners be avoided?
6 Can provision be made for proper tolerances in dimensions between planes and diameters when fixture quenching is to be used?
7 Can radical part combinations be avoided?
8 Can induction hardening be used? (Some parts can be hardened in no other way). This process often allows the use of lower priced steels.

9 Will special tooling be required for parts that are long and of small cross section or large in area and thin?

Each of these questions is now considered.

IS HEAT TREATMENT NECESSARY?

When strength alone (and/or hardness), without the toughness of a quenched and tempered microstructure, fulfills the minimum engineering requirements, the use of cold-finished bars made with extra heavy draft or elevated temperature drawing should receive consideration. In other words, why heat treat when it is unnecessary? Some typical steels and bar sizes in wide use are listed.

	Carbon Steels		
	1050	1144	1144 Mod.
Minimum yield point (psi)	100,000	100,000	125,000
Rounds (inches)	5/8–4	¼–4½	¼–3½
Hexagons (inches)	5/8–2	¼–2	¼–1½
	Alloy Steels		
	ETD 150*		ETD 180*
Minimum yield point (psi)	130,000		165,000
Rounds (inches)	7/16–3½		7/16–2

These materials should be particularly attractive to firms with no heat-treating facilities or with no commercial heat treater nearby. By adjustment of their composition and the degree to which they are cold- (or warm-) worked these steels can be made to have good machining characteristics.

CAN A LOWER CARBON STEEL BE USED?

Too many times steel selection is made without due attention to the affect of composition on heat-treating practices. By keeping in mind that increasing carbon content nearly always means increasing cracking tendencies a designer

*Trademarks of LaSalle Steel Company, Chicago, Illinois. Both are similar to 4145, with additions such as selenium and tellurium for improved machinability.

may avoid expensive scrap or inordinate heat-treating costs. As a general rule, steels with more than 0.35% C will require oil quenching to avoid cracking. This means that higher cost, higher alloy steels are required for adequate response to the slower oil quench when carbon content exceeds 0.35%. Further discussion of this subject is covered in Chapter 8.

DESIGN THE PART AS SYMMETRICALLY AS POSSIBLE

Perhaps the most difficult parts to quench without excessive distortion, at least when the entire part is austenitized, are those grossly asymmetrical. This is particularly true for carburized gears when tooth tolerances are of the order of 0.0005 in. per inch of face width. The classic example is shown in Fig. 3.5. The mechanism of this type of distortion is as follows: the thinner section below the web cools very rapidly in the quench, upsetting hot metal in the web-to-gear fillet. As the rest of the gear cools to quenchant temperature, it also contracts. The upsetting that has taken place, however, has altered the geometry of the gear. Consequently, when the entire part reaches quenchant temperature, the teeth will be tapered. An amount of web offset up to 0.15

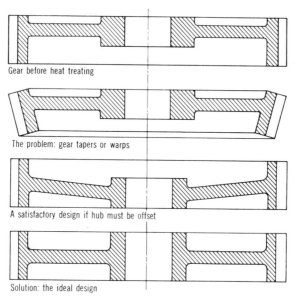

Gear before heat treating

The problem: gear tapers or warps

A satisfactory design if hub must be offset

Solution: the ideal design

Fig. 3.5 Asymmetrical and symmetrical design of a gear (from *Selecting Steels and Designing Parts for Heat Treatment,* American Society for Metals, 1969).

times the tooth width can usually be tolerated by the flow of oil and baffling in the quench press. By locating the web in the center of the 'gear OD and off-setting the hub, the taper was reduced (18 in. OD X 3 in. face width) from a range of 0.008 to 0.012 in. to one of 0.0015 to 0.002 in.

Induction hardening, especially when done tooth by tooth, has been very successful in hardening asymmetrical gears; however, the designer must recognize that the permissible contact stress for the same pitting life is approximately 10% lower than for carburized gears. Also the tendency to score is greater with the lower carbon surface of induction-hardened steel.

Another violation of the symmetry principle is the use of cluster gears (Fig. 3.6), especially when the ratio of pitch diameters between the two gears is greater than 3:2.5. The problem with cluster gears is twofold. First, one of the gears has to be cut on a shaper, which is not only another machine and setup (which introduces errors) but the shaping operation usually has less accuracy than hobbing or hobbing plus shaving. Second, it is, in effect, necessary to quench two gears. Whether it is on a plug or in a die, the quench has to be done

Fig. 3.6 A cluster gear that is difficult to heat-treat without excessive distortion (from *Selecting Steels and Designing Parts for Heat Treatment,* American Society for Metals, 1969).

simultaneously and the different configurations react differently; for example (Fig. 3.6) to obtain the proper case microstructure on the smaller gear, quench flows in the press of 100, 40, and 100 gal/min were necessary. This distorted the large gear beyond the permissible lead error of 0.0003 in. To solve this problem it was necessary to change to a more expensive steel (4320) and to reduce the vigor of the quench.

Another example of asymmetry that often causes difficulties in quenching occurs in shifter-fork grooves in gears. Depending on the configuration and the masses involved, these grooves will "bell-mouth" or pinch in slightly. The best design (Fig. 3.7) specifies the same section thickness of steel around the groove and the most liberal groove tolerances after heat treatment.

Symmetry is so important in "one-shot" induction hardening of gears that it is considered good practice to machine the underside of the rim in order to ensure consistency of rim thickness (which should be 1.5 to 2 times the whole depth).

In addition to dimensional symmetry in a part, it is usually desirable and often necessary to have what might be called metallurgical symmetry. This relates to changes in grain flow in forgings (e.g., at the flash line) and in castings in which gates and risers have been removed. This problem can be particularly troublesome when a forging calling for such high-hardenability steels as 4150 and 4340 is made on a hammer of insufficient size (Fig. 3.8). The thick flash on an axle forging made on a 25,000-lb hammer produced tears at the flash line and also exposed far too much "end grain" of the steel. As a result nearly

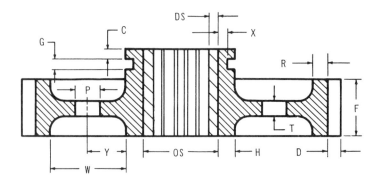

F = Face width of gear
D = Whole depth
R = (Carburizing and tooth by tooth induction)=D+0.15F
R = (One shot induction hardening)=1.5D+0.10F
T = R
DS = Whole depth of spline tooth

X = DS, C=X, G=C
OS = Major diameter of spline
H = 0.25 OS
W = Distance from hub to inside of rim
Y = W/2
P = W/3 max

Fig. 3.7   Recommended design of gears for carburizing and induction hardening.

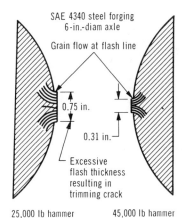

SAE 4340 steel forging
6-in.-diam axle

Grain flow at flash line

0.75 in.

0.31 in.

Excessive
flash thickness
resulting in
trimming crack

25,000 lb hammer   45,000 lb hammer

Fig. 3.8 Typical effect of inadequate hammer size in making a forging. Sections from 6-in. diameter axle, forged from SAE 4340 steel.

every shaft of some heats cracked in heat treating. By making the part on a 45,000-lb hammer it was possible to reduce the thickness of the flash from 0.75 to 0.312 in., which eliminated cracking. For castings or forgings the knowledge of a capable supplier is invaluable in designing parts for metallurgical symmetry.

HOLES, DEEP KEYWAYS, AND SPLINES

For the person involved with day-to-day production heat treating, holes, deep keyways, and splines perhaps give the most trouble. There are many opportunities for it to develop. First of all there may be a dimensional problem, either with the hole itself (including taper) or with its relation to other holes or surfaces. Second is cracking. Third, soot, heat-treating salt, or other debris may find its way into the holes to such an extent that the part becomes unacceptable. Holes, deep keyways, and splines are extremely troublesome when induction hardening is used (see Fig. 2.14). The ideal solution is to avoid holes entirely, at least in the heat-treated area, as shown at the right in Fig. 3.9. Holes in a pin can be eliminated in most instances by a change in the core to provide a grease reservoir if the mating part is a casting or by boring a reservoir into the female part of the joint if it is made of wrought steel. The result is not only a part that is much easier to harden but one that will have several times the fatigue strength of a cross-drilled pin.

A second and widely used solution is to drill after heat treatment. On carburized parts such as bevel gears this can be done after carburizing by copper plating all areas that are to be machined. For a direct hardening steel, such as that used in the shaft in Fig. 3.2, the design can often be made so that a

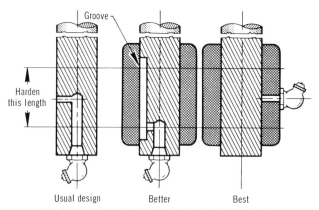

Fig. 3.9   Designs for induction hardening over holes.

machinable hardness is acceptable in the area to be drilled. The holes in this shaft are drilled through the flange, which is partly shielded from the quench.

A third alternative to drilling before heat treatment is to temper the drilling area selectively by induction, flame, or salt bath. A large installation for the selective induction tempering of the bores of crawler-tractor track links is shown in Fig. 3.10. On these parts both holes are tempered in approximately 20 sec, using 450,000-cycle current.

There is little that can be done in designs that require drilled or punched holes before heat treatment. Understanding the trouble-creating mechanism, however, is often helpful. As an austenitized part with a through-drilled hole enters the quench, cooling is proceeding at a rapid rate on all surfaces of the part. As quenchant tries to enter the hole, however, a steam or vapor bubble is continually maintained down to the boiling point of the quenchant. This bubble prevents quenchant from entering the hole. Consequently the inner surface of the hole cools much more slowly than the outer surface of the part. (This is especially true in water or vigorous oil quenching.) The cooling of the metal around the hole upsets the hole surface in compression. When the hole surface finally reaches quenchant temperature, the upsetting that has taken place causes high residual tensile stresses, and when these are of sufficient magnitude cracking occurs. Cracking will also appear at the periphery of the hole because of the stress concentration at this location. If the part does not crack and high hardness is required for endurance-limit fatigue strength, the residual surface tensile stresses in the hole will detract substantially from the fatigue strength in this area.

Fig. 3.10 Selective tempering of track link bores with induction heating for boring after hardening (courtesy of Allis-Chalmers Manufacturing Co.).

The use of oil, hot oil, or salt instead of a water quench to eliminate cracking is actually for the purpose of reducing the differential in cooling rate between the inside of the hole and the rest of the part. Cracking also can be eliminated by quenching in a fixture in which jets of quenchant prequench the holes before the major quench cools the part.

From a design standpoint a chamfer or radius at the entrance and exit of a hole will serve to distribute the tensile stress at the edge of the hole over more steel. In another solution the section in which the hole is drilled is thinned to accelerate its cooling in comparison with the rest of the part.

Maintenance of size on tapped holes can be effected during heat treatment by inserting SAE grade 8 set screws or bolts (see Chapter 13 for SAE grades). Taper of plain holes can usually be controlled by prequenching.

Distortion in hole spacing is a phenomenon that can be minimized by using steel of closely controlled hardenability, ensuring freedom from residual stress as heat treatment commences, and maintaining uniformity of quench from piece to piece. Heat treatments such as conventional oil quenching, hot oil quenching, and austempering usually provide lower average distortion and lower standard deviations.

SHARP CORNERS

Sharp corners on a part as it is quenched produce cracking and spalling, the usual mechanism being that shown in Fig. 2.1. This design feature is often not objectionable on pieces of bar stock cut to length and then heat treated because the stock removal specified will also remove the crack or spalled area. (The possibility of breaking tools and/or injuring a lathe operator must be considered.) Sharp corners are a hazard to manufacturing personnel handling the rough part. It should also be kept in mind that high-hardenability steels, particularly when oil quenched (Fig. 1.8), are residually stressed in tension on the surface. This condition could cause an expensive piece of steel to split along a minor surface imperfection, with the origin of the failure at a sharp corner.

The cracking and spalling we have described nearly always result in the scrapping of a semifinished or finish-machined part. Insurance against sharp corners calls at least for a chamfer or preferably a radius, the larger the better. The finish on the chamfer or radius and the faces leading to the corner is particularly important with high-hardenability steels of more than 0.45% max carbon (125 micro in. max preferred).

Sharp corners have a deleterious effect on the performance of a part in service; for example, it is common practice with sun and planet gears in a planetary design to grind to length after carburizing. This results in a razor-sharp edge on the end of each tooth with an almost infinite stress concentration factor. Grinding the ends of the teeth sometimes also results in grinding cracks which can lead to premature part failure. Grinding to length should be done perferably in the soft condition, followed by filing or milling a chamfer or radius, along the tooth profile (especially in the root radius) before heat treatment. Another alternative is to design a step on the end of the gear so that overall length can be ground in a zone around the bore, but grinding the ends of the teeth should be avoided (see Fig. 3.11).

TOLERANCES FOR FIXTURE QUENCHING

A quenching machine has two functions to perform in the hardening of a steel part:

1 To restore the geometry (flatness, roundness, and straightness) lost by a part in austenitization.
2 To quench the part vigorously, yet nonuniformly, if necessary, to produce a part that is acceptable both metallurgically and dimensionally.

It may come as a surprise to some that a quenching machine cannot "hold" a

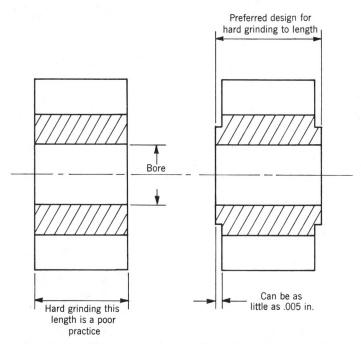

Preferred design for
hard grinding to length

Bore

Hard grinding this
length is a poor
practice

Can be as
little as .005 in.

Fig. 3.11 Gear design with a step so that overall length can be ground after carburizing without grinding the ends of the teeth.

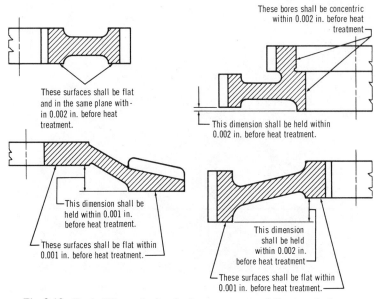

These surfaces shall be flat
and in the same plane within 0.002 in. before heat
treatment.

This dimension shall be
held within 0.001 in.
before heat treatment.

These surfaces shall be flat within
0.001 in. before heat treatment.

These bores shall be concentric
within 0.002 in. before heat
treatment

This dimension shall be held within
0.002 in. before heat treatment.

This dimension
shall be held
within 0.002 in.
before heat treatment

These surfaces shall be flat within
0.001 in. before heat treatment.

Fig. 3.12 Typical dimensioning for heat treatment of die-quenched gears.

43

part to the shape of a quench die. The machine will restore the geometry of the part at the austenitizing temperature but the quenching operation determines to a major extent the final shape of the part.

In order to quench a part in a die (or rollers), it is necessary that the dimensions of the surfaces on which the part is held and gaged be held to a close tolerance. This tolerance is usually one-third to one-half that expected on the finished part. Typical dimensions that should be held for gear quenching are shown in Fig. 3.12. For the quenching of shafts in a roller die (Fig. 3.13) the practice recommended by the Gogan firm is to hold three diameters (in a stepped shaft) to 0.002 in. max tolerance before hardening. On a plain round

Fig. 3.13 A shaft straightening and quenching machine (courtesy of Gogan Machine Corp.)

part two diameters are adequate. The power drive on the rollers will twist and distort the shaft if these dimensions are not held.

In die design it is important to remember the following:

1 Quenchant must get to all important surfaces of a part.
2 Arrangements must be made to prevent vapor bubbles from becoming trapped where they will restrict adequate flow of quenchant.
3 If possible, hot steel should not be quenched by hot quenchant. In other words, provision should be made to permit hot quenchant to escape so that the cold fluid can cool the austenitized part rapidly.

Vapor bubbles will result in soft spots, distortion, and even cracking due to the stresses set up when the hot spots cool last.

Design for die quenching should, whenever possible, include holes in the webs of gears to permit the residual quenchant to drain out of the part. This reduces the hazard of an employee's being burned by hot liquid remaining in the gear. Holes such as these (shown in Fig. 3.7) also sometimes assist in equalizing the quench severity of the part on both sides of the web. On large parts the holes also serve as a place to insert hooks for lifting out of the die.

RADICAL PART COMBINATIONS

Designers know that nothing is so efficient in keeping two or more parts together as making them in one piece. Production control departments also like this idea because there is only one part number to process and stock rather than two or three. For the machine shop, however, nearly perfect alignment must be held and for the heat treater many such combinations cause trouble.

A typical example is the cluster gear (Fig. 3.6). As already mentioned, there are problems in maintaining the machining accuracy required as well as in quenching the teeth. If the bore is an internal spline, there may be problems of taper.

Another example of a radical part combination is the gear and pinion shown in Fig. 3.14. The gear portion is approximately 15 in. diameter with 3.5 in. face. First of all because the combination shown on the left was in one piece it had to be made of the same grade of steel (in this example, SAE 8822). The two-piece design at right permits the use of a high-alloy steel in the pinion to ensure the proper microstructure and a lower alloy steel in the gear. Second, the pinion teeth on the combination design had to be shaped, whereas those of the two-piece design could be hobbed for greater accuracy. Third, the combination was too long to fit into a quench press; consequently distortion occurred on the 15-in. gear. The combination design also distorted out-of-straight on its longi-

Fig. 3.14 One- and two-piece designs for a gear and pinion. The latter (right) is easier to machine and heat-treat (from *Selecting Steels and Designing Parts for Heat Treatment,* American Society for Metals, 1969).

tudinal axis. The two-piece design is held together with a simple locking plate that bolts on the side of the gear teeth and spline of the pinion. Therefore, from the standpoint of both cost and quality, designs that use radical part combinations requiring heat treatment should be carefully scrutinized.

QUENCH EFFECTS IN INDUCTION HEATING

Most designs that call for induction heating for hardening utilize water or various aqueous solutions for quenching. The reasons for this are chiefly the following:

1 To permit the use of low-cost carbon steels.
2 Because exposed quenching is most common (as opposed to submerged quenching), oil would produce disagreeable working conditions as well as a fire hazard.

The use of aqueous quenches, often plus the self-quenching effect of cold steel beneath the induction-heated layer, can result in extremely drastic quenches. This means not only that the steel type must be carefully chosen and the quench, of excellent uniformity, but the design of the part must also prevent cracking as transformation occurs. The test results that follow show the high hardnesses obtained on SAE 1045 and several lean-alloy steels. (Samples for induction hardening were 1.25 in. round by 6 in. long. Travel speed was .086 in./sec, 175 kW 3000 cycle.) All steels were 0.42 to 0.43% carbon.

	End Quench Hardenability at 1/16 in.	Hardness at 1/16 in. Depth on Samples
SAE 1045	53.0 Rc	58.0 Rc
SAE 1045 + 1% Ni	55.8	61.5
SAE 1345	57.0	62.5
SAE 4042	55.5	60.0
SAE 5046	55.0	60.0

This indicates, as Tran and Osborn found, that induction hardening is capable of producing not only excellent surface hardness but also hardened depth beyond that of the end quench specimen.

Quenching steels to this hardness in simple round pins requires that the ends be prepared with radii to prevent spalling (when the hardness is required full length). Likewise, radial holes in this hardened length must have a proper radius. (For maximum bending or torsional endurance-limit fatigue the inside of these holes should also be manually peened with an air tool after induction hardening.)

The designer should recognize that the tolerance on a hardened length of a pin or shaft can be no closer than the hardened depth specified. Most progressive or scan-type induction-hardening procedures also utilize a spray quench built into the inductor coil that impinges at an angle of 30 degrees to the surface. Therefore hardening to the bottom of grooves and into the corners of all section changes is not always convenient (see Fig. 3.15). Preferably shafts and pins should be designed with center holes on both ends to ensure the accurate positioning of the part in the inductor coil. Parts requiring hardening for only sections of their length and not to one or both ends are an exception to this center-hole requirement because female locators can be used.

The quenching of induction-heated gears, especially those processed by the one-shot method, is a critical stage in this type of hardening. Because gears heat-treated in this manner shrink during quenching, it is imperative that the rim be centered over the web. If this is not done, excessive and erratic taper will result. When the rim is uniform in thickness on both sides of a centrally located web, the shrinkage in quenching will cause a desirable crown to develop on the faces of the gear teeth. To minimize the possibility of quench cracking (and excessive stress concentration) at the end of the teeth in the root area a generous deburring operation at the ends of the teeth should be specified.

For tooth-by-tooth induction hardening, especially with radio-frequency power, just about any configuration can be hardened with minimal distortion; however, the production rate per machine is approximately 15% of that obtained by the one-shot method. Also, in many instances the gear to be hardened tooth-by-tooth must receive prior heat treatment for adequate core strength.

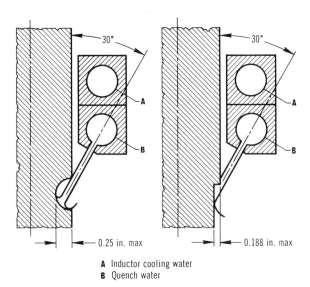

A Inductor cooling water
B Quench water

Fig. 3.15 Typical configuration limitations for induction hardening by scanning.

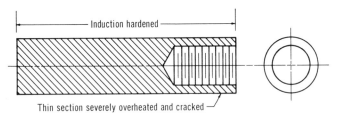

Fig. 3.16 Section showing cracking of a tapped pin (from *Selecting Steels and Designing Parts for Heat Treatment,* American Society for Metals, 1969).

For miscellaneous parts perhaps the most important design consideration for quenching is to provide generous fillets between section changes. This permits the quenchant to flow over the surface, do the required cooling, and escape; for example, the crawler tractor track roller in Fig. 2.15 was designed without a fillet between the flange and the hub. Evidence indicated that because quenchant could not consistently get into this corner, cracks and insufficient hardening resulted. A 3/32-in. radius solved the problem.

Finally, to control distortion and prevent cracking the part must be designed so that it can present reasonably uniform sections for quenching. The tapped pin shown in Fig. 3.16 is a simple example. Even after the problem of nonuniform temperature was solved, the part still cracked because of radical section differences. Drilling and tapping after induction hardening was the only solution. Nonuniformity of section in quenching induction-heated parts is worse than in furnace heating. The reason is that often the thin sections have been severely overheated; the result is grain coarsening, increase in hardenability, and loss of toughness.

It is suggested that part design, quenching medium, and quenching device be carefully studied for each new application of induction hardening.

LONG PARTS OF SMALL CROSS SECTION AND LARGE THIN PLATES

In Chapter 2 the austenitization of long thin parts and thin plates was described as difficult. This is especially true when parts are quenched without the aid of a roller die machine or press.

The definition of a long thin part is quite flexible because it depends on the straightness required and on the quenching facilities available. For free quenching a part whose length is more than 15 times its diameter is nearly always characterized as "long-thin," and the slightest nonuniformity in the quench will cause distortion. In the fastener industry most parts are free-quenched from belt furnaces into water, 5% caustic soda, or oil. The length limits used by one firm are listed:

Water or caustic quench	5 times diam max
Oil quench, martemper	8 times diam max
Austemper	10 times diam max

Beyond these lengths fixture quenching or straightening is considered necessary. Most plants lack quench fixtures, and parts must be straightened and/or ground after heat treatment. Press straightening is something of an art, therefore expensive; also of engineering importance is the fact that press straightening can decrease a part's surface compressive stress and thus its endurance-limit fatigue qualities. It has been reported that vertical direct-oil quenching of carburized transmission shafts from a continuous carburizer with the propellors off produced a straighter part than with the propellors running.* After a short initial quench the propellors were turned on to finish-cool the parts. These shafts were approximately 1 in. in diameter by 18 in. in length and the straightness tolerance was 0.003 in. TIR.

When only surface hardness is required, induction scanning can extend the ratio of length to cross section. If the part is a straight round, square, or hexagon or when machining of grooves and splines is done uniformly around the cir-

*J. Birnbaum, York Division, Borg Warner Corp., private communication, September 1973.

cumference, a scanning device can hold parts straight within 0.0025 in. max/12 in. of length.

Thin plates of large area, such as clutch discs, harrow disks, and thrust washers, are difficult to quench and to keep flat unless die equipment is available. With those that are case hardened, as by carbonitriding, it is not uncommon to grind off all the case to flatten the disk. The case depth can, of course, be increased, but because the parts distort more additional scrap is the result.

As a guide (assuming the part is flat and stress free, void of decarburization, and round or nearly square) limits are suggested for size and section that can be held flat within 0.001/in. of size when parts are racked and quenched edgewise.

Type of Quench	Ratio of Perimeter to Thickness
Water	30 max
Oil	80 max
Austemper	125 max
Gas	150 max

If the design exceeds these limits, either a quench press should be used or the parts may be purchased finished from firms specializing in their manufacture.

CHAPTER FOUR

COMMERCIAL ASPECTS OF
THE STEEL INDUSTRY

SECOND IN IMPORTANCE TO MEET-
ing the minimum engineering requirements in a part is its cost. Most products
manufactured from the constructional grades of steel are sold in a highly competitive market, whether the product is a wrench, automobile, farm tractor, or
mammoth mining shovel.

The recent recognition that global energy supply is limited was precipitated
by the OPEC oil embargo of 1973 and has greatly influenced the economy of
all industrial nations. In the United States, for example, the composite price of
steel rose from 6.5¢/lb in 1968 to an estimated 15.0¢/lb in 1977; the majority
of this increase has been effective since 1973 [1]. This 230% jump corresponds
to greater energy costs. Coal prices, for example, have climbed 400% since 1960,
oil approximately 300% in two years (1973-1974). The net effect is a shift of
economics for many products, from labor to material intensive. Today, typical
material costs for components made from constructional steels range from 40
to 70% of total cost, thus giving greater importance to prudent steel selection
and material control.

In a broad sense there are three aspects to the steel industry: (a) the technical operation, which includes research and product development; (b) the
manufacturing operation to produce the various steel products, qualities, and
grades; and (c) the commercial operation, which includes marketing of the
steel produced.

Manufacturers of steel products vary greatly in their technical backgrounds.
Most reputable firms, however, have at least an adequate technical background

by virtue of education, experience, or both. In any event, the common engineering characteristics of steels, both "as rolled" and "heat-treated," are generally readily accessible in various technical libraries and from the steel producers themselves.

Commercial details about the manufacture of steel are usually known only to a superficial degree and by only a small percentage of the steel user's technical staff. It could be argued rather convincingly that in many cases a materials engineer has little need to know the melting practice of steel company "A" at its No. 6 plant. However, this kind of knowledge would often enable him to make a more intelligent decision in guiding his purchasing department to particular sources for the best value. An example of this involving rejection rate of steel for heavy-duty crankshafts is shown in Fig. 4.1. This type of "know how" generally requires only minor capital investment and pays handsome dividends over the long pull. The most difficult part of the process consists of deciding what information is most significant and establishing a statistically sound method of analyzing the data.

Another weakness of many steel users is that rarely are their technical personnel familiar with the pricing structures of the various steel suppliers. We have known firms with hundreds of employees among whom not one could price a

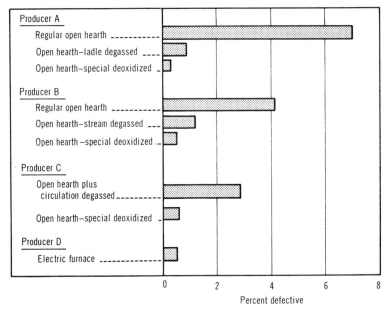

Fig. 4.1 Magnaflux rejection rate of crankshafts as a function of steel melting practice.

bar of SAE 1045 steel. This is incredible because some of these firms purchase and process as much as several hundred million dollars worth of steel a year.

To a technical employee learning to understand the commercial aspects of the steel industry may seem a dreary task, but it can provide a powerful tool for proper steel selection. Two sets of books are required to start this educational process: the Steel Products Manuals of the American Iron & Steel Institute [2] and a set of price books from a major steel producer. Both are free in limited quantity on request from the AISI and the producer by way of the company purchasing agent.

The foreword of a current issue of one typical manual in the set states:

It is the purpose of the Steel Products Manual to make available to purchasers, producers, consumers, and all others interested, information covering the major steel products. Each section deals with tolerances, methods of inspection, sampling and chemical analysis, definitions of technological terms, and other related subjects which have developed in the manufacture of iron and steel.

There is a separate manual for each of the following broad groups of steel products (a partial list):

Alloy steel: semifinished; hot rolled and cold finished bars
Carbon steel: semifinished for forging; hot-rolled and cold-finished bars; hot-rolled deformed concrete reinforcing bars
Alloy steel plates
Carbon sheet steel
Carbon steel: plates; structural sections; rolled floor plates; steel sheet piling
High-strength low-alloy steel and high-strength intermediate manganese steel
Railway track materials
Flat-rolled electrical steel
Wire and rods, alloy steel
Alloy steel sheets and strip
Tin mill products
Tool steels
Wrought steel wheels and forged railway axles
Wire and rods, carbon steel
Steel tubing specialties

It will become immediately obvious to the student that there are separate manuals for alloy and carbon steel products. Somewhat different manufacturing practices apply to these two broad types of steel, but to the user the main difference is that alloy steel is more expensive than carbon steel. For this reason carbon steel should be used whenever possible. Generally speaking, alloy steel does not possess any mysteriously superior qualities not contained in properly made carbon steel except that it hardens more deeply, or easily, as in oil, and that it imparts exceptional toughness, as in nickel-bearing steels. The principal distinctions between alloy and carbon steel are as follows [2].

ALLOY STEEL

When the maximum of the range given for the content of the alloying elements exceeds one or more of the following limits: manganese, 1.65%, silicon, 0.60%, copper, 0.60%, or when a definite range or a definite minimum quantity of any of the following elements is specified or required within the limits of the recognized field of constructional alloy steels: aluminum, chromium to 3.99%, cobalt, columbium, molybdenum, nickel, titanium, tungsten, vanadium, zirconium, or any other alloying element added to obtain a desired alloying effect.

CARBON STEEL

No minimum is specified or required for aluminum (except for deoxidation and or grain-size control), chromium, cobalt, columbium, molybdenum, nickel, titanium, tungsten, vanadium, zirconium, or any other element added to obtain a desired alloying effect, when the specified minimum for copper does not exceed 0.40%, or when the maximum content specified for any of the following elements does not exceed the percentages noted: manganese 1.65, silicon 0.60, copper 0.60. Boron may be added to carbon steels to improve hardenability. In all carbon steels small quantities of certain residual elements such as copper, nickel, molybdenum, and chromium are unavoidably retained from raw materials. These elements are considered as incidental.

The next step toward understanding the commercial aspects of the steel industry is to determine which of the AISI manuals is applicable. In other words, what is the correct trade designation of the grade and shape in question? To do this it is necessary to tabulate the dimensional limitations for each product; for example, if a piece of steel 0.50 in. thick by 7.0 in. wide is required, is it a bar, plate, sheet, or strip? Bars are defined as follows:

Rounds, 0.25 to 10 in. inclusive
Squares, 0.25 to 6 in. inclusive

Round-cornered squares, 0.375 to 8 in. inclusive

Hexagons, 0.375 to 4.06 in. inclusive

Flats, more than 0.203 in. thick and up to 6 in. inclusive in width

Flats, 0.230 in. and more in thickness, more than 6 to 8 in. inclusive in width

Bar size shapes — including angles, channels, tees, and zees when their greatest cross-sectional dimension is less than 3 in.

Ovals, half-ovals, and half-rounds

Special bar sections

As can be seen, the 0.50-in. thick by 7-in. wide section in question is a bar, and if the analysis under consideration meets the requirements of carbon steel the applicable AISI Steel Products Manual is "Carbon Steel: Semifinished for Forging; Hot-Rolled and Cold-Finished Bars." (This determination has a strong influence on steel cost, but the convergence of the dimensions of a number of flat-rolled steel products is confusing and is explained in Table 4.1.)

Table 4.1 Typical Product Classification, Flat Hot-Rolled Carbon and HSLA Steel by Size

Thickness (in.)	Specified Width (in.)					
	To 3.5	More than 3.5 to 6	More than 6 to 8	More than 8 to 12	More than 12 to 28	More than 48
0.230 and above	Bar[a]	Bar[a]	Bar[a]	Plate[a]	Plate[a]	Plate[a]
0.229 to 0.204	Bar[a]	Bar[a]	Strip	Strip	Sheet	Plate[a]
0.203 to 0.180	Strip	Strip	Strip	Strip	Sheet	Plate[a]
0.179 and below	Strip	Strip	Strip	Strip	Sheet[b]	Sheet[b]

[a]Subject to certain conditions, these dimensions are sold as carbon sheet or strip as well as bars or plate.
[b]These product classifications for hot-rolled sheet are based on the median point of the minimum thickness ordered plus full published thickness tolerances.

Once the appropriate AISI manual has been selected, the reader will find new commercial steel jargon. In the case of the 0.50- by 7-in. carbon steel bar there is a choice of two major qualities: "merchant" and "special." Subordinate qualities are then also available under "special quality":

1 Special internal soundness.
2 Special heat treating.

3 Nonmetallic inclusion.
4 Special discard.

The AISI manual also lists the tolerances for thickness and width, straightness, cold shearing limitations, and the check analysis tolerances permitted from the ladle analysis. Some useful excerpts from these tables are given in Appendixes III, IV, and V; for example, if the part to be made from the 0.50- by 7-in. flat bar of steel is a 6-in. diameter carbonitrided thrust washer, hot-rolled, special quality in a SAE 1524 grade is required. (See Chapter 5 for further discussion of steel qualities and types.)

PROCEDURES FOR PRICING

In the final step toward understanding the commercial side we use the pricing manual of a steel producer. These books have been prepared with considerable care and follow the same product and quality designations described in the AISI Steel Products Manual. Figure 4.2 is a typical price schedule for special-quality carbon steel bars. In general, steel pricing is not unlike pricing a new car in that the final price is the sum of a base price plus a series of optional extras. A correct price can be determined by following the index key shown in the pricing manual.

Note. All steel requirements should be inquired with suppliers for availability of grade and size and for confirmation of price and delivery.

As an example we shall determine the price of 15,000 lb of 7 X .50 in. hot-rolled steel bar. We assume that the engineer has specified SAE 1524 composition with silicon-killed, fine grain as additional requirements. The first item (1) in Fig. 4.2 is price bases. Table 4.2 (page 3 of the U.S. Steel manual) shows that on October 1, 1976, the base price of hot-rolled, special quality carbon steel bars was $15.10/cwt. Following the index key, the next increment of cost is the grade. As shown in Table 4.3, the grade extra for 1524 is $2.30/cwt. The next important price increment under "Metallurgical Requirements" is for silicon, as shown in Table 4.4. Since the silicon-killed and fine-grained extras are not cumulative, the charge for both is 65¢/cwt. The total cost at this point is as follows:

Base Price	$15.10
Grade (1524)	2.30
Silicon, fine grain	.65
	$18.05/cwt

INDEX KEY

Price Bases	Processing

Price Bases

Processing

Thermal Treatment
Surface Treatment

Metallurgical Requirements

Quantity

Grade
Chemistry
Mechanical Properties
Vacuum Degassing
Quality
Government Specifications
Inspection and Testing

Item Quantity

Pack, Mark, Load
Packing
Marking
Loading

Dimension

**General Information
and Tolerance**

Section
Length
Cutting
Tolerances

Metric Dimensions

USS COR-TEN, TRI-TEN, MAN-TEN, EX-TEN, MX, MACH 5
and Q-TEMP are registered trademarks of United States Steel
Corp.

Fig. 4.2. Index key from a typical steel pricing manual (courtesy of U.S. Steel Corporation)

Table 4.2 Step One in Steel Pricing (Base Price)[a]

Product	USS Price Bases	Effective Date
Hot rolled bars — special quality	15.10	10-1-76
Except Torrance, CA	15.60	10-1-76

[a]Courtesy U.S. Steel Corporation.

Special Quality Bars may be specified to standard or other than standard chemical specifications or to mechanical property specifications. Bars of this quality are produced from Rimmed, Capped, Semi-Killed and Killed Steels for hot forging, heat-treating, cold drawing, machining and other special purposes.

Special Quality Bars are manufactured to practices designed to minimize injurious internal irregularities and surface imperfections.

Table 4.3   Step Two in Steel Pricing (Determination of Grade Extra)[a]

AISI or SAE Number	Extras	Percent			
		Carbon	Manganese	Phosphorus Max[a]	Sulphur Max[a]
1513	1.35	.10–.16	1.10–1.40	.040	.050
1518	1.85	.15–.21	1.10–1.40	.040	.050
1522	1.85	.18–.24	1.10–1.40	.040	.050
→ 1524	2.30	.19–.25	1.35–1.65	.040	.050
1525	1.65	.23–.29	.80–1.10	.040	.050
1526	1.85	.22–.29	1.10–1.40	.040	.050
1527	2.30	.22–.29	1.20–1.50	.040	.050
1536	1.55	.30–.37	1.20–1.50	.040	.050
1541	1.55	.36–.44	1.35–1.65	.040	.050
1547	1.55	.43–.51	1.35–1.65	.040	.050
1548	1.25	.44–.52	1.10–1.40	.040	.050
1551	1.25	.45–.56	.85–1.15	.040	.050
1552	1.55	.47–.55	1.20–1.50	.040	.050
1561	1.05	.55–.65	.75–1.05	.040	.050
1566	1.15	.60–.71	.85–1.15	.040	.050
1572	1.15	.65–.76	1.00–1.30	.040	.050

[a]Courtesy U.S. Steel Corporation.

Following the index key (Fig. 4.2), the next price increment is for dimension. The 7 × 0.50-in. bar is considered as "flat," and its cost extra is $.80 cwt, as shown in Table 4.5. The other dimension extra is length, but if random lengths in the range of 15 to 20 ft inclusive are specified there is no extra.

The price increments for "Processing," "Quantity," and "Pack, Mark, Load" in the index key are as follows:

Processing	no special processing required
Quantity	$ 1/cwt
Packing	$ 0.35/cwt
Marking	none
Loading (open-top truck)	none

Total cost for the steel, F.O.B. mill:

	$18.05/cwt (previous increments)
Size	.80
Quantity	1.00
Packing	.35
	$20.20/cwt

Table 4.4   Step Three in Steel Pricing (Silicon and Grain Size)[a]

Silicon (Our standard ranges apply)		Extras (Applicable to All Grades and Sizes when Specified or Required)[b]
Silicon Minimum (%)	Silicon Maximum (%)	
.03 through .15	Over .10 through .30	.65
Over .15 through .40	Over .30 through .60	.90[d]
Killed Steel		.65[c]
Aluminum Killed		.65[c]
Fine Grain		.65[c]
Coarse Grain		.65[c]

[a]Courtesy U.S. Steel Corporation.
[b]Killed Steel Extra is generally applicable under any of the conditions listed below. It is not applicable to our listed Free Machining Steel Grades.
   Chemistry and Mechanical Properties
      Carbon Level exceeds .28% maximum.
      Manganese Level exceeds 1.00% maximum.
      Tensile Strength 70,000 PSI minimum and over.
      Yield Strength 40,000 PSI minimum and over.
   Size
      Rounds, Squares or Hexagons 3 in. and over in cross section. Flats 1½ in. thick and over.
[c]These extras are not cumulative even though more than one of the requirements may be specified to describe the desired type of steel and/or grain size.
[d]This extra applies in lieu of any other Killed Steel extras listed when such requirements are specified in conjunction with the higher Silicon levels.

Similar pricing arrangements are used for other commodities, such as plate, sheet, and structural shapes.

As we can see from the choices in the pricing of this bar, purchasing departments need more information than size and grade to avoid confusion and unnecessary cost. This leads to the matter of specifications.

Table 4.5 Step Four in Steel Pricing (Section Extra)[a]

Width (in.)	3 and Over	3 thru 2	2 thru 1	1 thru 3/4	3/4 thru 1/2	1/2 thru 3/8	3/8 thru 1/4	1/4 thru 13/64
3/8 through 31/64	−	−	−	−	−	− [c]	− [c]	− [c]
1/2 through 3/4	−	−	−	−	3.25 [c]	3.75 [c]	− [c]	− [c]
Over 3/4 through 1	−	−	−	2.25	2.50	3.00 [c]	3.50 [c]	− [c]
Over 1 through 1¼	−	−	.95	1.10	1.25	1.50	1.75 [c]	3.50 [c]
Over 1¼ through 2	−	−	.90	.85	.85	1.00	1.30	2.50 [c]
Over 2 through 3	−	.90	.85	.80	.80	.90	1.20	2.00
Over 3 through 6	1.15	.85	.80	.80	.80	.85	1.00	1.50
Over 6 through 8	1.15	.90	.85	.80	.80	1.15	1.70	1.80 [d]

Header span: Thickness (in.) — ↓ Under

[a]Courtesy U.S. Steel Corporation.
[b]Carbon .75% minimum and under; carbon over .75% minimum .20 plus size extras shown in table above.
[c]Flats under 1 lb. per lineal ft – inquire.
[d]Minimum thickness for flats over 6 in. wide is .230 in.

SPECIFICATION: ENGINEERING OR PURCHASING?

A manufacturing firm that uses any significant amount of steel — say more than 1000 tons a year — should have inhouse specifications that accurately describe what is required in commercial terms. There are many arguments in favor of these specifications and few against them except the cost of their preparation, maintenance, and distribution. Specifications not only state exactly what is wanted but also prevent drawings from being cluttered with materials-engineering jargon (merchant quality, silicon-killed, vacuum-degassed, etc.). When specification numbers are used, the material in hundreds or even thousands of different parts can be changed by altering only a few specification pages. (Individual drawing changes cost $100 to as much as $1000, depending largely on the size of the firm.) When the use of a certain steel composition is more than approximately 1000 tons a year, nonstandard grades are often justified, and inhouse specifications are the ideal method to document the requirements and provide the flexibility to optimize cost/benefit relations. Finally, specification numbers lend themselves well to modern data processing systems for such functions as inventory control and purchasing in contrast to long notes on drawings. In many cases ASTM or SAE specifications will suffice, but most often they are inadequate for a firm's complete requirements and an inhouse numbering system must be established.

When writing specifications and setting up the necessary control systems, it is well to keep the following points in mind:

1 If the specification is to be used to purchase steel, it becomes part of a purchasing contract and must define what is wanted in commercial steel terminology.

2 It is generally good practice to define the intended use for the steel as well as the expected qualities and mechanical properties. If it is a grade for heat treatment, a knowledge of mechanical properties that can be expected in different sizes is useful engineering information.

It is common practice to mix these two requirements (engineering and purchasing) in a specification, but confusion and excess cost are frequently the result. It is suggested, therefore, that engineering and purchasing requirements be physically separated within the specification or, preferably, that the specification be written for purchasing only, putting engineering characteristics, capabilities, and suggested usages in a separate document for internal use only.

The following is a typical specification format:

Article	Covering
1 Scope	(a) General type of product, such as hot-rolled carbon steel bars (b) Subordinate specification references
2 Application	Broad anticipated usage, such as diesel engine piston pins and hardened and ground bushings.
3 Form and quality	e.g., hot-rolled bars — special quality, or cold-finished bars — standard quality.
4 Composition	(a) Heat analysis (b) Chemical ranges
5 Grain Size	(a) ASTM — E112 — fine (b) Coarse grain practice (c) No grain size specified.
6 Hardenability (if appropriate)	(a) End quench Distance ___ Rockwell C ___ (b) SAC test (for carbon steels) ___ to ___ Rockwell inches

Article	Covering
7 Methods of test	(a) Per SAE ____
	(b) Per ASTM ____
	(c) Per AISI recommended practice
8 Thermal treatment	(a) As-rolled, annealed, etc.
9 Tolerances	(a) Per appropriate AISI Manual
	(b) Per appropriate AISI Manual except as follows:
10 Pack, mark, load	(a) Size of lifts
	(b) Painting or stamping
	(c) Type of loading: in gondola cars, etc.
11 Reports	Shipping notices showing results of metallurgical tests to be sent to:

All specifications should carry a change number, a change date, and the number of pages in the specification. It is also advisable that each book of specifications bear a serial number and that a record be maintained of its holder. Whenever possible, specifications should be sent not to an individual but to a position such as sales manager because of possible personnel changes. Also, each change should carry an acknowledgment form, to ensure that suppliers have received the new specification and applicable book index change. (The latter should show the latest change number and its release date.)

CONTROLLING STEEL COSTS AT THE SPECIFICATION LEVEL

It is usually productive to control costs at the specification level. When purchasing steel, the following are some typical items to check carefully:

1 Carbon Bars
(a) Can merchant quality be used instead of special quality? Savings: $23/ton.
(b) Can modified merchant grades be used instead of standard grades? Savings: $1/ton.
(c) Can carbon steel be used instead of alloy steel? Savings: $19.50/ton in base price alone.
(d) Are chemical composition ranges standard? Each 0.01% restriction in carbon costs $1/ton in carbon steel bars, for example.

(e) Is annealing or normalizing really necessary? Depending on size, omission of these processes could save $43 to $88/ton.

2 Alloy Bars

(a) Is the lowest cost grade that will meet engineering requirements being used and is it readily available? The potential savings are $30/ton or more.

(b) Are chemical composition and hardenability ranges standard? Restrictions cost 10% of base price, plus open hearth or BOP grade.

(c) Are special quality extras really necessary? Axle shaft quality, for example, costs $6/ton.

A good way to detect any unnecessary cost increments is to check steel-supplier invoices.

ECONOMIC ORDER QUANTITY

Design and materials engineers should recognize that steel, like other parts and materials, has to be purchased in an economic order quantity (EOQ). In instances of a nonstandard chemical composition this means buying in heat lots that will weigh 50 to 300 tons, depending on the supplier. Special rolled sections, which may require many hours of downtime in making a mill setup, usually require minimum rollings of at least 100 tons. Accordingly, in working with special sections the question of obsolescence must be kept in mind. The initial investment in a set of mill rolls is generally $100,000 or more and sometimes a dimensional change as small as 0.03 in. can make half the set obsolete.

The alternative to ordering large quantities from a producing mill is buying them from a warehouse or service center. The advantage of this source of material is almost immediate delivery in small quantities or full truck loads. The disadvantages are price (approximately 20 to 40% more than mill) and severe limitations in sizes and grades available. During periods of abundant supply one may even be able to obtain common grades from service centers at or below mill price.

Several of the major service centers in the United States stock the following grades:

Carbon Bars (round)

Hot-rolled, merchant quality	M1020, M1044
Hot-rolled, special quality	1018, 1020, 1117, 1035, 1040, 1045, 1141, 1144, 1095

Cold-finished, standard quality	1018, 10L18, 1117, 11L17, 1045, 1141, 11L41, 1144, 1213, 1215, 12L15
Drawn, ground, and polished	1213, 1215, 1141
Turned, ground, and polished	1018, 1141, 1144 heavy draft, 1025, 1035, 1043, 1045, 1046,
Cold-drawn, extra heavy draft	1144

Alloy bars (round)

Hot-rolled, BOH or BOP quality	4615/20, 4140, 41L40, 41L47, 4340, 8620, 86L20
Hot-rolled, heat-treated, BOH	4140, 41L40, 41L50
Hot-rolled, aircraft quality	4340, Nitralloy 135 Mod.*
Cold-finished, standard quality	4140, 41L40, 41L47, 4620, 8620, 8640, 86L20
Cold-finished, heat-treated BOH	4140, 41L40
Cold-finished, aircraft quality	E4130, E4140, E4340, Nitralloy 135 Mod.*, E8740
Heat-treated, turned and polished	4140
Heat-treated, turned, ground, and polished	4140
Elevated-temperature-drawn	ETD-150,* ETD-180*

Carbon bars (flats)

Hot-rolled, merchant quality	M1020, M1044
Hot-rolled, special quality	1020, 1045, 1095
Cold-drawn, standard quality	1018, 11L17

Alloy bars (flats)

Hot-rolled, BOH or BOP quality	4130, 4140
Hot-rolled, aircraft quality	E4130
Cold-drawn, standard quality	1018, 11L17

Carbon plates	1015, 1020, 1025, ASTM-A36, 0.40–0.50% carbon, ASTM-A283, Grade C.

*Proprietary trade names of various producers.

High-strength, low-alloy	Hi Steel,* Van 80,* INX 50–60 and 70,* Cor-Ten*
Alloy Plates	
Hot-rolled	4140, 4147, 8620
Hot-rolled, heat-treated	TI,* TIA,* TIB,* RQ–100A*
Sheets	
Hot-rolled	Commercial quality, commercial quality pickled and oiled
	P & O, Hi Steel,* Cor-Ten,* 0.40–0.50% carbon
Cold-rolled	Commercial quality, drawing quality, aluminum-killed drawing quality, stretcher-leveled, vitreous enameling
Bar-Size Shapes	M1020
Structural Shapes	ASTM–A7, ASTM–A36

IMPORTED STEEL

The spread of industrialization has raised worldwide challenges to all American industries. The steel industry is no exception and imported steel is an increasingly significant market factor. In 1970 13.8 million tons of steel, representing about 15% of all domestic shipments, were imported. In 1974 imported steel increased to 16.7 million tons despite near record domestic shipments of 109.5 million tons [3,4]. Imported alloy constructional grades, representing less than 0.5% (47,757 tons in 1974) of the total steel imports [4], are still somewhat rare.

It is recommended that caution be exercised when using imported alloy grades for heat-treated components. Although steel of excellent quality is available from foreign sources, it is also recommended that it not be used for highly stressed components unless it is made to inhouse specifications far more restrictive than standard ASTM or SAE documents. Failure to provide such specifications and the necessary quality control to ensure that they are being met could

*Proprietary trade names of various producers.

easily jeopardize any saving that might be generated by using a foreign source. Purchases from overseas mills generally means that the manufacturer is faced with long lead times to incorporate necessary changes in steel chemistry or manufacture. Another consideration when using offshore suppliers for critical components is that the manufacturer could stand alone in the event of product-liability litigation stemming from defective material.

STOCK LISTS

Small steel consumers (those who use substantially less than 1000 tons of steel a year) may find it advantageous to provide a stock list of steels currently obtainable so that their designers and materials engineers will not introduce a proliferation of grades and sizes. Even in large firms a stock list is needed for flat bars which often account for an inordinate portion of inventory costs.

The abuse of stock lists occurs when the origin of a drawing calling for a certain material grade, size, finish, and heat treatment considered necessary to meet the minimum engineering requirements becomes obscure. For example, when a designer calls for 4140 heat-treated to 32-38 Rc on a drawing, the use of a carbon steel at a later date even for reasons of expediency sometimes becomes difficult. The 4140 may have been called for initially because this grade was carried in stock for a number of parts and not because the part really required heat-treated alloy steel. A new engineer in charge of design will hesitate to use a lower cost steel when there is no record of the reason for specifying the original steel.

Firms with international manufacturing operations may encounter problems when the stock list is used at widely scattered design facilities. For example, suppose that at the American plant 1-in.-diameter cold-drawn SAE 1045 bar stock is specified because the requirement is 100 pieces a day of a part and the steel is machined on an automatic screw machine. The part requires only 30,000 psi yield point, but the cold-finished 1045 on the stock list is of the right size; therefore it is called for on the drawing. The Hong Kong subsidiary uses 100 pieces a year of this part, but they are made on an engine lathe. The Hong Kong manufacturing department assumes that the 1045 cold-drawn steel is specified because of its 75,000 psi yield point. It is then ordered from the United States or Japan at an exorbitant cost and delay, although the hot-rolled 1045 on hand would have been adequate. If the Hong Kong plant makes several hundred different parts, the specification of American stock list steels with little if any regard for engineering requirements creates a difficult situation indeed. It is important, therefore, that the designer and materials engineer in a multiplant firm specify steels and heat treatments on the basis of universal availability and minimum engineering requirements so that the manufacture of parts can then be moved from plant to plant with a minimum of difficulty. Of equal importance are good records that reflect accurately the reasons for materials or process changes as they are made.

REFERENCES

1 Steel Price Hikes Provoke Marketplace Grumbles *Iron Age,* May 23, 1977, p. 28.
2 AISI Steel Products Manual for Alloy, Carbon, and High-Strength/Low-Alloy Steels: Semi-finished for forging; hot rolled bars; cold finished bars, etc. American Iron and Steel Institute, Washington, D.C., copyright, owner, 1977.
3 *Minerals Yearbook 1974,* Vol. I, Metals, Minerals, and Fuels, Bureau of Mines, U.S. Department of the Interior, 1974.
4 *Metal Bulletin Handbook 1976,* 9th ed., Metal Bulletin London, 1976.

CHAPTER FIVE

CHARACTERISTICS OF CONSTRUCTIONAL STEELS

THE CONSTRUCTIONAL STEELS have numerous characteristics that are important in their selection for heat-treated parts, in addition to their wide range of chemical compositions and hardenabilities.

It may be said that any steel can be heat-treated. However, the result may sometimes be of questionable engineering usefulness; for example, internal unsoundness and chemical segregation can contribute to quench cracking and highly nonuniform engineering characteristics. Also, coarse-grained steels are much more prone to quench cracking and brittleness than fine-grained steels.

Accordingly, the selection of steel for heat-treated parts should include consideration of the following:

1 Internal soundness.
2 Uniformity of composition.
3 Grain size (deoxidation practice).
4 Surface condition.
5 Ease with which the grade can be made to a high degree of cleanliness.
6 Idiosyncrasies in heat treatment.
7 Engineering potential.

STEELMAKING CHARACTERISTICS

Internal soundness, uniformity of composition, and grain size are functions of grade, steelmaking practice, and ingot type. The three types of ingot mold in standard use are shown in Fig. 5.1. Steels to be used in the "as rolled" or normalized condition, including steels for drawing-quality sheet, are usually cast into the big-end-down mold, either open-topped or capped (Fig. 5.1a and b). In the rimmed steels (which are cast unkilled) the evolution of gases from the reaction of FeO with carbon and also nitrogen causes extreme segregation in the ingot. As solidification takes place carbon and other elements, as well as nonmetallic inclusions, are rejected from the liquid/solid interface and migrate inward. This results in an outside layer of nearly pure iron or "rim" on the surfaces of the ingot. When rolled down to a sheet, this nearly pure iron layer provides a smooth, ductile surface. A micrograph of a rimmed steel sheet is

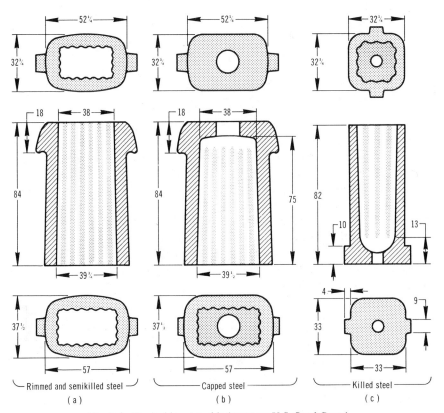

Fig. 5.1 Typical ingot molds (courtesy U.S. Steel Corp.).

shown in Fig. 5.2. The composition of the rim and core is typically as follows:

	Rim (%)	Core (%)
Carbon	0.04	0.09
Manganese	0.36	0.38
Silicon	0.01	0.01
Sulfur	0.029	0.039
Phosphorus	0.010	0.010

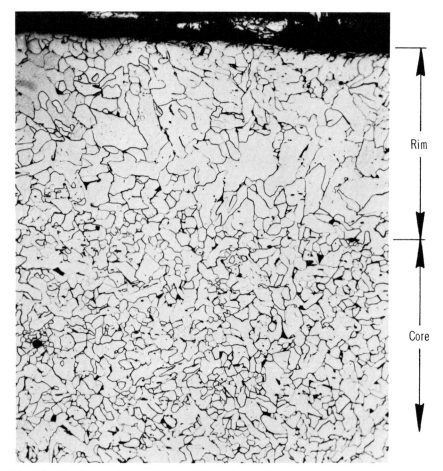

Rim

Core

Fig. 5.2 Rim (low carbon) and core of rimmed steel. 2% nital etch; 250 × (courtesy U.S. Steel Corp.).

Capped steels are tapped with somewhat lower FeO contents and are permitted to "rim" in the ingot mold for only a few minutes. The rimming action, or gas evolution, it stopped by placing a heavy iron or steel cap on the mold. By reducing the time for gas evolution capping the ingot also reduces the amount of segregation. Low manganese and silicon contents are necessary to achieve the desired rimming action, but they add little hardening capability to the steel. Carbonitriding and similar processes by which nitrogen is added are the only applicable heat treatments.

Semikilled steel with higher silicon content (approximately 0.10% max) is also cast into the big-end-down, open-top ingot (Fig. 5.1a) and is amenable to simple heat treatments. Grades with maximum ladle carbon content of 0.22% and adequate manganeses and residual alloy content which allows them to be quenched in water to 80% min martensite can be tempered to a yield point of 80,000 psi. Substantial cost can be avoided by employing semikilled steels for some applications that require heat treatment. The user should tell the supplier what he intends to do with the steel and also be prepared to divert some substandard material simply because of the lack of uniformity and predictability typical of semikilled steel. Carbonitriding or other processes that employ oil or gas quenches can be used for hardening. Semikilled steels of more than 0.22% ladle carbon are not generally suitable for heat treating to high hardness levels because the high quenched hardness may lead to splitting and cracking. Brittleness may also result from the usual coarse-grain, chemical nonuniformity and relatively weak center condition.

The recommended practice for steel to be heat-treated is to increase the silicon content and cast into a big-end-up mold with a hot top, as shown in Fig. 5.1c. Aluminum may be added to refine the grain (to ASTM grain size No. 5 or finer). The silicon content is usually 0.15 to 0.30% for both carbon and alloy steels. The hot top is made from an insulating, and sometimes even exothermic, refractory that keeps the metal in the top of the mold liquified so that molten metal can feed down into the ingot as it shrinks and freezes.

Silicon-killed steel made without the addition of aluminum provides the best machinability. This is due to coarser grain size and absence of very hard and brittle A_2O_3 inclusions, which are deterimental to tool life. These steels, especially when cast into hot-top molds, are suitable for some heat treatments such as induction hardening. The higher hardenability induced by the coarse grain and the attendant brittleness of the coarse martensite usually require the use of soluble oil/water emulsions or synthetic quenchants of similar severity. Most free-machining, hardenable grades of steel, such as 1137, 1141, and 1144, are made in this manner, although the manganese and sulfur segregations add to the difficulty of heat treating without cracking.

Aluminum-killed steels without the extra silicon, such as cold-rolled drawing-quality sheet or strip, present a special problem in heat treating. The condition manifests itself in the micrographs of carburized samples in Fig. 5.3. The aluminum-killed steel in Fig. 5.3*a* shows a structure of cementite and ferrite that will not austenitize properly. Even with a drastic quench, such as brine, this steel will not harden uniformly to more than 50 Rc. The microstructure in Fig. 5.3*b* is made of a silicon-aluminum-killed steel, carburized at the same time, which can readily be fully hardened with ease to higher than 60 Rc.

A summary of melting and ingot practices for various carbon steel products is given in Table 5.1.

CONTINUOUS CAST STEEL

In 1975 approximately 9% of all steel produced in the United States was continuously cast through a water-colled copper mold directly to bloom or slab rather than statically cast in an ingot mold. This percentage is expected to rise sharply in the future because approximately 15% of the nearly 9 million BTUs required to finish a ton of raw steel on a blooming mill can be saved with this process. Some of the other important advantages of continuous or "strand" casting are the following:

1   Substantially increased yield from melt to shipping weight (little or no crop loss).
2   Lower rolling cost to reduce the steel to the specified size because the primary hot reduction in a blooming or slab mill is bypassed.

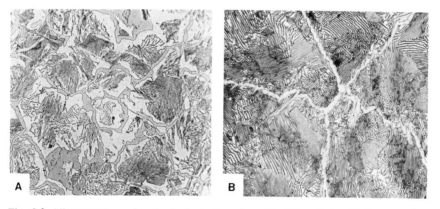

Fig. 5.3 Microstructure of carburized steels made with different deoxidation practices (courtesy U.S. Steel Corp.). (*a*) Aluminum-killed steel; 300 X (*b*) Silicon-aluminum-killed steel. 2% nital etch; 1000 X.

Table 5.1 Some Cost Relationships Between Quality and Melting and Casting Practice (Typical Carbon Steel Products)

Product	Quality Designation	Base Price	Usual Practice	Price Addition
Semifinished	Merchant	$226.00/ton	Rimmed, capped, semikilled	No extra
	Forging	$264.00/ton	Semikilled, killed, killed fine-grain	None to $13/ton
Hot-rolled bars	Merchant	$ 11.75/cwt	Rimmed, capped, semikilled	No extra
	Special	$ 15.10/cwt	Rimmed, capped, semikilled, killed, fine-grain	None to 65¢/cwt
	Special shell A	$ 15.10/cwt	Killed fine-grain	$1.15/cwt
Plates	Regular	$ 14.10/cwt	Rimmed, semikilled, killed, fine-grain	To 95¢/cwt
	Structural	$ 14.10/cwt	Semikilled, killed, fine-grain	To $1.05/cwt
	Cold pressing	$ 14.10/cwt	Rimmed, semikilled, killed	To $1/cwt
	Forging	$ 14.10/cwt	Killed, fine-grain	85¢ to $1.15/cwt
	Marine	$ 14.10/cwt	Killed, fine-grain	$1.75/cwt

Note. Prices shown (1976) do not include many other additions, such as grade extras or size extras. Producers ordinarily will not accept an order that specifies melting and casting practice, but only the quality, grade and silicon content (or silicon content and/or grain size).

3 Usually superior surface quality.
4 Much greater uniformity of chemical composition throughout the cross section due to much more rapid cooling rates during solidification.

However, unless this continuously cast steel is reduced in cross section by a ratio of at least 10 to 1, the center may be unsound. When maximum uniformity of mechanical properties is required from surface to center, as, for example, in a high-strength heat-treated bolt, continuously cast steel should be avoided unless the reduction is at least 20 to 1. This also applies to high-strength forgings, particularly when there is minimal hot working and the flash is thick or it occurs in a highly stressed area of the part. Manufacturers of cold-drawn bars also avoid the use of hot bars from continuously cast billets unless the aforementioned reduction is assured.

Some closed die forgings, such as crankshafts, are made from "as-cast" or slightly strand-reduced continuously cast billets. Forgings from continuously cast steel may exhibit somewhat lower and erratic fatigue strength than those manufactured from conventional ingot-cast steel when treated to hardnesses above 40 Rc. Induction-hardened pins and shafts subjected to wear and bending where a slightly weak center is unimportant may be made from bars with less than a 10 to 1 reduction.

STEEL "QUALITIES"

For most heat-treated parts for which the constructional grades of steel are specified basic open-hearth or basic oxygen-process quality is adequate. Because there is a range of characteristics in these qualities, it is considered good practice to inform the supplier of the kind of part for which the steel is to be used. Most suppliers will recommend metallurgical qualities particularly suited to the intended use. Some severe applications, however, require special melting practices and surface conditioning. These special manufacturing practices are sometimes related to processing and at other times to engineering requirements. Examples are "cold heading," "bearing," or "aircraft quality" steels.

Obviously, the need for cold-heading quality is appropriate when a part is to be upset, forged, or extruded at room temperature. The seams and nonmetallic stringers present in wire or bars of ordinary qualities would result in splitting and excessive scrap.

The extra cost of bearing quality can be money well spent when the surface of a carburized or induction-hardened shaft is to be used as the inner race of a highly loaded rolling bearing. Magnetic particle inspection of the raceway surface is advisable, although rejection from such inspection is not always honored as a claim by the steel supplier. For these types of surface an upset forging operation is sometimes desirable to break up nonmetallic inclusions and change their orientation (e.g., in the bore and on the teeth of planet gears).

For parts in which a nonmetallic stringer could bring about catastrophic failure the use of magnetic particle testing and "magnaflux quality" steel is appropriate. When calling for this quality, the buyer must negotiate a mutually agreeable test specification with the producer.

Some purchasers include a vacuum degassing requirement in steel specifications, whereas others go so far as to demand carbon deoxidization. Vacuum degassing will not clean up an improperly made heat of steel. Also, especially in the Dortmund-Horder (D-H) process, how extensive should vacuum degassing be? Accordingly, it is suggested that rather than tell the producer how to make the steel the specification writer should relate what is needed in terms of cleanliness, postinspection standards, and other requirements and leave the details of steel production to the steel supplier.

EFFECTS OF RESIDUAL ALLOY CONTENT IN CARBON STEELS

It is common knowledge that the various grades of constructional steel, both carbon and alloy, have different compositions and hardenabilities (see Appendix VI). Some important characteristics of these steels are less well known and should be considered in making selections for specific parts.

The first is the residual alloy content of carbon steels, often neglected in specifications. As a result the heat-treatment response can vary greatly and a state of constant turmoil in the heat-treating operation will ensue. The residual alloy content depends mainly on four factors:

1 Percentage of heats made by the producer as alloy steel.
2 Percentage of melting-furnace charges that are purchased scrap.
3 Percentage of furnace charge as hot metal or pig iron.
4 Grades of alloy steels made.

This may appear to be a complicated set of circumstances and becomes just that when a producer can ship from one of several mills in his company. In practice, however, it usually develops that a producer falls into one of three categories:

1 One whose output is primarily carbon steel with low residuals because of extensive use of hot metal.
2 One whose output consists of a substantial percentage of alloy steel heats but who may or may not use large amounts of hot metal in his charges.
3 One who uses 100% scrap for his charge and makes a high percentage of alloy steel.

The respective residual alloy content can be predicted to be low, erratic, or high. The amounts of alloy found by experience with both low- and high-residual producers are as follows:

	Low (%)	High (%)
Nickel	0.01–0.05	0.10–0.20
Chromium	0.02–0.06	0.10–0.20
Molybdenum	0.00–0.02	0.03–0.08
Copper	0.02–0.05	0.10–0.20

Using the Grossmann multiplying factors (see Chapter 6 and Appendix V) for a comparison, the multiplying factor for D_I (ideal critical diameter) of the high residuals is 1.562. In a 0.45% carbon steel of grain size 7 with 0.75% Mn the heats would have calculated D_I's of 0.828 for low residuals and 1.235 for high, which can be significant in many heat-treating processes.

Often it is preferable to use low-residual steel because its response to heat treatment is usually more consistent. When maximum hardened depth is required (e.g., for components requiring abrasive wear resistance), high-residual steel is distinctly advantageous. Firms that heat-treat substantial tonnages of steel by different methods and/or for different reasons may find it advantageous to have two specifications for the carbon steels they use, with a minimum and maximum chemical factor for each. The chemical factor specification is described in Chapter 6.

HEAT-TREATING CHARACTERISTICS OF
SOME CARBON STEELS

The following comments and recommendations are general in nature and based largely on our own experience:

1015, 1017, 1020, 1023, 1025. Not recommended for heat treatment except when produced by a high-residual mill because of the low manganese content.

1016, 1018, 1019, 1022. Recommended for carburizing and water quenching or carbonitriding. The 0.70% minimum Mn grades are preferred.

1033, 1035, 1037 low-residual. Suitable for induction hardening with water quenching of parts that are somewhat nonuniform in section and irregular in shape; for example, the rear axle shaft of a passenger car.

1040, 1042, 1045 low-residual. Suitable for induction hardening with water quenching of parts of simple regular shape and uniform cross section. For hardened depths greater than 0.125 in. high-residual grades with soluble oil or synthetic quench are suggested. Low-residual grades are ideal for hot-caustic quenching in a roller die machine.

1046, 1049, 1050. Low-residual steel is suitable only for induction heating, with water quenching in regular cylindrical shapes. High-residual material may require soluble oil or synthetic quench.

1055 to 1075. Limited use in heat-treated parts because of danger of quench cracking in aqueous quenches; hardenability, however, is inadequate to respond properly in oil.

1080, 1085, 1090. High-residual material is preferred to austempering parts such as tools and cutting devices. Hardenability is generally inadequate for oil quenching; will usually crack in water.

1078, 1086, 1095. Suitable for shallow hardening applications to high-surface hardnesses that employ water or, preferably, brine or caustic quenching. Holes, grooves, and notches present cracking problems.

1108, 1109, 1115, 1120. Adequate hardenability only for cyanide hardening with water quenching or carbonitriding. Rockwell 15-N hardness of 85 min can be achieved.

1116, 1117, 1118, 1119. All grades in high- and low-residual type suitable for carbonitriding up to 1.25 in. diameter to a Rockwell 15-N surface hardness of 88 min. For carburizing 1118 is a popular grade, especially the high-residual variety, when purchased as silicon-killed, fine-grained. Sulfide inclusions, however, strongly detract from the transverse toughness.

1132, 1137 low-residual. Suitable for water quenching and tempering to machinable hardnesses. Also suitable for induction hardening with water quenching.

1141, 1144. Especially when made with low residuals, these grades in sections up to 2.00 in. diam. require water quenching at the low side of the composition range, but will crack with heats on the high side. With high residuals, the high-side composition only will respond properly in oil quenching. Suitable for induction hardening; however, cracking is often a serious problem especially with 1144, which is particularly difficult to heat treat.

1524. Suitable for lightly loaded carburized gears, especially when produced with high residuals. Poor machinability is a problem with expensive cutting tools such as hobs.

1536, 1541. Suitable for direct hardening; however, high manganese tends to produce microsegregation. The result is poor fracture toughness. Low-side heats (especially 1541) require a vigorous quench, such as water; high-side heats will crack if water-quenched. Difficult to heat treat.

HEAT TREATING CHARACTERISTICS OF SOME ALLOY STEELS

1300 series. Difficult steels to process because of poor machinability, wide range in hardenability, and poor fracture toughness. The tendency of manganese to float in the melt can create serious manufacturing problems. Generally a poor selection for a heat-treated part.

4000 series up to 4027. An economical direct-quench carburizing steel for gears. Either low or high residual should be specified and should not be mixed in order to narrow the hardenability band as much as possible. Tendency to microcrack is very low.

4032 to 4047. Little need for using these grades because more economical steels can be substituted.

4100 series. In spite of their wide use, this series is erratic in heat-treating response. The reason seems to be the banding of carbon-molybdenum-rich areas in the steel. In a normal heat of 4140, for example, an electron-microprobe traverse through a 1-in. diameter bar showed several areas containing up to 1.5% carbon and 1% molybdenum as opposed to the nominal composition of 0.40% C and 0.20% Mo. Some producers also have had difficulty in making these grades with an acceptable surface and internal cleanliness. These steels also have a reputation of quench cracking when the nominal carbon content exceeds 0.40%. Therefore care should be taken in recommending them for a complex heat-treated part. Exceptions are parts for which mild oil quenches can be used (50% martensite acceptable) or parts to be nitrided.

4320 and 4320 modified (with 0.70 to 0.90 Mn). These grades are excellent steels for heavily loaded carburized gears with contact stress over 250,000 psi. Can be made very clean and are popular rolling bearing grades. These steels can be carburized with great freedom from microcracking and have good short- and long-cycle fatigue properties. A major disadvantage is cost. Comparative current grade extras are as follows*:

4023	3.15 ¢/lb
4118	3.10
4320	9.60
4620	9.20
4820	15.55
8620	5.50
8822	6.45
9310	19.40

4340. Outstanding as an engineering material, especially in 2-to-5-in. sections. Can be melted clean, heat-treats with a minimum of difficulty, but, as in the case of 4320, is very expensive. The beneficial effect of the nickel content on short-cycle fatigue properties can be an important advantage.

4400 series. These carburizing steels contain substantial amounts of molybdenum and are prone to microcracking when quenched directly from the carburizing furnace.

*Source. U.S. Steel Alloy Steel Bar Price Book — 1976.

4600 series. These carburizing grades machine and heat-treat uniformly with good case hardenability; however, the core hardenability, with the possible exception of 4626, is low. Because 4626 has only 0.70 to 1% Ni, it exhibits much lower case hardenability than the other grades in the series. (In general, 4320 provides better value than 4626, even though it is priced slightly higher.)

4815, 4817, 4820. These steels have exceptionally high case hardenability and excellent resistance to short- and long-cycle fatigue. Excellent for heavily loaded carburized gears. The high case hardenability requires that case carbon content must be held between 0.65 and 0.80% to prevent excessive austenite formation when direct quenching is required.

5015 and 5046. Little need for these grades in heat-treated condition. Carbon steel with 0.80 to 1.10% Mn obtained from high-residual producers can be used at substantial cost saving.

5115 through 5160, 6118, 6150, 8115. Little need for these as well. Suggested substitute is C-Mn-B steels with appropriate source selection for residuals.

8615 through 8627. Excellent carburizing steels for fine- and medium-pitch gears. Case carbon should be controlled to 0.95% max to prevent microcracking with direct quenching when used in six-pitch or finer gears.

8630. An excellent steel for abrasive wear resistance when water quenched and tempered at 400 F.

8637. Suitable for water quenching in sections up to 2-in. round and tempering to a machinable hardness.

8640. Suitable for oil quenching in thin sections.

8642 to 8645. Suitable for oil-quenching sections to 2-in. diameter or equivalent in flats.

8655 to 8660. Excellent steel for hot-wound, quenched, and tempered coil springs and torsion bars to 2-in. diameter and leaf springs to 1.50 in. thick.

8700 series. The usefulness of these steels is seriously questioned. The $3/ton grade premium over the 8600 series can seldom be justified from an engineering standpoint.

8822. An excellent steel for heavy-duty carburized gearing but will usually microcrack in direct quenching and requires vigorous oil quenches when reheated for hardening to prevent bainite formation.

9254, 9255, 9260. These steels are rather difficult to produce clean and roll with a good surface because of the high silicon content. When used for valve coil springs, these grades have a reputation for developing transverse cracks. A chromium-vanadium valve spring wire is often preferred. For heavy springs the 50B60 or 51B60 grades are often a better value.

50B44, 50B46, 50B50. Can be replaced with C-Mn-B steels at a lower price.

50B60 and 51B60. Outstanding values in steels for heavy springs.

81B45. This steel can usually be replaced with a high-residual C-Mn-B grade.

86B45. An excellent steel for heavy shafts and forgings at a substantial saving over 4340.

94B17. Although not widely used today, this grade, or a modification of it, could be a popular carburizing steel in the future.

94B30. This steel can often be replaced with a high-residual C-Mn-B grade at a lower cost.

As noted, many of the constructional grades of alloy steel are, in our opinion, of limited value. In later chapters this is further explained and justified.

EFFECT OF PHOSPHORUS AND SULFUR

Phosphorus content in carbon steel is normally specified as 0.040% max and in alloy steels, as 0.035% max. Steels with phosphorus content at these levels will tend to be brittle when quenched and tempered to hardnesses above Rockwell C40. Constructional steels made today usually have less than 0.025% P; a preferable maximum is 0.015%. When maximum toughness is required, steels can be made consistently with sulfur as low as .005% max. There is an extra cost for this low sulfur content, but in certain applications, it may be justifiable.

EFFECT OF Ms TEMPERATURE

An important characteristic of constructional steels which is often overlooked is the temperature at the start of martensite transformation (Ms) and the temperature range through which transformation occurs. Whenever possible, a steel should be selected with the highest possible Ms to minimize the tendency to quench-crack. The Ms can be calculated from the formula developed by R. A. Grange and H. M. Stuart [1]:

$$Ms\,(^{\circ}F) = 1000 - (650 \times \%\,C) - (70 \times \%\,Mn) - (35 \times \%\,Ni)$$
$$- (70 \times \%\,Cr) - (50 \times \%\,Mo)$$

Note. Boron has an insignificant effect on the Ms temperature.

The martensite transformation temperature range is particularly important when decarburization is involved. Leeper [2] used an example of 8660 decarburized to 0.30% C on the surface. Cracking occurred because the 0.30% C layer had completely transformed to 90% martensite before the transformation in the interior of the steel, containing 0.60% C, had started. The expansion of the 0.60% C martensite beneath the transformed 0.30% C surface initiated cracks. This same phenomena, in reverse, contributes to the compressive residual stress found in carburized cases in which the expansion due to transformation in the case is restrained by the previously transformed core.

REFERENCES

1 R. A. Grange and H. M. Stuart, *Metals Tech.,* American Institute of Mining, Metallurgical, and Petroleum Engineers, **167,** 467, June 1946.
2 W. A. Leeper, letter to *Metal Progr.,* **100,** No. 5, 15 (1971).

CHAPTER SIX

HARDENABILITY AND
TEMPERING PARAMETER

HARDENABILITY IS COMMONLY
defined as the property of a ferrous alloy that determines the depth and distribution of hardness induced by quenching. This definition is not always adequate because, with the possible exception of tensile strength, nearly all of the important engineering characteristics of steel are more dependent on microstructure than hardness. A more precise measure of hardenability would be "that property in a ferrous alloy which determines the ease with which a completely martensitic microstructure can be obtained." No matter which definition applies, a thorough understanding of this property and its measurement is essential for proper steel selection. This chaper discusses the history, procedures, limitations, and application of the principal hardenability tests in wide use today.

THE SAC TEST

Late in the 1930s the demands that carbon steel have better uniformity from heat to heat were such that in 1938 Burns, Moore, and Archer announced their breakthrough work on quantitative hardenability [1]. Their test, known today as the SAC (surface-area-center) test, consists of fixture quenching 1-in. rounds in water and measuring the area beneath the curve on a graph of the diametral hardness traverse. The hardenability index A (area beneath the curve) is measured in Rockwell inches, now usually determined in accordance with SAE Standard J406a [2] (see Figs. 6.1, 6.2, and 6.3). The S value is the average sur-

Fig. 6.1 Rockwell machine index fixture for SAC hardenability test traverse (courtesy Republic Steel Corp.).

face hardness of the specimen and C is the average center hardness. The hardenability index in the SAC test actually consists of a set of three numbers that represent surface hardness, Rockwell-inch area, and center hardness; for example, although the numbers 52–43–37 might be registered at the bottom of the test report (Fig. 6.3), the value for A (i.e., 43) is the most significant.

At the outset, it was possible to purchase steel with controlled hardenability to a range of 10 Rockwell inches but at extra cost. The installation of spectrographs at the producer's melting floors resulted in a narrowing of the ladle composition range and, because manganese additions could be adjusted on the basis of residual alloy content of a heat, hardenability limits as narrow as eight Rockwell inches became common. Even narrower limits are now subject to negotiation with some suppliers. Substantial tonnages of carbon steel are purchased typically to the following composition and hardenability ranges excerpted from a purchase specification:

Composition and Hardenability

Carbon	(see below)
Manganese	0.55–0.90%
Silicon	0.15–0.30%
Sulfur	0.050% max
Phosphorus	0.040% max

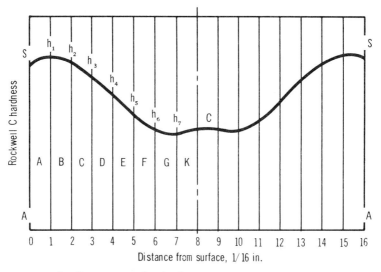

Let S = average surface hardness
     h_1, h_2, h_3, and so forth = average hardness at depths indicated
     C = average center hardness

Then: area of A $= \dfrac{S + h_1}{2} \times \dfrac{1}{16}$ area of B $= \dfrac{h_1 + h_2}{2} \times \dfrac{1}{16}$

total area $= 2 (A + B + C + D + E + F + G + K)$

$$= \frac{1}{8}\left(\frac{S}{2} + h_1 + h_2 + h_3 + h_4 + h_5 + h_6 + h_7 + \frac{C}{2}\right)$$

Fig. 6.2 Calculating A in the SAC hardenability test (from *SAE Handbook*, 1972, p. 20).

Grade	Hardenability[a] (Rockwell inches)	Carbon[b] (%)
1	33–41	0.31–0.45
2	35–43	0.33–0.47
3	37–45	0.35–0.49
4	42–50	0.37–0.51

[a]From tests at bottom of first ingot and top of last ingot representing product supplied.
[b]All billet or bloom checks shall be within this range.

Another typical specification calls for 0.43 to 0.50% ladle C, 0.60% min Mn, and hardenability of 43 to 51 Rockwell inches.

Part of the work of Burns, Moore, and Archer was the development of a method for determining the chemical factor of a heat of steel. As used today,

SAC TEST REPORT

DATE_____ TEST NO._____

LABORATORY_____ SOURCE_____

GRADE	HEAT NO.	GRAIN	ANALYSIS									
			C	Mn	S	P	Si	Ni	Cr	Mo		

NORMALIZING TEMP (F)_____ QUENCHING TEMP (F)_____

REMARKS _____

_____

_____

_____

	INDIVIDUAL READINGS				AVERAGE ORDINATES			INDIVIDUAL READINGS				AVERAGE ORDINATES	
S					$S/2$		S					$S/2$	
1					1		1					1	
2					2		2					2	
3					3		3					3	
4					4		4					4	
5					5		5					5	
6					6		6					6	
7					7		7					7	
C					$C/2$		C					$C/2$	
TOTAL ORDINATES							TOTAL ORDINATES						
AREA = TOTAL ORDINATES DIVIDED BY 8							AREA = TOTAL ORDINATES DIVIDED BY 8						
HARDENABILITY RATING					S	A	C	HARDENABILITY RATING				S	A	C

Fig. 6.3 Typical SAC hardenability test report form (from *SAE Handbook,* 1972, p. 20).

85

this formula is as follows:

$$\text{Chemical factor} = 1000\ (\%\ C) + 500\ (\%\ Mn) + 400\ (\%\ Cr) + 100\ (\%\ Ni)$$
$$+ 25\ (\%\ Cu) + 1000\ (\%\ Mo) \qquad (6.1)$$

A good relationship was found between chemical factor and Rockwell-inch hardenability, as shown in Fig. 6.4.

This test method has helped to control cracking in parts made of carbon steel, either induction-heated or furance-heated and water-quenched. One such part was a crawler-tractor track-link forging with section thicknesses of 0.25 to 0.875 in. that was furnace-heated and quenched in 5% caustic solution. Before the SAC test was adopted it was believed that carbon content was the controlling factor in quench cracking, but data for hundreds of heats showed no correlation. Manganese was also suspected but, likewise, no clear relationship could be established. Finally, a plot of percentage of parts cracked versus Rockwell-inch hardenability showed hardenability to be the controlling factor (see Fig. 6.5).

The SAC test has three serious shortcomings:

1 It is suitable only for direct-hardenable carbon steel with 0.30 to 0.50% C.

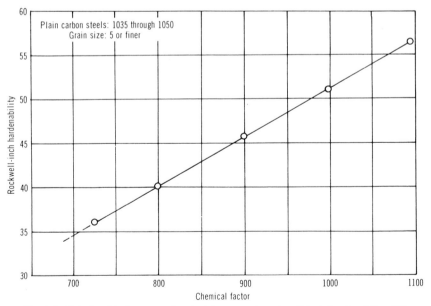

Fig. 6.4 Relationship between chemical factor and Rockwell-inch hardenability [1].

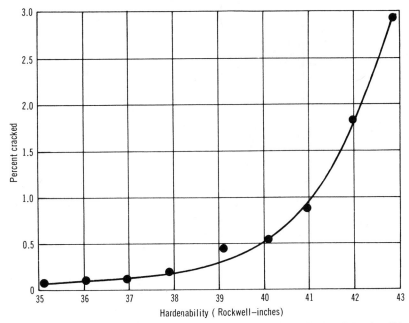

Fig. 6.5 Relationship between percentage of parts quench cracked and Rockwell-inch hardenability.

2 The Rockwell-inch value provides no means of predicting accurately the hardness that would be obtained in a part with various quench-cooling rates.

3 In trying to meet specifications for a certain hardened depth, the SAC test may give misleading data because the specified Rockwell-inch value can be obtained with curves of different shape (see Fig. 6.6).

The last-mentioned problem is serious because on an induction-hardened gear, for example, the depth of 50 Rc would differ substantially for the two steels shown in Fig. 6.6 but would not be predictable from the SAC hardenability rating, which is nearly the same for these steels. This led to specifications in which a minimum and maximum Rockwell C hardness at some depth from the surface on the SAC traverse is required. A typical specification may state

45 Rc min @ 3/16 in. depth
40 Rc max @ 7/16 in. depth

In order to meet this specification, factors for composition of a low-residual steel can be developed (see Table 6.1). Hardness depth equals the sum of the fac-

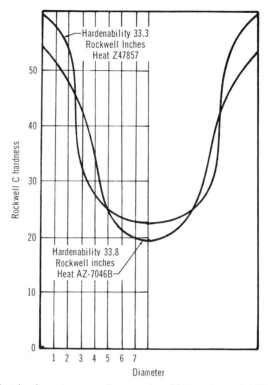

Fig. 6.6 SAC Test hardness traverses. Part requires 40 R_C min. at 0.125-in. depth, a location having a quench-cooling rate of 3/16 in. on SAC traverse. Heat AZ-7046B will meet specification; heat Z47857 will not.

tors given for percentages of carbon, manganese, sulfur, and residuals (residuals are the sum of nickel and chromium content minus molybdenum). This method has proved to be successful for steels with the following ladle composition limits:

Carbon	0.37 –0.51
Manganese	0.56 –1.32
Phosphorus	0.004–0.042
Sulfur	0.014–0.044
Silicon	0.15 –0.30
Copper	0.00 –0.11
Nickel	0.00 –0.09
Chromium	0.00 –0.08
Molybdenum	0.00 –0.03

Table 6.1 Factors for Estimating Rockwell C at 3/16-in. Depth in SAC Test

Contents (%)	0	1	2	3	4	5	6	7	8	9
					Carbon					
0.3						0.73	1.60	2.47	3.35	4.22
0.4	5.10	5.98	6.86	7.73	8.61	9.49	10.37	11.24	12.12	13.00
0.5	13.88	14.75								
					Manganese					
0.5							4.83	5.18	5.54	5.89
0.6	6.25	6.60	6.96	7.31	7.66	8.02	8.37	8.73	9.08	9.43
0.7	9.79	10.14	10.50	10.85	11.21	11.56	11.91	12.27	12.62	12.98
0.8	13.33	13.69	14.04	14.39	14.74	15.10	15.46	15.81	16.16	16.52
0.9	16.87	17.23	17.58	17.93	18.29	18.64	19.00	19.35	19.71	20.06
1.0	20.41	20.77	21.12	21.48	21.83	22.19	22.54	22.89	23.25	23.60
1.1	23.96	24.31	24.67	25.02	25.37	25.73	26.08	26.44	26.79	27.14
1.2	27.50	27.85	28.21	28.56	28.92	29.27	29.62	29.98	30.33	30.69
1.3	31.04	31.40	31.75	32.10	32.45	32.80	33.16	33.51		
					Sulfur					
0.01						1.68	1.81	1.93	2.05	2.17	2.29
0.02	2.41	2.53	2.65	2.77	2.89	3.02	3.14	3.26	3.37	3.50
0.03	3.62	3.74	3.86	3.98	4.10	4.22	4.34	4.46	4.58	4.70
0.04	4.82	4.95	5.07	5.19	5.31					
				Residual: (Nickel + Chromium − Molybdenum)						
0.0		17.57	18.09	18.61	19.13	19.65	20.17	20.69	21.21	21.73
0.1	22.25	22.77	23.29	23.81	24.33	24.85	25.37	25.89	26.41	

By specifying the hardness values at different depths on the SAC traverse the ability of a steel to harden at given cooling rates is established. *Note.* The center of the SAC specimen cools at the same rate as a point 3/16 in. from the quenched end of a Jominy specimen.

The SAC test is useful for controlling the hardenability of plain carbon steels with 0.35 to 0.50% C and 1.20% Mn max. Susceptibility to cracking on one hand and the need to meet a substantial depth requirement on the other represent a problem that can be solved at minimum cost with the SAC test. The end-quench test, discussed next, has not been very useful in the control of these grades of steel.

THE END-QUENCH TEST

About the same time the SAC test was announced W. E. Jominy and A. L. Boegehold developed a test for hardenability, originally intended for carburizing steels [3]. This is called the Jominy or end-quench test and is widely used today; the details are also covered in SAE Standard J406a [2].

In this test the cooling rates at various positions are determined (see Fig. 6.7). The hardening response of a steel at each specific cooling rate in the end-quench test is known as the end-quench curve, and steel can be purchased to minimum and/or maximum hardnesses at specified distances from the quenched end or to hardenability bands, as shown in Appendix VI and published in the SAE Handbook [4, 5] and the AISI Steel Products Manual [6]. For specification purposes the customer has the option of calling for several variations in hardenability requirements (see Fig. 6.8). It should be noted that when the hardenability combination requirement is specified, as shown by A, B, C, D, or E, no other part of the band may be specified without extra charge except the maximum and minimum hardness at 1/16 in. When the full H band is specified, the hardenability can be described without cost penalty by hardness values at 1, 2, 4, 8, 12, 16, 20, 24, 28, and 32 sixteenths from the quenched end. In this

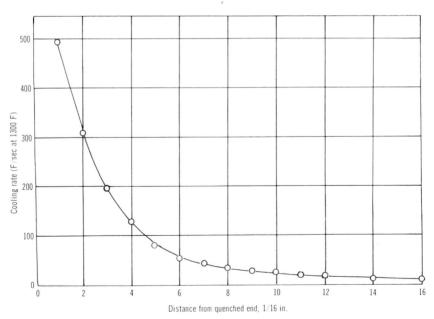

Fig. 6.7 Cooling rates at various positions on Jominy end-quench test specimen.

HARDENABILITY BAND 8630 H

C	Mn	Si	Ni	Cr	Mo	
.27 / .33	.60 / .95	.20 / .35	.35 / .75	.35 / .65	.15 / .25	

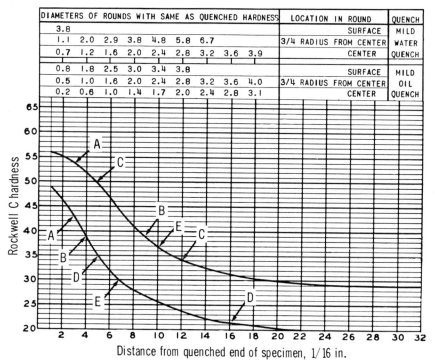

DIAMETERS OF ROUNDS WITH SAME AS QUENCHED HARDNESS									LOCATION IN ROUND	QUENCH
3.8									SURFACE	MILD
1.1	2.0	2.9	3.8	4.8	5.8	6.7			3/4 RADIUS FROM CENTER	WATER
0.7	1.2	1.6	2.0	2.4	2.8	3.2	3.6	3.9	CENTER	QUENCH
0.8	1.8	2.5	3.0	3.4	3.8				SURFACE	MILD
0.5	1.0	1.6	2.0	2.4	2.8	3.2	3.6	4.0	3/4 RADIUS FROM CENTER	OIL
0.2	0.6	1.0	1.4	1.7	2.0	2.4	2.8	3.1	CENTER	QUENCH

Distance from quenched end of specimen, 1/16 in.

Fig. 6.8 Specification of hardenability requirements using H bands (from *AISI Steel Products Manual for Alloy, Carbon, and High Strength, Low Alloy Steels Semifinished for Forging; Hot Rolled Bars; Cold Finished Bars,* Copyright AISI, 1977).

instance it is customary to accept a tolerance of two points Rockwell C over a small portion of the upper or lower curve.

Alloy steels can be purchased to restricted hardenability (usually 50% of the published spread) for an extra 10% of the base price plus the BOH or BOP grade extra and rounded to the nearest $.05/cwt. The purchase of restricted hardenability can *sometimes* be a satisfactory cost-reduction measure (see Fig. 6.9) where restricted hardenability 8720H met the minimum requirement at substantially lower cost than the closest standard steel (4820H).

The end-quench test has made it possible to predict, with a fair degree of accuracy, the mechanical properties in a heat treated steel part. This is pos-

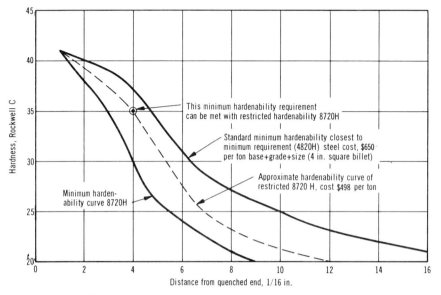

Fig. 6.9 Cost reduction by purchasing a restricted-hardenability steel.

sible through the use of equivalent cooling rates in terms of Jominy distance (J), as shown in Figs. 6.10 through 6.14. It should be kept in mind that these cooling rates are for scale-free steel, furnace-heated in a controlled atmosphere. For hot-rolled steel products and uncleaned forgings, or for steel heated in an oxidizing atmosphere, the cooling rate will be not only lower than shown but very erratic, depending on the tightness of the scale.

Figure 6.15 shows how to determine quench-cooling rates at important locations in a part:

1 Obtain at least two test parts made from the same heat of steel.
2 Machine the parts to the condition in which they will be hardened. Process parts by heat-treating operations for times estimated to be approximately those for the production part. Quench part No. 1 in a manner as close to production as possible (no temper).
3 Cut, grind, and polish hardened sections from part No. 1 so that hardness readings may be taken (see Fig. 6.15).
4 Machine end-quench hardenability test specimens from part No. 2. Test location on hardenability specimens should correspond to depth D below surface for maximum accuracy. Harden end-quench specimens from the same temperature as part No. 1. Sample test results are as follows:

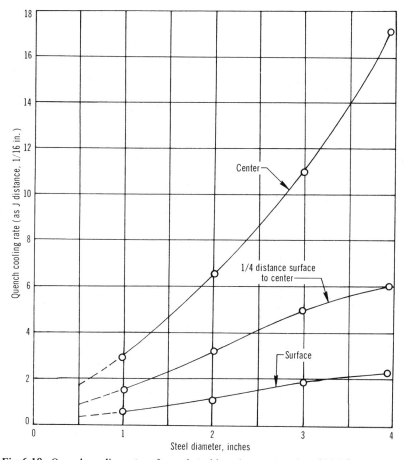

Fig. 6.10   Quench-cooling rates of round steel bars (water quench at 200 ft/min flow).

Distance from quenched end,
1/16 in.   1   2   3   4   5   6   8
Hardness; Rc   56   55   55   54   52   48   43

5 By comparing hardness results obtained at reference location on Step 3
   (Rc 42.7) to end-quench results (Step 4) it can be seen that this hardness
   occurs at 8/16 in. on the end-quench curve. The quench-cooling rate at
   the reference point is approximately equal to 8/16 in. (J 8) in the end-
   quench-test.
6 Confirm the cooling rate subsequently on a number of different heats or

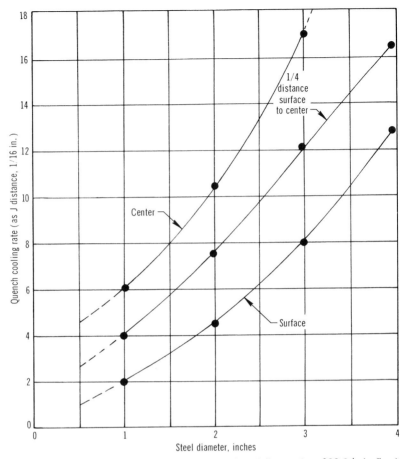

Fig. 6.11   Quench-cooling rates of round steel bars (oil quench at 200 ft/min flow).

production parts and adjust material or heat treatment or both to obtain the engineering requirements consistently.

With this method the quench-cooling rate for parts of irregular shapes can be determined quite accurately if due care is exercised. It must be remembered that the total tolerance on a series of operations is the sum of the individual tolerances for each step; for example, if the hardness tester is erratic by one or two points or if the sample was burned slightly in cutting (use nital etch to check), the estimate of cooling rate could vary significantly. If the quench cooling rate in the core of a carburized gear of 8620H, for example, varies from 4/16 to 6/16 J, the estimated hardness would be expected to range from 21 to 41 Rc.

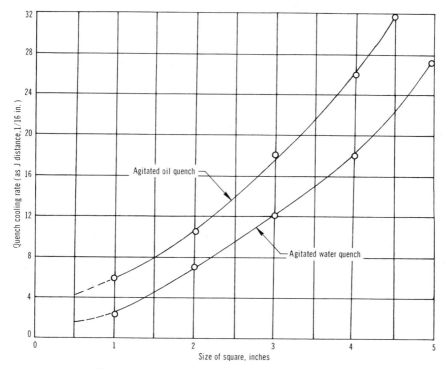

Fig. 6.12 Quench-cooling rates at center of square steel bars.

The surface of the trepanned Jominy specimen on which the hardnesses are taken should represent, if at all possible, the same position that will be used for the "as-quenched" hardness readings. This is required for maximum accuracy, for, due to segregation, the hardenability throughout a section is not completely uniform. This can be confirmed by carefully cutting a quenched Jominy specimen transversely at a selected position and comparing the surface readings with those across a diameter.

In some applications it may be desirable to make the test parts from a steel with a definitive end-quench curve shape. The 5100 series of steels has been used for this purpose because its sloping Jominy curve makes cooling-rate determinations more accurate. Also, when a drastic change in material is contemplated for a forging in which a hammer setup is costly and the steel in the forgings on hand is unsuitable for accurate cooling-rate determinations, a steel casting of more suitable grade can be made by using the forging for the pattern.

Quite often the material on which it is necessary to determine the hardenability is too thin to make a standard end-quench specimen. In this situation a

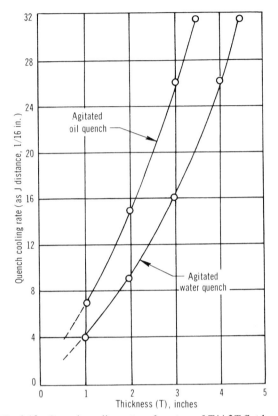

Fig. 6.13 Quench cooling rates of centers of $T \times 2T$ flat bars.

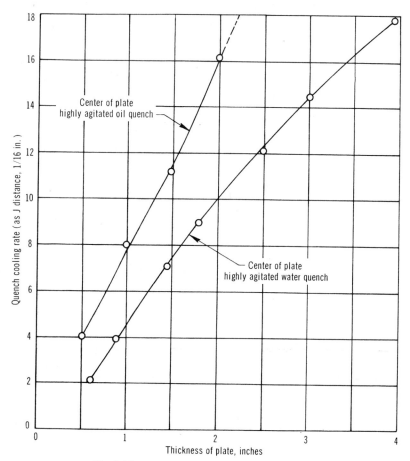

Fig. 6.14   Quench-cooling rates of steel plates.

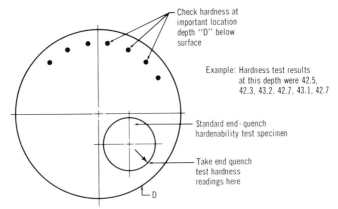

Fig. 6.15 Determination of quench-cooling rate in a part (from *Selecting Steels and Designing Parts for Heat Treatment,* Copyright ASM, 1969).

fabricated specimen can be prepared (see Fig. 6.16). After welding the composite should be normalized or annealed for machining along with the parts to be hardened. The Jominy quench temperature should always be the same as the quench temperature for the part. In determining the quench-cooling rates, tests

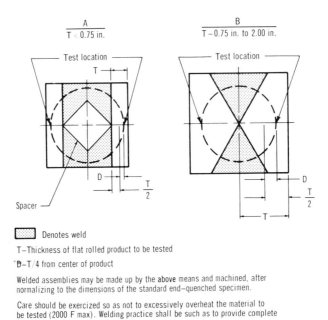

Fig. 6.16 Welded construction of basic 1.00-in. round end-quench hardenability specimen from flat-rolled products.

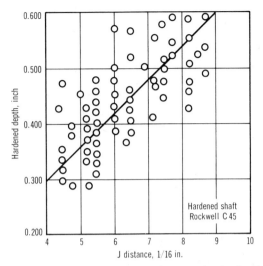

Fig. 6.17 Typical scatter when attempting to correlate case depth of an induction-hardened shaft to producer's end-quench data.

should be run, if at all possible, with at least two and preferably three different heats of steel and the results, averaged.

Using steel producers' Jominy data usually lead to inaccuracies because of segregation and the probability that the austenitizing temperature will not be the same as that employed in production. The chart in Fig. 6.17 illustrates this phenomenon for identical induction-hardened shafts heat-treated by the same procedure. Each point for this curve represents a shaft from a different heat of steel. The hardness found at the specified depth was located on the end-quench curve of that heat and its J distance, plotted. The scatter in this approach made it impossible to apply the data to any useful purpose.

CALCULATED HARDENABILITY

Shortly after the announcement of the end-quench test it became apparent that a method was needed to caculate the Jominy hardenability from composition and grain size. Grossmann [9], Field [10], and Boyd and Field [11] did just that. The method involved the use of multiplying factors known as Grossmann factors. Tables of these factors, along with methods of calculation, appear in Appendix VII. This appendix also includes methods developed by Climax Molybdenum Co. [12, 13], Crafts and Lamont [14], and Jatczak and Girardi [15, 16].

A procedure recently developed by Grange of the Research Department of the U.S. Steel Corp. is also described in Appendix VII. The hardenability is expressed as the diameter at which a steel will harden to 90% martensite at the center. This procedure is particularly useful in selecting steels for parts of simple shapes such as fasteners, but it can also be adapted for complex shapes that require at least 90% martensite through critical sections.

Using the tables in the appendix, hardenability can be calculated and the boron factor of a boron-treated steel, determined. As noted in the appendix, the *estimated* boron factor is given by

$$B_f = 1 + 1.60(1.01 - \%C) \tag{6.2}$$

The *actual* boron factor is determined from the equation

$$B_f = \frac{\text{actual } D_I \text{ of a heat}}{\text{calculated } D_I \text{ of the heat without boron}} \tag{6.3}$$

In order to make this calculation, the composition, grain size, and end-quench hardness data of a heat must be known. (Ideal diameter, D_I, is defined as the diameter of a steel that will harden to a microstructure of 50% martensite at the center with a quench of infinite severity. The D_I is also a useful means of expressing hardenability in a single number.) The hardness at 50% martensite can be obtained from Table VII.3 (from Hodge and Orehoski [17]). The distance at which 50% martensite occurs determines the actual D_I, shown in Table VII.4 (J distance in sixteenths of an inch from quenched end); for example, assume the following composition and grain size:

Carbon	0.40%
Manganese	0.85%
Silicon	0.25%
Phosphorus	0.030%
Chromium	0.10%
Nickel	0.15%
Molybdenum	0.03%
Boron	0.0015%
Copper	0.15%
Vanadium	0.02%
Zirconium	0.05%
Grain size	ASTM 8

The Rc hardnesses at different J distances on the end-quench bar were

R_c	J 1	2	3	4	5	6	7	8	10	12
	56	56	55	53	50	48	43	39	36	33

From Table VII.3 it can be seen that the hardness at 50% martensite in a 0.40% carbon steel is 39.0 Rc. For the example heat it will be noted from the end-quench data that this hardness occurs at 8/16 in. From Table VII.4 we find that a heat with 50% martensite at a J distance of 8 has an actual DI of 2.97 in.

Using the published Grossmann and Crafts-Lamont values, we calculated hardenability of the example heat *without boron* and with a grain size of ASTM 8 as follows:

Carbon	0.40%	0.1976
Manganese	0.85%	3.833
Silicon	0.25%	1.175
Sulfur	0.035%	0.973
Phosphorus	0.030%	1.078
Chromium	0.10%	1.216
Nickel	0.15%	1.055
Molybdenum	0.03%	1.090
Copper	0.15%	1.000
Vanadium	0.02%	1.030
Zirconium	0.05%	1.120

Calculated D_I (without boron) = 0.1976 X 3.833 X 1.175 X 0.973 X 1.078 X 1.216 X 1.055 X 1.090 X 1.00 X 1.030 X 1.120 = 1.50. The actual boron factor (B_f), then, is 2.97 ÷ 1.50 = 1.980. In other words, the presence of approximately 0.03 lb of boron in a ton of this steel nearly doubles hardenability. The actual boron factor compares well with the estimated boron factor of 1.976 from (6.2); therefore this is properly produced boron steel that can be expected to exhibit good fracture toughness and notch tensile strength as discussed in Chapter 15.

DEFICIENCIES OF THE JOMINY END-QUENCH TEST

The end-quench test has three serious weaknesses:

1 The cooling rates up to 5/16 in. from the quenched end decrease far too rapidly and rates 6/16 in. and greater, too slowly (see Fig. 6.7.). This makes the test difficult to use on somewhat shallow hardening steels and

to correlate with heat-treating results on deep hardening steels. Use of different types of specimen for the lower hardenability steels has been suggested, but this idea has not yet gained significant acceptance; even the SAE lists a 1045H steel based on the standard end-quench specimen. Modified quench procedures to increase cooling rates, shown in Fig. 6.18, have seen only limited usage. Another method employs a simple round, quenched in the same manner as the SAC specimen, and utilizes the surface-to-center traverse in 2-mm steps as the hardenability curve.

2 As a result of the unusual cooling rates in the test, it has been difficult to devise an accurate means of calculating hardenability from composition. Accuracy of the results obtained with the calculation methods proposed by Grossman et al. should not be expected to be better than ±10% of the actual hardenability. Several years ago G. Walters of International Harvester Co., and Kern proposed the regression analysis of hardenability data by computer for each 1/16-in. position on the end-quench specimen. Because of the few accurate hardenability test results of various compositions, the usefulness of this method so far is limited to narrow composition ranges; for example, Kern's work, based on 338 sets of data, is useful for steels within the following ranges of composition:

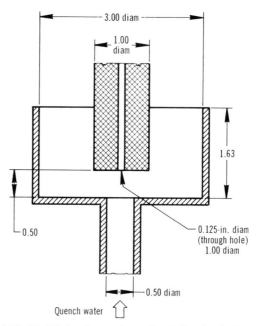

Fig. 6.18 Modified end-quench test for shallow hardening steels.

Carbon	0.28	-0.46%
Manganese	0.80	-1.40%
Silicon	0.13	-0.39%
Nickel	0.00	-0.28%
Chromium	0.05	-0.25%
Molybdenum	0.01	-0.06%
Copper	0.08	-0.22%
Boron	0.0001	-0.0019%

Grain size was not included, but was within the range of ASTM 8–12.

Most of the heats were of 35 to 175 tons, made by single slag practice in electric arc furnaces. The formulas for hardness as a function of heat composition for J distances of 1 through 20 are shown in Table 6.2.

The inability to calculate hardenability accurately means that a steel producer will know the hardenability of a heat only after it is cast. Accordingly, the H bands must, for economic reasons, remain excessively

Table 6.2 Equations for Hardness as a Function of Composition Based on Regression Analysis of 338 Heats

Jominy Distance (1/16ths)	Hardness (Rc)
1	$204.C + 4.3Si + 8.32Cu - 241.3C^2 + 11.03$
2	$207.9C + 7.06Cu - 246.3C^2 + 400MnB + 9.94$
3	$226.3C + 2.28Mn + 6.15Cu - 281.7C^2 + 7.43 \times 10^3 C^2 B + 4.176$
4	$7.02Ni - 13.07Cr + 23.9 \times 10^3 CB - 9.01 \times 10^6 CB^2 + 47.76$
5	$17.88Ni - 11.76Cr + 33.8 \times 10^3 CB - 19.0 \times 10^6 CB^2 + 5.29$ $\times 10^3 MnB + 39.8$
6	$41.73Ni - 80.32MnS + 23.5 \times 10^3 CB - 23.1 \times 10^6 CB^2 + 10.27$ $\times 10^3 MnB + 32.9$
7	$8.46Mn - 115.6S + 64.4Ni + 24.7Cr - 17.4 \times 10^6 CB^2 + 12.47$ $\times 10^3 Mn^2 B + 18.1$
8	$14.34Mn - 80.34S + 68.77Ni + 36.84Cr - 16.13 \times 10^6 CB^2 + 9.89$ $\times 10^3 Mn^2 B + 7.7$
9	$27.15Mn + 136.3P + 69.07Ni + 33.6Cr + 1.715Mn^2 B - 9.329$
12	$14.01Mn + 87.59P + 31.33Ni + 21.17Cr + 70.76Mo + 5.49$
16	$22.93C + 9.173Mn + 50.54P + 16.36Ni + 13.29Cr + 57.44Mo$ $+ 1.696$
20	$29.11C + 10.41Mn + 10.2Ni + 12.71Cr + 50.43No - 2.93$

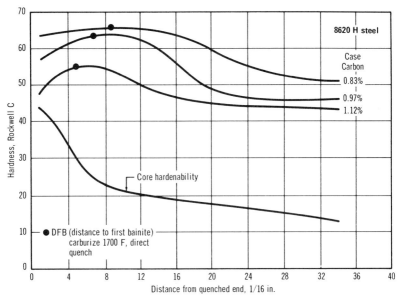

Fig. 6.19 Typical microstructural capability of 8620H steel (composition 0.17 C, 0.70 Mn, 0.19 Si, 0.53 Cr, 0.41 Ni, 0.17 Mo, 0.18 Cu).

wide; this, however, reduces their usefulness to the engineer and heat treater.

3 Most of the important engineering characteristics of heat-treated steel are dependent on microstructure rather than hardness (see Table 1.1). The percentage of martensite in the "as-quenched" microstructure can be estimated from the hardness if carbon content is known but in some cases with only disastrously low accuracy. This is illustrated in Fig. 6.19. Hardnesses higher than 60 Rc with 0.83 and 0.97% c are usually assumed to be 100% martensitic; yet in the example shown for 8620H bainite occurs at 0.97% C, as seen by the DFB (Jominy distance to first bainite) at a cooling rate of 6.5 sixteenths; with 0.83% C bainite is present at 9.5 sixteenths. This undesirable transformation product exists even though the hardness is approximately 65 Rc.

IMPORTANCE OF MICROSTRUCTURAL CAPABILITY

In our opinion the microstructural capability of the constructional steels is the ultimate measure of hardenability. As specific engineering demands become higher and higher, the need to know this capability assumes greater importance. Applications in which hardenability, in terms of microstructure, is of utmost importance are in parts like heavily loaded carburized gears and rolling bearings. Traditionally, selection of steels to obtain optimum microstructure has been largely by trial and error. More recently, rough screening, using the Jominy test on a carburized specimen, has provided helpful data for the selection process.

In 1972 [18] Kern described a method based on a factor known as DFB (J distance to first bainite, sometimes referred to as "quenching pearlite"). The purpose of this test was twofold: first, to determine the lowest quench-cooling rate a steel could tolerate without forming any upper bainite, and, second, to determing the specific effect on this capability of each alloying element. It was found that this distance varied not only with composition but with heat-treating practice. Different case carbon contents would also obviously have a marked effect (see Fig. 6.19). Limited testing resulted in the data shown in Table 6.3 for steels carburized to 1% C. A regression analysis of this data gave formulas for DFB for two methods of heat treatment:

1   Carburize at 1700 F, furnace-cool to 1500 F, direct quench:

$$\text{DFB (in.)} = 4.038 \, (\% \, Si) - 0.448 \, (\% \, Mn) + 0.319 \, (\%Ni)$$
$$-0.813 \, (\% \, Cr) + 3.942 \, (\% \, Mo) - 0.107 \, (\text{if B treated})$$
$$- 12.804 \, (\% \, S) - 56.\,414 \, (\% \, P). \qquad (6.4)$$

2   Carburize at 1700 F, furnace-cool to room temperature, reheat to

1500 F, quench:
$$\text{DFB (in.)} = 0.215 \, (\% \, Mn) - 0.494 \, (\% \, Si) + 0.586 \, (\% \, Ni)$$
$$+ 0.583 \, (\% \, Cr) - 2.094 \, (\% \, Mo) + 0.54 \, (\text{if B treated})$$
$$- 1.343 \, (\% \, S) + 2.789 \, (\% \, P) \qquad (6.5)$$

Although these results were obtained on the steels shown in Table 6.3, the small amount of data and its analysis may only be an indication of tendencies. Hodge and Orehoski [17] reported J distances to 99.9% martensite. With additional data it should be possible to predict the effect of each alloying element on microstructure to permit actual design of carburizing steels for their microstructural capabilities in terms of J distance to first bainite. Also, in hypoeutectoid steels it may be possible to establish the J distance to first ferrite, which is important in some applications.

In selecting steels for heat-treated parts, it is often necessary to know the capability of developing the desired final microstructure, at least as estimated from hardness; for example, in choosing a steel for a high-strength fastener, it is desirable to have a microstructure of 99.9% martensite completely to the

Table 6.3 Typical Bainite Sensitivity of Carburizing Steels (Distance to First Bainite in Inches From Water-Quenched End of Jominy Specimen Carburized to 1% C)

Steel Grade[a]	Carburize at 1700 F, Cool to 1500 F, Direct Quench	Carburize at 1700 F, Cool to Room Temperature Reheat to 1500 F, Quench
10B16 mod	. . .	0.122
1018	0.075	0.055, 0.080
10B22	. . .	0.105
10B24	0.122	0.116
1117	0.122	0.116
1213	0.122	0.118
3310	2.000+	2.000+
4118	. . .	0.085
4120[a]	. . .	0.114
41B16 mod		0.186
4320	0.960[b]	. . .
4620	. . .	0.272
4620 (0.40 Mo)	. . .	0.250
	2.000[b]	2.000+
5120	. . .	0.080
EX-15	. . .	0.116
EX-24	0.385[b]	. . .
EX-29	0.760[b]	. . .
EX-31	2.000[b]	. . .
8620	0.232[b]	0.108
8720	. . .	0.132
8822 low chem	. . .	0.189
8822 med chem	1.270	0.300
X9115	. . .	0.104
9120	. . .	0.084
94B17	. . .	0.173

[a] 0.20% C, 0.80% Mn, 1% Cr, 0.05% Ni, 0.25% Mo.
[b] Test results courtesy of Climax Molydenum Co.

Table 6.4   Estimated Minimum Microstructural Capabilities of Constructional Steels (Maximum J Distance in Relation to Percentage of Martensite)

Grade	99.9% (min)	95.0% (min)	90.0% (min)	Grade	99.9% (min)	95.0% (min)	90.0% (min)
1330H	2	3+	4	5147H	2	5	6+
1335H	1	3	4	5150H	3	4+	5+
1340H	1	3+	5	5155H	2	4+	5+
1345H	3	5	5+	5160H	1–	5	6
4027H	1+	2	2+	8617H	1	2	2+
4037H	1	2+	2+	8620H	1	2	2+
4047H	1+	2+	3	8622H	1	2+	3
4118H	1	2	2+	8625H	1+	2+	3
4130H	2	3	3+	8627H	2	2+	3
4137H	2	6+	8	8630H	2	3+	4
4140H	1	6	8	8637H	2	4+	5+
4142H	3	8	10	8640H	1+	4+	6
4145H	3	9	12	8642H	2	5	6+
4147H	4	11	12	8645H	3	5+	6+
4150H	7	11	12+	8655H	3	6	8
4161H	1–	14	16	8720H	1	2+	3
4320H	1	2+	3	8740H	2	5	6
4340H	7	13	18	8822H	2	3+	3+
4620H	1	1+	2	9260H	2	3	3+
4621H	1	2+	3	50B44H	4	6+	7
4815H	1	3	3+	50B46H	1	3	4
4820H	1	4+	5	50B50H	4	6+	7
5120H	1–	1+	2	50B60H	5	6+	7+
5130H	1+	3	3+	51B60H	5	7	8+
5132H	2	3	3+	81B45H	5	8	9
5135H	1+	3	4	86B45H	3	10	14
5140H	1	3	4+	94B17H	2	5	5+
5145H	2	4	5	94B30H	5	8	8+

center of the part. From the curves in Fig. 6.10 or 6.11 the quench-cooling rate can be estimated in the center of round sections. Suppose, for example, that a manufacturer of studs wishes to select steel for a 1-in. diameter part using a water quench for hardening. Figure 6.10 indicates that the cooling rate in the water quench at the center of a 1-in. round corresponds to that at 3/16 in. from the end of the end-quench specimen. For maximum freedom from quench cracking a steel with nominal carbon content of 0.30% max should be used. Referring to Table 6.4, we find that the only standard steel meeting the requirements of 99.9% martensite min at a J distance of 3/16 in. min with 0.30% nominal C is 94B30H. This grade is not widely used and therefore has limited availability, so it may be necessary to go to a higher carbon steel and oil quench. To meet the microstructure requirements with oil quenching we must have a steel that shows 99.9% martensite at 6/16 in. (from Fig. 6.11). Table 6.4 shows only two steels that have this level of hardenability: 4150H and 4340H. The excessively high carbon content of 4150H, as well as other factors, make it an undesirable choice. Therefore 4340H would be the best choice for oil quenching.

An examination of Table 6.4 reveals some interesting facts:

1 With the exception of 4340H and the boron-treated steels, few have any significant capability to harden fully; that is, to 99.9% martensite.

2 The extremely high cooling rates necessary to achieve this quality of microstructure (99.9% martensite) indicate that for sections of any significant size water quenching must be employed.

3 Although not determined for all the steels shown, the microstructural capability of most of the steels varies widely from low side to high side in composition and H-band limits. The following are three examples:

	Maximum J Distance in Relation to Percentage of Martensite		
	99.9%	95.0%	90.0%
4140H (low side)	1	6	8
4140H (high side)	11	18	20+
8620H (low side)	1	2	2+
8620H (high side)	3	4+	4+
94B30H (low side)	5	8	8+
94B30H (high side)	12	15+	18

For 4140H the microstructural capability from high to low side for 99.9% martensite is 11:1, for 8620H, 3:1; for 94B30H, 2.4:1. At other percentages of martensite the range is of the order of 2:1 to 3:1. There is little wonder that we encounter cracking and size control difficulty in heat treating!

As the microstructural capability of steels is examined in Table 6.4, the true value in terms of engineering characteristics becomes of interest. Expecting microstructures of 99.9% martensite in production, however, is, in general, unrealistic, and only the severest applications should require this quality of structure. Nevertheless, on page 191 of The American Society for Metals (ASM) Metals Handbook, Vol. I, the advice is properly given that in most dynamically loaded parts the surface and some percentage of the cross section should have an as-quenched microstructure of at least 90% martensite.

TEMPERING PARAMETER

Once a steel is selected to harden fully to the proper depth, the next problem is to determine the tempering cycle (temperature and time) to achieve the desired final hardness. The temperature-time parameter method of Holloman and Jaffe [7], as modified by Grange and Baughman [8], makes it possible to estimate an appropriate tempering cycle from chemical composition as long as tempering occurs in the range 650 to 1200 F. This method provides a reasonably reliable estimation of the hardness of the resulting tempered martensite (generally about ±1 Rockwell C units or ±10 diamond pyramid hardness units) for AISI-SAE carbon and alloy steels containing 0.2 to 0.85% C and less than 5% total alloying elements.

The combined effect of temperature and time of tempering is expressed by Holloman and Jaffe [7] in the parameter $T (c + \log t) \times 10^{-3}$, where T is the absolute temperature equal to degrees $F + 460$, t is time in hours, and c is a constant for a particular steel. The use of a single value of 18 for c, suggested by Grange and Baughman [8], gives the same parameter for all steels so that the curves of different steels are directly comparable. Figure 6.20 shows the effect

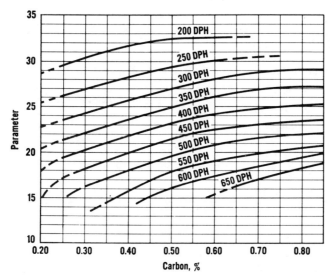

Fig. 6.20 Variation in parameter value with carbon content at each indicated hardness level [8].

of carbon content in plain carbon steels on the parameter at a number of hardness levels. Table 6.5 lists factors for predicting the increment of hardness added to carbon steels by a number of alloying elements. Figure 6.21 shows the relationship of the parameter, $T(18 + \log t) \times 10^{-3}$ to temperature and time.

The following example illustrates the construction of an estimated hardness curve. The composition of AISI-SAE 4340 steel in weight percent is

C	Mn	Si	Ni	Cr	Mo
0.42	0.78	0.24	1.85	0.81	0.27

Step 1. Construct on a parameter chart (Fig. 6.22) a tempering curve for a base plain carbon steel of identical carbon content (0.42%), using values from Fig. 6.20.

Step 2. Adjust the 1042 curve for effects of alloying elements from values in Table 6.5 for a series of parameters; for example, at parameter 20

$$5 \times \% \text{ Ni} + 50 \times \% \text{ Cr} + 20 \times \% \text{ Mo} = 55 \text{ dph}$$

Repeat for parameters 22, 24, 26, 28, and 30 until the 4340 estimated curve is developed.

If, for example, we wish to know the hardness after tempering a specimen of fully hardened 4340 steel for five hours at 1000 F, the parameter for this tempering treatment would be 27.3 (see Fig. 6.21); according to the estimated curve in Fig. 6.22 the hardness would be 368 dph or 37.5 Rc. Incidentally, this

Table 6.5   Factors for Predicting the Hardness of Tempered Martensite[a]

Element	Range	Factor at Indicated Parameter Value					
		20	22	24	26	28	30
Manganese	0.85–2.1%	35	25	30	30	30	25
Silicon	0.3–2.2%	65	60	30	30	30	30
Nickel	Up to 4%	5	3	6	8	8	6
Chromium	Up to 1.2%	50	55	55	55	55	55
Molybdenum	Up to 0.35%	40	90	160	220	240	210
		(20) (b)	(45) (b)	(80) (b)	(110) (b)	(120) (b)	(105) (b
Vanadium(c)	Up to 0.2%	0	30	85	150	210	150

[a]Grange and Baughman [8].
[b]If 0.5–1.2% Cr is also present, use this factor.
[c]For AISI-SAE chromium-vanadium steels; may not apply when vanadium is the only carbide formerly present.
Note: Boron factor is 0.

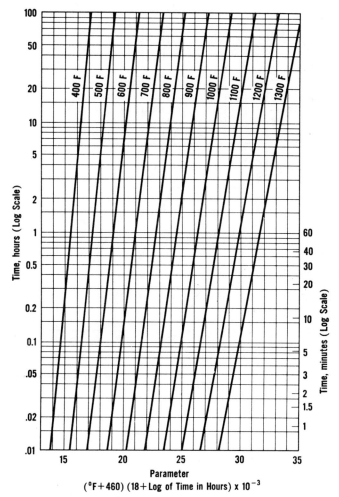

Fig. 6.21 Time-temperature parameter chart with $C = 18$ to temperature and time (after Grange and Baughman [8]).

is within 0.6 Rc of the measured value. Similarly, 10 hours at 1100 F gives a parameter of 29.65 and a hardness of 32 Rc.

In another example we wish to know the tempering time at 1200F to give the same hardness, 32 Rc, obtained in 10 hr at 1100F. The time indicated in Fig. 6.21 by the intersection of parameter = 29.65 and the 1200F curve is 0.7 hr or 42 min. The same result may be calculated with the parameter relation

$$T_1 (18 + \log t_1) = T_2 (18 + \log t_2)$$

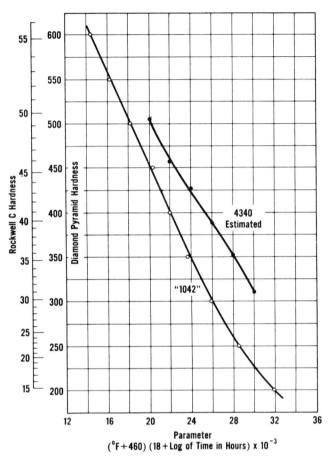

Fig. 6.22 Estimated hardness of tempered martensite of a 4340 steel (after Grange and Baughman [8]).

where T_1 = 1100 + 460 F, t_1 = 10 hr., T_2 = 1200 + 460 F and t_2 is the time to be found. Substituting and solving gives t_2 = 0.7 hr or 42 min as above. Areas in a part not fully hardened will temper somewhat softer than the tempered martensite. Whenever possible, tempering schedules should be selected to avoid tempering in the range of 475 to 750 F inclusive because of possible temper embrittlement (see Chapter 1).

REFERENCES

1 J. L. Burns, T. L. Moore, and R. S. Archer, *ASM Trans.*, **26**, No. 1, 1 (1938).

2 Methods of Determining Hardenability of Steels — SAE J406a, *SAE Handbook,* New York, 1972, p. 15.

3 W. E. Jominy and A. L. Boegehold, *ASM Trans.*, **26,** 574 (1938).
4 Hardenability Bands for Carbon H Steels − SAE J776d, *SAE Handbook,* New York, 1974, p. 21.
5 Hardenability Bands for Alloy H Steels − SAE J407c, *SAE Handbook,* New York, 1974, p. 24.
6 *AISI Steel Products Manuals,* American Iron and Steel Institute, New York.
7 J. H. Hollomon and L. D. Jaffe, *Trans. AIME,* **162,** 223 (1945).
8 R. A. Grange and R. W. Baughman, *ASM Trans.,* **48,** 165 (1956).
9 M. A. Grossmann, *Trans. AIME,* **150,** 227 (1942).
10 J. Field, *Metal Progr.,* **43,** No. 3, 402 (1943).
11 L. C. Boyd and J. Field, Calculation of the Standard End-Quench Hardenability Curve From Chemical Composition and Grain Size, *Contributions to the Metallurgy of Steel,* No. 12, American Iron and Steel Institute, New York.
12 A. F. deRetana and D. V. Doane, *Metal Progr.,* **100,** No. 3, 65 (1971).
13 I. R. Kramer, S. Siegel and J. G. Brooks, *Trans. AIME,* **167,** 670 (1946).
14 W. Crafts and L. Lamont, *Trans. AIME,* **158,** 157 (1944).
15 C. F. Jatczak and D. J. Girardi, *ASM Trans.,* **51,** 335 (1959).
16 C. F. Jatczak, *Met Trans.,* **4,** 2267 (1973).
17 J. M. Hodge and M. A. Orehoski, *Trans. AIME,* **167,** 627 (1946).
18 R. F. Kern, *Metal Progr.,* **102,** No. 4, 127 (1972)

CHAPTER SEVEN

COMPROMISES IN
STEEL SELECTION

To ACHIEVE THE GOAL OF SELECT-
ing steels to meet the minimum engineering requirements of the part at the
lowest total cost compromises are usually necessary for one or more of the
following reasons:

1 Similar-parts materials engineering.
2 Availability.
3 Stock standardization.
4 Weldability.
5 Machinability.
6 Availability of proper heat treating equipment.
7 Formability (forging, cold-heading, extruding).
8 Ease of heat treatment.

SIMILAR-PARTS MATERIALS ENGINEERING

In most firms the design engineer has full responsibility for ensuring satisfactory
performance of a part. This includes specifying steel and heat-treating requirements. The materials engineer is called on only if there is a new and unusual
design or if severe processing or field problems are involved. In relatively few

firms are design and materials-processing standards available to enable the designer to make optimum steel selections by himself. Consequently, steel selection for heat-treated parts is usually made by the designer from experience with similar parts. This is known as similar-parts materials engineering. For example, if the design engineer has had success with SAE 8620 in a five-pitch carburized gear of a certain size and function, he will select this grade for a new part for similar use. If this gear is part of a transmission that will be subjected to dynamometer and field tests before production, the profit-oriented firm will take advantage of this opportunity to test new lower cost steels such as EX-15. This sort of change would effect a saving of approximately 1¢/lb *based on the slug weight or raw stock required to make a part.*

Similar-parts materials engineering has retained its popularity because of the following:

1 The designer feels safer with a tried and tested steel.
2 In some cases it contributes to stock standardization.
3 The processing characteristics of the steel are well known.
4 Steel availability is usually assured.
5 Often a large percentage of parts in a design is made from low-cost, unheat-treated steels that would waste a materials engineer's time to study.

Sometimes this kind of materials engineering leads to problems other than excessive costs; for example, a large manufacturer of speed reducers had success in using SAE 3310 steel in its gears to compensate for erratic heat-treat control. This steel then became nonstandard and the firm was forced to procure it in heat lots at a substantially inflated price. Another manufacturer used 8617 steel for all carburized gears up to five pitch. These gears were reheated for hardening. An equipment-modernization program was introduced that included a change to direct quenching. Microcracking then occurred in some of the finer pitch gears, followed by a rash of premature breakage. Changing to SAE 4023 finally solved the problem and reduced material cost, but failure to recognize initially all the effects of changing the process resulted in financial loss.

Similar-parts materials engineering is sometimes carried on to the extent that a firm will change from a reasonably successful steel, or one tailored to meet its own engineering requirements, to one used by a competitor. It seldom occurs to the designer that the competitor may be having severe shop or field problems with the steel he is using.

One of the greatest difficulties a materials engineer faces is the technically unfounded preference or dislike for some steel grades by the designer and/or his superiors. At the present time there seems to be some apprehension about specifying boron-treated steels because of these rumors:

1 They cannot be made clean.
2 They cannot be rolled or forged with a smooth surface.
3 They are erratic in heat-treat response.

All of these allegations are generally untrue. Boron steels developed a bad reputation during World War II because substitutions were made for high-alloy steels such as 4817 with grades such as 94B17 on the basis of core hardenability. Carburized case microstructural capability was explored only after serious epidemics of field failures. Careful applications of boron steels today not only provide cost reduction but may actually improve performance as discussed in Chapter 15.

In summary, the prolonged use of similar-parts materials engineering can result in the loss of a firm's technical edge over competition that will eventually erode profitability. Many firms in the United States are 10 to 20 years behind the leaders in materials engineering largely because they have relied on this method of steel selection.

AVAILABILITY

Almost any constructional steel composition is available from a steel producer in heat lots. Lots can vary from 20 tons at the so-called minimills to several hundred tons from the major producers. Most users cannot afford to buy that much steel at one time, especially on a trial basis; consequently, they must order parts of heats of the grades already being made in substantial quantities. The tonnage produced of the various grades of carbon steels for heat treatment are listed in Table 7.1 as percentages of total production of a major supplier; figures for alloy grades are given in Table 7.2.

Users of very small quantities of steel (less than three tons) must usually rely on steel service centers (warehouses) as their source. This seriously restricts availability of grades and sizes as shown in Chapter 4. If a firm wishes to use a nonstandard grade on an ongoing basis, some service centers will stock it but generally at a substantial premium over steel purchased directly from a producer.

When the need for a particular grade amounts to more than 100 tons per quarter, it is often possible to tailor the steel for the application and to buy it in heat lots. It must be remembered, however, that not until a mill has made a considerable number of heats will they guarantee hardenability. Composition will be guaranteed, but hardenability on a "best effort" basis only. Tailored steels are effective in cost reduction when a firm makes a series of parts similar in proportion but of different section thicknesses, depending on size; for ex-

Table 7.1 Shipments of Heat-Treating Grades of Carbon Steel by a Major U.S. Producer — Second Quarter, 1973

Grade	Percentage of Production	Grade	Percentage of Production
1018	16.88	1527	1.50
1045	9.09	1117	1.45
1020	7.24	1043	1.35
1038	5.97	1141	1.30
1022	4.93	1046	1.25
1035	4.15	1069	1.08
1541	3.63	1050	1.05
1040	3.38	1042	1.03
1215	3.36	1118	1.02
1030	2.85	1080	1.01
1039	2.60	1065	0.93
1026	2.58	1049	0.80
1541H	2.57	1551	0.79
1060	2.35	1524	0.77
1019	1.81	1053	0.76
12L14	1.80	1146	0.54
1566	1.66	1038H	0.52
1561	1.60	1137	0.50
1144	1.55	1084	0.26

Table 7.2 Shipments of Heat-Treating Grades of Standard Alloy Steels by a Major U.S. Producer — Second Quarter, 1973

Grade	Percentage of Production	Grade	Percentage of Production
5160	24.40	4340H	1.20
5160H	12.64	8637	1.17
8637H	7.68	1340	1.15
5132H	6.68	4118H	1.13
4140H	4.90	8622H	0.90
9260H	4.88	5140H	0.85
8620H	3.16	4140H	0.73
8620	2.92	8640	0.65
4027H	2.33	8742	0.65
51B60H	1.96	9260	0.54
1335	1.86	4037H	0.51
50B60H	1.70	4820	0.64
4137H	1.32	4820H	0.27
4037	1.25		

ample, the maximum section thickness of a series of diesel-engine connecting rods may be 1.5 in. for the largest and then range down to 1.0, 0.75, 0.50, and 0.25 in. It is not uncommon that the steel used for the 1.5 in. rod will also be used for all other rods, and it may be an expensive product such as 4340. It should be possible to save money on the smaller rods by changing to a cheaper, leaner alloy grade. If no lower cost standard steel is suitable, a special grade could be designed and substantial money would still be saved without sacrificing quality.

To design a special steel, we suggest calculating the required composition limits by using the chemical factors or SAC traverse method for carbon steel described in Chapter 6; for carbon-boron steel, the calculations suggested in Chapter 6 and Appendix VII; for alloy steel, the Climax Molybdenum factors up to 0.25% max C, Grossman values for 0.25 to 0.60% C, except vanadium, for which the Crafts and LaMont factors are preferred; for case hardenability, the factors developed by Jatczak and Girardi. All of these factors are given in the tables in Appendix VII. Carbon steel should be considered whenever possible to avoid paying the base-price premium for alloy steel by following these principles:

1 Select carbon content and range according to the standard set forth in the appropriate AISI Steel Products Manual. Keep carbon content as low possible; 0.45% should be the top limit for most applications requiring direct-hardening (except for springs for which 0.60% nominal carbon is preferred). For carburized gears it is preferable to restrict carbon to 0.25% max. For applications that are designed on yield strength Fig. 7.1 may be used to establish the minimum % C. This nomograph relates minimum yield strength to hardness for through-hardened (95% martensite) sections. In making use of this figure, it is recommended that the "as-quenched" hardness be at least five points Rc higher than the final hardness.

2 For additional hardenability start with 0.80 to 1.10% Mn and avoid higher than 1.10 to 1.40% Mn in carbon steels. For alloy steels, 0.75 to 1.00% Mn is preferable; do not exceed 1.30% Mn.

3 Set silicon content at 0.15 to 0.30% for carbon and alloy grades.

4 Sulfur content should be .050% max for carbon steel and 0.040% max for alloy grades.

5 Hold phosphorus to 0.040% max for carbon steel and 0.035% max for alloy grades.

6 Boron content, when specified, should be 0.0005 to 0.003%.

If an alloy steel must be designed, the following procedures will provide the

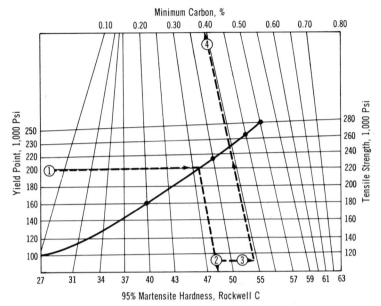

Fig. 7.1 Determination of minimum carbon content and necessary hardness when required yield point or tensile strength is known [1].

best quality-cost combination based on current price books:

1 Start with molybdenum at 0.13 to 0.20%.
2 If short-cycle fatigue strength is an engineering requirement, call for at least 0.40 to 0.70% Ni.
3 If additional hardenability is required, call for chromium at 0.25 to 0.40% or 0.45 to 0.65%.
4 If still higher hardenability is required, increase molybdenum to 0.15 to 0.25%, 0.20 to 0.30%, or 0.30 to 0.40%. *Note.* For carburizing with direct quench hardening of gears do not use more than 0.30% Mo; change heat-treating practice to reheat for hardening.
5 If further increased hardenability is required, increase chromium to 0.70 to 0.90%.
6 For maximum hardenability and/or short-cycle fatigue strength increase nickel to 0.80 to 1.00%, 1.20 to 1.50%, or 1.65 to 2.00%.

After designing the special steel composition submit it, along with hardenability requirements, to several producers for their comments. If acceptable to at least two producers, order a small heat (25 tons min) and have it bloomed. After the hardenability and microstructural capability has been checked, have some of the blooms rolled into sections suitable for the parts. Process the steel into finished parts and evaluate the results. It may be found necessary to make adjustments in composition to meet the minimum engineering requirements because of the inaccuracies inherent with calculated hardenability and with variations in the heat.

If endurance-limit fatigue strength is a major engineering requirement, it will be necessary to develop a composition that can be readily melted by the producer to a high degree of cleanliness, keeping in mind that some steels such as 4100 and those containing more than 1.00 % Si are difficult to melt clean.

It is often desirable to buy steel in a special rolled shape (or section), as typically shown in Fig. 7.2. These sections, however, present a number of problems in availability:

1 The purchaser must buy the first set of mill rolls, which may cost $25,000 to $250,000.

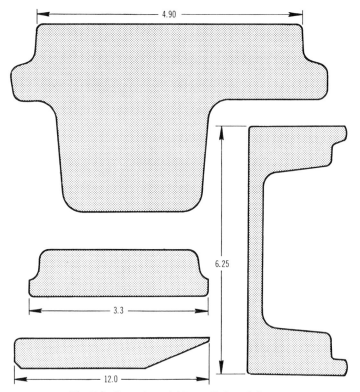

Fig. 7.2 Several special hot-rolled-steel shapes.

2 The purchaser must be in a position to order and accept delivery of minimum rollings of at least 100 tons and at some mills as many as 500 tons.

3 Because of the roll cost and, in some instances, the size of the section that would require a certain size of rolling mill, the purchaser may have only one source. A major breakdown, wildcat strike, or other interruption in production could create a difficult procurement situation.

Prices, tolerances, and minimum rollings on all special sections must be negotiated with the steel producer.

A factor that must be considered in the availability of steel is the reliability of sources of alloying metals. As shown in Figure 7.3, the United States depends substantially on imported alloying elements. Even short of war, the international

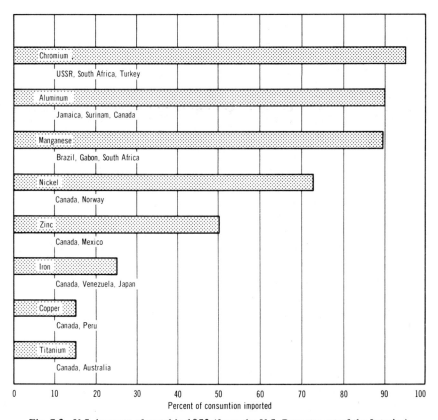

Fig. 7.3 U.S. imports of metal in 1972 (from the U.S. Department of the Interior).

political situation can create shortages. We are especially susceptible to chromium shortages because this metal must come from great distances. Although most of our manganese is imported, a substantial portion comes from the western hemisphere; there are also sizable deposits of low-grade manganese ore in the United States. We are self-sufficient in requirements of silicon, boron, and molybdenum. Canadian nickel is in good supply, but it has in the past been subject to shortages.

STOCK STANDARDIZATION

Another factor that must often be contended with in the selection of steel for a heat-treated part is the pressure to standardize on the minimum number of grades and sizes. Because no further price reduction is realized when more than 40,000 lb of a steel is ordered at one time, the only possible cost saving by standardization is to eliminate the cost of stocking. This varies from $250 to $1000 a year, depending on the firm's clerical costs, and would seem to indicate that to pay its way a new grade would have to produce a minimum saving of this amount. This, however, is not always the case because the elimination of 10 steel types would not result in the need for one fewer employee. The real problem in inventory is often the cost, as shown in Fig. 7.4, where, for example, only 500 lb or less of an item must be purchased and stocked.

Most firms find it necessary to buy small quantities of steel (less than 1000 lbs), at least on occasion. The best that can be done with these purchases is to minimize the expense of processing the necessary paperwork to use a grade stocked by a warehouse. For firms that write their own specifications (instead of using those of ASTM, SAE, etc.), a shortcut will show the alternates permitted, at least one of which is warehouse-stocked. Inventory control departments can cut down purchasing costs by combining alternates.

When firms use SAE or similar specifications, suitable alternates should whenever possible be shown on part drawings. In the use of alternates procedures must be set up so that manufacturing departments, especially heat treating, will know what grade they are processing. Also, grades on a given shop order should not be mixed.

Another method that multiplant firms sometimes employ to minimize procurement and stocking costs (Fig. 7.4) centralizes steel purchasing, at least for high-cost proprietary items.

The most important contribution to standardization that can be made by design engineering personnel is to make sure that the drawing shows the minimum engineering requirements. The materials engineering department, working with the manufacturing engineering department can then select steels to meet these requirements from alternates in the specifications or from materials standards.

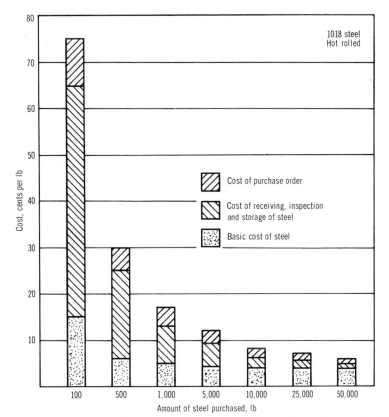

Fig. 7.4 Typical cost of steel versus quantity purchased (from *Metals Handbook,* Vol. 1, 284, Fig. 8).

Standardization for the purpose of increasing the quantity to at least a minimum mill order is a desirable objective because of the potential saving (35 to 50%). Ordering from a mill will also sometimes permit a firm to select a less expensive grade than is available at a warehouse.

Another way in which engineering personnel can be helpful in standardization is in what might be called "standard sizes"; for example, a designer ideally may want a shaft to be 1.062 in. in diameter. By using a somewhat higher strength steel — such as 1-in. round 1045 instead of 1.062-in. round 1018 — equivalent resistance to yielding could be provided; however, the design would be subject to slightly more deflection under load. Standard tolerances for size, as shown in Appendix III, must be considered in making a selection. Producers make odd sizes; delivery is often slow, however, because they attempt to accumulate orders for sufficient tonnage to justify rolling. One way in which an all-out effort should be made to standardize on certain sizes is in flat bars. Oversized rounds can usually be turned down at minimum expense, but flat bars are expensive to mill or plane down to narrower or thinner sections. Another item to watch closely is tubing, because of the countless combinations of sizes, grades, and manufacturing processes.

WELDABILITY

Welding of high-strength steel should be avoided whenever possible. Strength and weldability are nearly always conflicting characteristics of steel so that compromises must be made between the ideal material for an unwelded part and a steel that can be joined with freedom from underbead cracks or other defects; for example, a recent problem involved welding a strain-tempered SAE 1144 stud to a low-carbon steel tube. This grade is a popular stud steel, but in this application (1.25-in. diameter) it produced severe underbead cracking because of its high carbon and manganese contents. Also, its high sulfur content made the welds porous. Changing the studs to 8620 steel, water quenched and tempered to 28-33 Rc, resolved this problem but increased material cost.

Nearly any constructional grade of steel can be welded; however, the precautions and techniques required to produce an acceptable weld are often prohibitive. According to most structural welding codes, steels with carbon equivalents of more than 0.475%, as defined by (7.1) [2], require preheating to prevent hydrogen-induced cracking for arc welds on sections more than 0.75-in. thick.

$$CE = C\% + \frac{Mn\%}{6} + \frac{Ni\%}{20} + \frac{Cr\% + Mo\%}{10} + \frac{Cu\%}{40} \qquad (7.1)$$

The main function of preheat is to reduce the cooling rate in the heat-affected zone of the weld, thus to produce a more ductile microstructure that is less susceptible to cracking, and to provide more time for hydrogen to diffuse out of the weld.

A more sophisticated definition of cracking susceptibility is given in Fig. 7.5; it involves a similar formula for carbon equivalent but also recognizes the major contribution of carbon content to this tendency [3]. Zone I steels have high hardenability, but carbon content is so low that even the hardest microstructure is not susceptible to cracking. Steels in Zone II are shallow hardening but can develop sensitive microstructures because of their increased carbon content. Preheating of these steels is an effective means of reducing cracking tendency. Steels in Zone III combine high hardenability with high carbon content, compounding the problem considerably. In this instance extremely careful procedures, which may include special precautions to minimize hydrogen ingestion, controlled cooling, and postheat treatment, may be required.

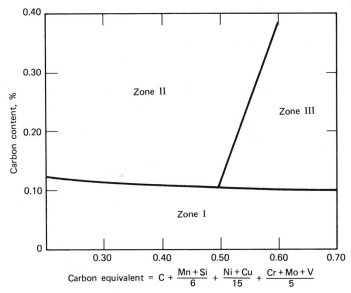

Fig. 7.5 Cold cracking tendency of steels as a function of carbon content and carbon equivalent [3].

The reader is referred to the work by Linnert [4] and the Welding Handbook [5] for more detailed discussions of the weldability of all types of steel.

Sulfur and lead are also elements to be avoided when specifying steels that are to be arc welded. Sulfur contents substantially higher than 0.05% can lead to "hot cracking" of the weld due to the formation of low-melting iron sulfide in the weld. Arc-welding leaded steels can present a health hazard to the welder unless special precautions are taken to protect him from the fumes.

MACHINABILITY

The materials engineer must frequently compromise his optimum steel selection to provide good machinability. This is particularly true when very expensive cutting tools such as hobs, broaches, and bevel-gear cutters are used. In one instance, when machining a worm gear with approximately 5-in. pitch diameter, the hob cost per piece was more than $2.50 (the hob cost thus being by far the largest increment of factory cost). Generally speaking, the following steps can be taken to ensure good machinability:

1   Keep the hardness as low as practical. Hardness is the characteristic most directly related to machinability.

2 Avoid high-manganese steels such as SAE 1524, 1320, 1340. Medium carbon 4100 series (e.g., 4140, 4150) can also be difficult to machine, especially in the quenched and tempered condition.

3 If equipment is available, call for a suitable isothermal anneal for machining before hardening.

4 Use steel of special cleanliness only when absolutely necessary. (The extremely low sulfur content is deleterious to machinability.)

5 Utilize mildly resulfurized grades such as 4024.

6 In extreme cases consider the use of steels containing 0.15 to 0.35% lead, keeping in mind that the fatigue strength of leaded steel may be reduced at high hardness [6]. (See Fig. 7.6)

Steels such as 1524 and 1320 would be much more widely used in automotive ring gears, for example, if their machinability were such that the cutter cost per piece were reduced. Lead (0.15 to 0.35%) is used quite extensively in carburizing grades such as 8620; however, its cost is high ($20 to $30/ton). Proper annealing or normalizing usually pays its way in improved tool life and is an absolute necessity on highly alloyed grades such as 9310 to 9317.

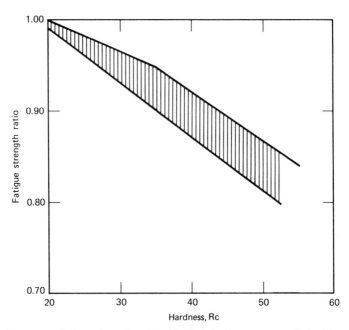

Fig. 7.6 Transverse fatigue strength ratio of leaded steel versus same steel without lead at various hardnesses (from Simon [6]).

Modern steel-making practice sometimes produces steels so low in sulfur that in order to get good machinability it may be necessary to specify grades such as 4024 with 0.035 to 0.050% S added, although this increases steel cost and can be detrimental to mechanical properties. Exceptionally low sulfur is also a characteristic of medium-carbon steels produced to special cleanliness. Carbon-manganese-boron steels for heat-treated line pipe are being made to 0.005% max S for exceptional toughness, but the extremely poor machinability can cause difficulty in weld-joint preparation.

A difficult machining problem exists on parts such as close tolerance 4140 gears heat-treated before machining to a high hardness. In one example changing to 41L40 increased the number of pieces per sharpening of the gear hob from 75 to 100 pieces up to 130 to 165 pieces. Single-tooth bending fatigue tests of the nitrided gears did not show any deleterious effect caused by the presence of lead in the microstructure.

FORMABILITY AND FORGEABILITY

Sometimes the ideal steel to meet minimum engineering requirements must be bypassed in order to facilitate cold forming; for example, a hardened and ground bushing might be made from 52100 steel tubing. However, if quantities are large (50,000 to 100,000 pieces a year), it may be more advantageous to use a steel such as 10B16, cold extrude the part, and carburize it.

Steels containing more than 1.0% nickel have a reputation for developing a tight scale in heating for forging. If a forging has large flat surfaces that must be smooth, nickel-containing steels may be a questionable choice unless those areas are machined.

Steels containing more than 0.50% Cu or more than 0.13% S tend to be hot short and to crack in hot forging. This also applies to boron steels if the boron content exceeds 0.008%.

AVAILABILITY OF PROPER HEAT
TREATING EQUIPMENT

An important factor contributing to a compromise in the ideal steel for a part is the availability of suitable heat-treating equipment; for example, the ideal for a 3-in. diameter by 36 in. long axle might be 1040 steel, shell hardened in a roller-die-quench machine. Few firms have these machines, however; as a matter of fact, we are aware of only one commercial heat treater that uses this equipment. Consequently, most firms will make the shaft from an alloy steel such as 4140, oil quench and temper it to a machinable hardness, and then finish-machine the part. This results in an excess cost per axle of as much as 10%. With the 4140 shaft, nearly all machining must be done after heat treatment, whereas with the roller-die-quenched 1040 only the bearing diameters need to be ground after hardening. Also, the endurance-limit fatigue strength of the 4140 axle will be only a fraction of the shell-hardened 1040 part. A possible alternative to using 4140 steel in the axle would be to induction-harden a machined 1040 shaft over its entire length. This heat treatment would cost between $7.00 and $10.00 a piece, which would probably make the cost of the 1040 part more than that of the 4140, but it would provide a shaft with improved endurance-limit fatigue life.

The firm with no heat-treat facilities must rely on commercial heat treaters; however, it is also possible to subcontract parts made of the most appropriate steel to firms that have proper heat-treating equipment. For a firm that maintains only direct-fired hardening furnaces, it should be remembered that parts can be pack-carburized in them and then properly hardened with the same equipment.

EASE OF HEAT TREATMENT

Often the steel that is ideal from an engineering standpoint is difficult to heat treat without excessive distortion and/or cracking. An example is the crawler track shoe shown in Fig. 7.7. Ideally, this part should be made of a 10B35

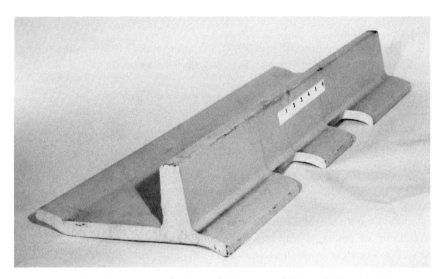

Fig. 7.7 Crawler tractor track shoe made from a special hot-rolled-steel section.

modified steel, water-quenched. With heats containing 0.36 to 0.38% C, however, severe cracking occurred. Consequently, the mean carbon was reduced to 0.30%. The manganese level than had to be increased to offset the loss in hardenability at the lower carbon level, thereby increasing cost $8/ton. Grades such as 1340H, 1345H, 4140H, 4145H, and 4150H have adequate hardenability for many parts to be oil quenched, but they are prone to quench cracking. Therefore, grades such as 86B45 or 4340 are preferred for intricate parts even though they are more expensive.

Carburizing grades have idiosyncrasies that are often the cause of serious trouble when overlooked or ignored:

1 Steels such as the 4100, 4400, 8700, and 8800 series are quite susceptible to microcracking when direct-quenched for hardening. Therefore case carbon must be held below 0.85% in gears of six pitch or finer. These grades, as well as the 5100 series, also require close control of case carbon to prevent excessive amounts of carbide.

2 The high-nickel grades, such as the 4300, 4600, and especially the 4800 series, require close control of case carbon content in order to prevent the retention of excessive austenite. (Suggested maximum is 35%.)

A firm that has a large, trained metallurgical staff, along with sophisticated control equipment, by exercising constant vigilance in processing can use steels that would be impractical for a small firm. The capable materials engineer recognizes this and makes selections that can be heat-treated and controlled by the sources available to him with only a minimum amount of attention. Only by careful consideration of the numerous factors described herein can he be assured of an uninterrupted flow of reliable parts at the lowest total cost.

REFERENCES

1 Roy F. Kern, Selecting Steels and Designing Parts for Heat Treatment, Materials and Progress Engineering Bookshelf, American Society for Metals, Metals Park, Ohio 1969.

2 AWS D14.3-77 Specification for Welding Earthmoving and Construction Equipment, American Welding Society, Miami, Florida, 1977, pp. 21-22.

3 B. A. Graville, Conference on Welding of HSLA (Microalloyed) Structural Steels, November 9-12, 1976, Rome, Italy. Reported by A. B. Rothwell, Weldability of HSLA Structural Steels, *Metal Progr.*, **III**, June 1977.

4 G. E. Linnert, *Welding Metallurgy*, Vol. 2, American Welding Society, Miami, Florida, 1967.

5 *Metals and Their Weldability*, Section 4., *Welding Handbook*, 6th ed., American Welding Society, Miami, Florida, 1972.

6 W. Simon, Fatigue, Impact and Tensile Properties of High Strength, Leaded and Non-Leaded SAE 4140 Steel, *Metal Progr.*, July 1962, p. 166.

CHAPTER EIGHT

GENERAL PRINCIPLES
FOR THE SELECTION OF
DIRECT-HARDENING STEELS

DIRECT HARDENING IS A HEAT treatment that does not change the composition of the surface except when necessary to restore the surface carbon content to that of the base steel. Direct-hardening steels are used when the loading condition is one or a combination of the following:

1 Pure tension
2 Shear.
3 Bending.
4 Torsion.
5 Impact.
6 Abrasive wear.

Tension, shear, bending, and torsional loading is most often cyclic. When the cyclic strain amplitude is less than 0.1%, it is assumed that endurance-limit (long-life) fatigue conditions exist under which the life is usually better than 200,000 cycles. When the strain approaches 0.5%, it is assumed that short-cycle (short-life) fatigue conditions exist in which the life is less than 200,000 cycles. Because fatigue failure can occur at stresses substantially below the tensile yield strength, the engineering requirement must be known not only in terms of maximum stress but also of the required life under dynamic loading.

It is not uncommon that designers hesitate — and sometimes even refuse

— to specify the minimum engineering requirements accurately. Their reasons are either that it requires too much effort and time to make even an educated guess or they do not know how to make a determination. It must be remembered, however, that if the designer does not define the minimum part requirements, someone usually less qualified must guess what they are.

When the stresses in complex structures are too difficult to predict, experimental units are often tested with strain measuring devices attached, (Fig. 8.1). The information obtained from this test shows the validity of previous estimates, calculations of stress amplitude, frequency, and direction. The more precisely the minimum engineering requirements can be determined, the more accurate the selection of the steel and its heat treatment.

HARDNESS AND CARBON CONTENT

Providing the minimum engineering requirements with constructional steels requires a knowledge of the interrelation of strength, carbon content, hardness, and microstructure. Hardness and microstructure in heat-treated steel are functions of the composition and the severity of the quench. The necessary hardness and the suggested minimum carbon content to attain the required strength can be determined from Fig. 7.1. The relationships described are based on microstructures of at least 95% tempered martensite. This type of microstructure is perhaps the practical optimum for day-to-day heat treating and

Fig. 8.1 Strain gages attached to operating machinery to determine stress level and direction. Electronic equipment in the van processes signals transmitted from strain gages attached to parts under test.

provides an excellent combination of strength and toughness. It is recommended that at least in the critical areas of parts it be called for on drawings and in heat-treating specifications.

Once the hardness and microstructure requirements are known, along with the minimum percentage of carbon, the materials engineer must determine how best to obtain these qualities. The type of quench used (based on quench cooling rate) determines the necessary alloy for hardenability.

SELECTION OF QUENCHING MEDIA

Considerations in the selection of quenching media were discussed in Chapter 3 and are summarized as follows:

1 Utilization of quench facilities available.
2 Freedom from quench cracks.
3 Freedom from excessive distortion.
4 Production of the required residual-stress pattern.

Deciding how a part can be hardened is largely an art, gained only by extensive experience, but the following guidelines may be helpful:

1 Generally speaking, if the mean carbon content in carbon or alloy steels can be held to 0.30% max, the part can be water-quenched without cracking. Steels with mean carbon content of 0.30 to 0.38% can be water-quenched in simple shapes such as round bars. When the mean carbon content must be more than 0.38%, oil quenching should be employed. Exceptions are carbon steels of low residual alloy content with manganese of 1.00% max. Carbon steels containing 0.95% nominal carbon and 0.30 to 0.50% Mn can be water-quenched in simple shapes (containing no drilled or punched holes).

2 If the part has wide variations in section sizes (a ratio of more than 3 to 1) or if it has holes, keyways, or snap-ring grooves machined in it, water quenching may result in cracks, regardless of carbon content. Providing generous fillets at these discontinuities is an effective means of solving the problem.

3 When distortion must be as low and as consistent as possible (e.g., gears) oil or salt quenching should be employed. Exceptions can be made when localized heating by flame or induction can be used.

4 When the microstructure requirement is for essentially 100% bainite, austempering in molten salt should be done. In order to be sure that austempered parts do not contain any retained austenite, which may

transform in service to untempered martensite with attendant brittleness, a final temper slightly below the austempering temperature is recommended.

5 When the engineering requirements call for a substantial level of residual compressive stresses in the surface of a part, water quenching must be employed in direct-hardening steels (when through heated). Oil and molten-salt quenches create little or no residual stress. Some steels, such as 8660, will develop detrimental tensile residual stresses in the surface when oil-quenched (see Fig. 1.8).

STRESSES ENCOUNTERED WITH
VARIOUS TYPES OF LOADING

When the type of quench to be employed has been determined, the quench-cooling rate can be estimated or actually measured in significant part locations, as described in Chapter 6. The various kinds of loading encountered at significant locations are typically as follows:

1 Pure tension and shear loading: for maximum strength through-harden the part to at least 90% martensite. In extreme cases, at required strength levels above the 175,000 psi yield point, 99% martensite throughout is suggested.

2 Bending and torsion: for these types of loading the stress is at a maximum at the surface; it decreases to zero at the neutral axis according to the following typical equations:

Bending stress of a cantilever beam = PL/Z in psi   (8.1)

P = load in pounds
L = length from anchor point to point of load applications in inches
$Z = I/c$, where

for rounds

$I = 0.0491D^4$, D = diameter of round in inches
c = distance from the center of the round in inches

for Plates

$I = BH^3/12$, where
B = plate width in inches

H = plate thickness in inches
c = distance from center of plate in inches

Torsional stress in a round bar = Tc/J in psi   (8.2)

T = torque in inch-pounds
c = distance from the center in inches
$J = 0.0981D^4$, where
D = diameter of round in inches

Formulas for other types of loading and for other shapes are available in numerous engineering handbooks. It is significant to recognize that stress can be drastically reduced by only small increases in diameter or thickness. Also, most designs have stress concentrations at grooves, fillets, and holes that must be accounted for. A recommended reference for such stress concentrations is *Stress Concentration Factors* by R. E. Petersen [1].

To determine the level of hardenability necessary to provide the required strength several procedures may be followed. Product function, volume, and reliability will dictate the method of selection to be used:

1 When the quench-cooling rate is known in significant locations, the steel grade can be selected from published hardenability bands, as in Appendix VI. The minimum hardness of the band should be at least five Rc points above the final hardness needed in the significant locations. *Note:* Be sure that the hardness and carbon content at the minimum of the band will provide the minimum percentage of martensite required (see Table 6.4).

2 From Fig. 8.2 determine the minimum ideal diameter (D_I) and from Table 8.1 select the grade.

3 For a quick estimate of grade required when the quench-cooling rates are not otherwise available use Fig. 8.3. The grade selected may be somewhat higher in hardenability because the quench severities from which the curves were derived were mild.

4 If the requirement is very large and cost must be held to a minimum, tailor a special steel by using the hardenability guidance in Chapter 6 and the composition suggestions in Chapter 7.

In all cases the lowest possible carbon content should be selected to minimize the possibility of quench cracking and still provide the required microstructure. The need for various carbon levels is discussed later in detail; however, the following are major uses of direct-hardening steels with certain carbon

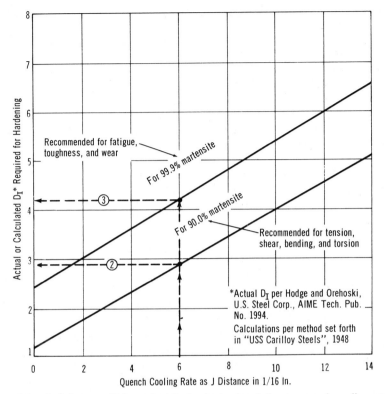

Fig. 8.2 Ideal diameter (D_I) required for hardening in relation to quench-cooling rate.

levels:

1 The 0.20% carbon steels are used mainly for welded structures or when maximum toughness is required.

2 The 0.30% level is widely used when combinations of excellent toughness and high hardness are in demand for wear resistance. Typical parts are scarifier teeth, cold chisels, and wrenches.

3 The 0.40% level is for general use in high-strength applications.

4 The 0.50% carbon level is used both for general high-strength applications and for parts to be induction- or flame-hardened for wear resistance.

5 The 0.60% level is most popular for heat-treated springs.

6 Steels with carbon contents above 0.60% are for specialized applications such as austempered cutting tools that must hold a sharp edge and support rolling bearings.

Table 8.1 Minimum D_I and Cost for Typical Constructional Steels as of January 1, 1974

Grade	Minimum D_I	Cost[a]	Grade	Minimum D_I	Cost[a]
1018	0.47	$10.675	4621H	1.75	16.250
1022	0.57	10.675	4718H	2.12	16.700
1524	1.02	11.425	4720H	1.37	15.400
1035	0.66	10.325	4815H	2.12	19.850
1536	1.06	11.025	4817H	2.30	19.850
1038H	0.80	10.325	4820H	2.58	19.700
1040	0.71	10.325	5046H	1.47	12.050
1541H	1.75	11.025	5120H	1.16	12.600
1045H	1.00	10.325	5130H	2.29	12.550
1050	0.81	10.325	5140H	2.45	12.350
1552	1.34	11.025	51B60H	3.64	12.600
1060	0.87	10.325	6118H	1.37	13.300
1080	1.01	10.325	8617H	1.37	14.100
1117	0.61	11.425	8620H	1.57	14.100
1118	0.77	11.575	8622H	1.75	14.100
1141	1.27	11.525	8625H	1.85	14.100
1144	1.14	11.875	8627H	2.12	14.000
4027H	1.35	12.700	8630H	2.20	14.000
4028H	1.35	12.950	8640H	2.07	14.000
4032H	1.37	12.700	8645H	3.32	14.000
4042H	1.75	12.700	86B45H	5.00	14.250
4118H	1.16	12.750	8650H	3.64	14.000
4130H	2.12	13.050	8660H	6.20	14.000
4140H	4.00	13.250	8720H	1.75	14.250
4150H	5.10	13.250	8822H	1.93	14.700
4161H	5.40	13.400	9260H	2.12	12.550
4320H	2.12	16.450	9810[b]	5.00	22.100
4340H	7.00	16.650	94B15H	2.58	14.400
4419H	1.16	13.650	94B17H	2.58	14.400
4620H	1.16	16.150	94B30H	3.64	13.950

[a] Cost is in dollars per hundredweight for hot-rolled bars, special quality to 12 sq in cross section in carbon steels and BOH or BOP quality in alloys except when indicated otherwise.
[b] Electric-furnace quality and grade.

136

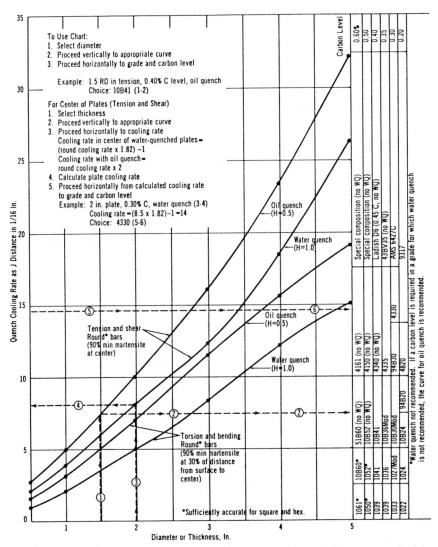

Fig. 8.3 Development of optimum properties in heat-treatable steels. Illustrates method for quick estimate of grade required (from R. F. Kern [4]).

137

SELECTION FOR TENSION
AND SHEAR LOADING

In the selection of steels for pure tension and shear loading these additional factors are important:

1 If absolutely no yielding can be allowed, the relevant mechanical property is the proportional limit, which is 75 to 90% of the tensile yield strength, depending on the quality of microstructure required (100% tempered martensite is ideal). When a small amount of permanent deformation is permissible, the conventional tensile yield point determined by the 0.2% offset method can be used satisfactorily.

2 Notch tensile strength is important for parts that are loaded in tension (usually threaded). Steels containing effective boron* exhibit higher notch tensile properties than boron-free steels. Because boron performs no solid solution strengthening, steels containing this element for hardenability can often be cold-headed, without the full spheroidize anneal required for boron-free steels of equal hardenability.

3 To provide for maximum resistance to sudden brittle failure, steels loaded in tension should exhibit a high degree of toughness at the temperature at which the part is to be used. This toughness, or ductility, is required to accommodate the misalignment and out-of-squareness that often occur in an assembly or structure. For best results the carbon content should be kept below 0.40%.

4 The room-temperature mechanical properties of tempered martensite are mainly a function of the carbon content. They are affected only slightly by the small amounts of alloy present in constructional steels. Exceptions are nickel and boron, but for tension and shear applications at room temperature and down to –20F steel selection can be made on the basis of hardenability.

5 Fixture quenching, in which the quenchant impinges on the hot steel, is preferred because it provides these advantages:

   (*a*) Maximum hardened depth is obtained for a given composition, which permits the use of the lowest cost steel.
   (*b*) Maximum microstructural quality can be consistently maintained.
   (*c*) With water quenching high levels of residual compressive stress are

---

*The term effective boron means that the steelmaking practice was such that the boron increased the hardenability by a Grossmann factor of approximately $1 + 1.6 (1.01 - \% \text{C})$. "Ineffective" boron not only does not increase hardenability but the benefits to notch tensile strength are also absent.

developed in the part, thus enhancing the endurance-limit fatigue strength (see Fig. 1.8).

(*d*) The uniformity of cooling minimizes distortion and quench cracking. Table 8.2 compares the uniformity of results in free quenching versus impingement fixture quenching of SAE 1045 bars, 2 ft long, from the same heat of steel. High-pressure impingement quenching, at least with water, should only be used on simple symmetrical parts such as studs, pins, and shafts, unless the quenching equipment can be carefully designed to provide controllable heat-removal rates from sections of varying thickness.

6 Parts, such as fasteners, that are subject to tension loading should be as straight as possible to ensure freedom from bending stresses that might cause premature failure. When parts are more than 2 in. in cross section, it is good design practice to use spherical nuts and abutment to facilitate self-alignment.

SELECTION FOR BENDING OR TORSIONAL LOADING

A steel chosen to provide bending or torsional strength must have a yield strength when heat-treated that is high enough to exceed the service load at any point throughout the cross section. Heat treating to a microstructure of 50% min martensite at the center and 90% min martensite to a depth of at least 30% of the distance from the surface to the center may be substantially more than is actually required (except for shell-hardened* or induction-hardened parts), but proper performance will be ensured. When selecting steels that will harden to only 10, 15, or 20% of the distance from surface to center, a superbly uniform quench must be available. Unless this is so the variations that exist in most

Table 8.2 Advantages of Impingement Fixture Quenching — 1045 Bars

Bar Diameter (in.)	Surface Hardness (Rc)		Straightness TIR (in.)	
	Free	Fixture	Free	Fixture
1	55–57	59–60	0.038	0.008
2	50–57	58–59	0.026	0.006
3	41–56	56–58	0.017	0.004
4	35–52	55–58	0.015	0.003

production quenches will result in soft spots (see data in Table 8.2 for 1045 steel bars). The curves for torsion and bending applications in Fig. 8.3 are used by many firms. As for parts subject to tension and shear, the most precise method of steel selection is to determine the quench cooling rates in significant locations for the available quench and then select a suitable grade from published hardenability bands.

SELECTION FOR CYCLIC LOADING (FATIGUE)

In most parts premature failures are a result of fatigue. The design of parts and selection of steels and processing methods to prevent fatigue failure constitute an extremely broad subject on which volumes have been written. One of the most useful books on design is the *Fatigue Design Handbook,* published by the Society of Automotive Engineers. [2]. The cited reference points out that fatigue results from cyclic straining that can be described by a plastic and elastic component:

$$\frac{\Delta\epsilon}{2} = \frac{\Delta\epsilon_p}{2} + \frac{\Delta\epsilon_e}{2} \tag{8.3}$$

$$\frac{\Delta\epsilon}{2} = \text{total strain amplitude}$$

$$\frac{\Delta\epsilon_p}{2} = \text{plastic strain amplitude}$$

$$\frac{\Delta\epsilon_e}{2} = \text{elastic strain amplitude}$$

For high-stress conditions in which the plastic strain predominates the general equation for fatigue life can be approximated:

$$2N_f = \left(\frac{\Delta\epsilon_p}{2\epsilon'_f}\right)^{1/c} \tag{8.4}$$

where

$$2N_f = \text{fatigue life in cycles}$$

$$\epsilon'_f = \ln\left(\frac{100}{100 - \%RA}\right)$$

C = fatigue ductility exponent (for heat treated steel
$-0.70 \leqslant c \leqslant -0.50$)

%RA = reduction of area determined by standard tensile
test

It is important to note that this equation predicts that toughness measured in %RA is of primary importance to low-cycle fatigue strength.

For conditions of low nominal strain amplitude in which the elastic component $\Delta\epsilon_e$ is predominant

$$\frac{\Delta\epsilon}{2} \approx \frac{S_a}{E}$$

$$2N_f = \left(\frac{S_a}{\sigma'_f}\right)^{1/b} \tag{8.5}$$

where

S_a = nominal stress amplitude

E = modulus of elasticity

σ'_f = fatigue strength coefficient (for hardness less than
BHN 500 $\sigma'_f \approx S_u + 50,000$ psi where S_u
= ultimate tensile strength)

b = fatigue strength exponent (approx
= –0.085 for hardened steel)

This relation indicates that ultimate tensile strength is of prime importance under long life conditions. For intermediate conditions both elastic and plastic strain must be considered and the life relation becomes

$$\frac{\Delta\epsilon}{2} = \epsilon'_f \ (2N_f)^c + \frac{\sigma'_f}{E} \ (2N_f)^b \tag{8.6}$$

These relations are useful for general analysis but become very complex for components containing discontinuities where stresses and strains concentrate.

When selecting the proper steel for complex parts such as crankshafts, welded assemblies, and housings, highly sophisticated stress analysis methods are often necessary to determine the strain history. Sometimes a test part is made in a manner as close to production configuration and processing as possible. This test part is coated with brittle lacquer and under closely controlled conditions of temperature and humidity is stressed as it will be in service. The crack pattern of the brittle lacquer locates the highly stressed areas. The load is removed and strain gages are attached in these areas to measure precisely the strain from which the fatigue life can be calculated. These strain histories can be verified in actual operation (Fig. 8.1). The very slow and costly process of field-fatigue testing can sometimes be replaced or supplemented by full-size fatigue tests of components (Fig. 8.4).

Once the true operating strain history and areas of concentration have been established, selection of the proper steel and processing is greatly simplified. If abnormally high strain amplitudes are found in the birttle-lacquer test and strain-gage study, the designer can often solve the problem much more easily by a change in configuration than the materials engineer can by changing the material and /or processing. An example is shown in Fig. 8.5 in which a change in fillet radius reduced the stress concentration sufficiently to solve the failure problem in a large hub.

When space, weight, or shape limitations are inflexible, the materials engineer can resort to a number of alternatives, other than a composition change, which can be used either singly or in combination to improve fatigue properties. All rely on the control of defects and/or processing:

1. In any important application involving fatigue the materials engineer should take steps to ensure that the cleanest possible steel is used, consistent with cost limitations; he should also call for magnetic-particle inspection of the parts. It must be remembered that it is the largest nonmetallic inclusion at or

Fig. 8.4   Fatigue testing of a large, welded machine component.

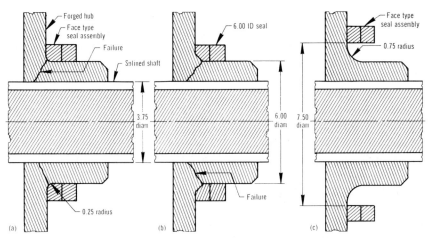

Fig. 8.5 Example showing solution of a fatigue-failure problem by a simple change in configuration.

slightly beneath the surface of a hgihly stressed area that most often causes premature fatigue failure. A successful way to minimize the inclusion problem is to obtain a carbon deoxidized vacuum degassed steel. The drawing should show critical areas and state, for example, "no stringers permitted as determined by magnetic particle inspection in this area." If parts are identified by the producer and his heat number, the quality control department can assist, on the basis of percentage of rejections, in establishing the best steel sources. It may be possible to avoid the degassing extras when the mill capabilities are well known.

Steel makers use several types of melting practice to improve steel cleanliness. In order of increasing cleanliness, steel made by the various processes is known as

(*a*) vacuum degassed,
(*b*) carbon deoxidized, vacuum degassed,
(*c*) carbon deoxidized, vacuum degassed, induction stirred,
(*d*) vacuum induction melted,
(*e*) vacuum arc remelted.

These processes in themselves may not necessarily produce clean steel, and each step in manufacturing must be properly performed. A comparison of the degree of cleanliness of steel produced by different methods is based on the step test for total stringer length.

	Total Stringer Length (in.)	Approximate Relative Cost[a]
Basic open hearth melt	4.000	1.00
Basic oxygen melt	3.500	1.00
Basic electric melt	2.500	1.10
Basic eletric melt + carbon deoxidize, vacuum degas	1.500	1.15
Vacuum induction melt	0.500	1.71
Vacuum arc remelt	0.125	2.57

[a]Based on 1045 hot-rolled bars, special quality.

Microcleanliness testing is of questionable value for most parts for two reasons: first, only a miniscule portion of a heat of steel is tested; second, inclusions of a microscopic size are not ordinarily the ones that promote premature fatigue (except at hardnesses above 55 Rc in applications such as rolling bearings or carburized gears). Nevertheless, microcleanliness specifications, based on SAE Recommended Practice J422 [3], that are acceptable to most producers are as follows:

	Maximum Ladle Carbon Content (%)		
	To 0.18 Inclusive	More than 0.18 to 0.33 Inclusive	More than 0.33
Basic oxygen process or basic open hearth quality alloy steels or RRA or MRR carbon steels	7–O, 5–S	6–O, 5–S	6–O, 5–S
Electric furnace quality steels	6–O, 5–S	5–O, 4–S	5–O, 4–S
Electric furnace bearing or aircraft quality steels	4–O, 4–S	4–O, 4–S	4–O, 4–S

A specimen is classified 7–O (oxide) and 5–S (silicate) to indicate that the longest oxide included was comparable to that in micrograph 7 of the J422 specification and the longest silicate inclusion noted was comparable to that in micrograph 5 of the same specification.

2. When stresses are such that a short-life fatigue situation exists (i.e., the cyclic strain amplitude approaches 0.5%), use strong and tough steels. One of the most relevant demonstrations of this type of loading is given in Fig. 1.5. Toughness can be optimized by keeping the carbon content below 0.35% nominal and the phosphorus and sulfur, below 0.010%, by using a boron steel and by using nickel as a major alloying element. Preferably the microstructure should be 95% min martensite and tempering in the range of 475 to 750F must be avoided.

3. For maximum long-life fatigue strength (more than 200,000 cycles) specify a material and heat-treatment combination (such as shell or induction hardening) that will develop high levels of residual compressive stresses in the surface of the part. Shot peening, cold or warm rolling, and nitriding (when applicable) are also effective in increasing long-life fatigue strength. A disadvantage of surface rolling is that unless the design is such that the amount of cold work can be measured, control is difficult. Processes such as shot peening and nitriding can be controlled more closely with process test samples. The effects of surface processing on fatigue life are shown in Fig. 8.6 as well as in Fig. 1.8.

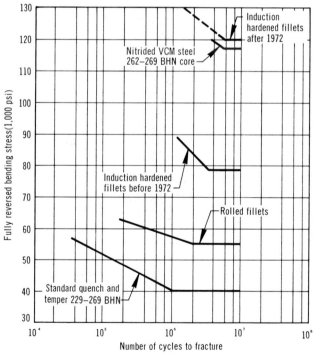

Fig. 8.6 Effect of surface treatments on fatigue life of a crankshaft (courtesy Allison Detroit Diesel Division, General Motors Corp.).

Table 8.3 shows the substantial increases in long-life fatigue strength that can be obtained with shot peening; however, unless a firm has had considerable experience with shot peening, the work is best subcontracted to those who are in this business. Figure 8.7 shows the beneficial effects of nitriding on long-life fatigue strength. This process is limited to steels that contain substantial amounts of nitride formers (e.g., Cr, Mo, Al) and is relatively costly.

Table 8.3   Typical Successful Applications of Shot Peening[a]

Part	Steel Grade	Hardness or Condition	Stress, (ksi)	Fatigue Life, Cycles Unpeened, (min)	Fatigue Life, Cycles Shot Peened, (min)
Automotive leaf spring	5160	40 to 48 Rc	170	12×10^3	80×10^3
			150	50×10^3	10×10^4
			140	80×10^3	30×10^4
			120	50×10^3	40×10^4
Latch spring	NR[b]	42 to 48 Rc	NR	6.5×10^3	12.5×10^{3c}
					17.5×10^{3d}
					24×10^{3e}
Shaft	NR	Quenched and tempered	30	2×10^6	. . .
			43	. . .	5×10^7
Connecting rod	NR	Quenched and tempered	NR	2×10^5	2×10^6
Gear	NR	Carburized	80	2×10^5	3×10^7
Landing-gear cylinder	NR	Quenched and tempered, CR-plated, and baked	60	10×10^4	. . .
		Quenched and tempered, shot-peened, CR-plated, and baked	100	. . .	10×10^4
Flat steel	4340	NR	90	5.3×10^4	. . .
		NR	90	. . .	20×10^4
		Electroless Ni-plated	90	. . .	14×10^4

[a] Data from Metal Improvement Company, subsidiary Curtis-Wright Corp., Hacksensack, N.J.
[b] Not reported.
[c] Shot-peened on outside surface only.
[d] Shot-peened all over.
[e] Strain-peened all over.

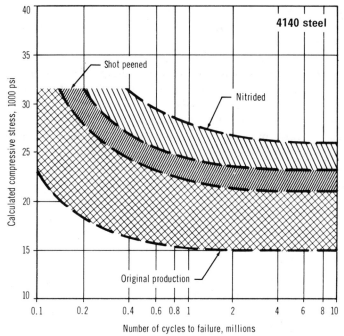

Fig. 8.7 Typical effect of nitriding on the long-life fatigue strength of steel shown by tests on a nitrided aircraft-engine crankshaft [6].

4. Control the surface finish in critical areas to 32 μin. max. to maximize the fatigue strength of a part. The processing to achieve these fine finishes must be properly selected; for example, Dr. Michael Field of Metcut Research Associates has found that steel harder than 40 Rc is nearly impossible to machine (with carbide or high-speed tools) without severely damaging the surface and thereby detracting from its fatigue strength. The detrimental effects of grinding on surface compressive stresses, as well as the danger of burning and even cracking, are well documented but sometimes overlooked. Manufacturers of grinding machines and wheels usually have a competent staff to give advice on the choice of proper wheels and grinding techniques to ensure the preservation of good surface qualities.

5. Select a steel and heat-treating practice that will eliminate the need for straightening after heat treatment. Many of the failure epidemics in passenger car axles in the 1920s were traced to cold straightening after heat treatment. These shafts were straight as ground, but after a few load cycles they returned to their original warped shape which resulted in wheel wobble and premature fatigue failure. One solution was to straighten hot (up to the tempering temperature)

and in some cases to temper after straightening. Either hot or cold straightening, however, may remove the desirable residual compressive stresses from the part's critically loaded areas.

6. Fatigue failure entailing corrosion can often be prevented by conventional nickel plating. A short-cycle fatigue failure (often fewer than 100 cycles) can occur as a result of stress corrosion. This is a problem in steels such as 4140 or 4340 heat-treated to strength levels above a 200,000-psi yield point. The corrosion can be caused by water or mildly corrosive solutions of other chemicals, such as chloride salts. When parts are to be heat-treated to this strength level and there is a possibility of stress   corrosion, the following expedients are suggested:

(a) Use 300M steel (essentially 4340 with 1.5% Si) to permit a higher tempering temperature. Whereas regular 4340 must be tempered at 425 to 475 F to obtain 200,000-psi yield strength, 300M can be tempered at 575 to 600 F.
(b) Make sure that all nicks and other surface defects are removed from the part. Maintain finish at 125 in. max.
(c) Make sure that the part is not carbon enriched during austenitization if atmosphere control is less than perfect.
(d) Shot-peen all accessible surfaces.
(e) Cadmium-plate and bake or cadmium-plate in vacuum.
(f) Apply a polyurethane or other nonmetallic protective coating to all exposed surfaces.

Hard electroless nickel plating reduces fatigue strength of heat-treated steel 10 to 40% and therefore should be avoided when maximum fatigue properties are required. Because of its precracked surface, hard chromium is also deleterious to fatigue strength of heat-treated steel. A step that is often necessary when plating steels harder than 30 Rc is baking to remove the hydrogen absorbed during plating. Parts should always be baked within 15 min after plating. The following shedule is suggested:

Tempering Temperature of Steel ($^\circ$F)	Suggested Baking Cycle
300 to 350	300 F, 6 hr
350 to 400	350 F, 4 hr
400 to 750	400 F, 3 hr
More than 750	500 F, 1 hr

One of the examples in Table 8.3 shows how the deleterious effects of plating can be largely eliminated by shot peening before plating.

7. The fatigue strength of a problem area that is only a small portion of the total part area can be increased by local hardening with flame or induction heating. This gain stems from induced residual compressive stresses, plus the higher endurance limit of the hardened steel. It is important to select the proper depth of hardening because insufficient hardened depth may induce surface residual tensile stress. Figure 8.8 shows how the critical fillet in a forged-steel wheel spindle was induction-hardened for additional resistance to fatigue failure (the small diameter adjacent to the fillet is 2.5 in.) induction hardening of this fillet increased its bending fatigue life by a factor of eight. This type of hardening is very effective in improving the torsional and bending fatigue strength of heavy-duty forged-steel crankshafts (Fig. 8.6). It should be remembered, however, that there is a zone at the end of an induction- or flame-hardened area that is residually stressed in tension. Provision therefore must be made to have the hardened area extend well beyond the highly stressed area.

8. For parts that are used without stock removal, such as many forgings and spring wire, surface quality is important to long-life fatigue strength. Surface defects that adversely affect fatigue resistance are decarburization, seams, laps, and cracks. A decarburized layer not only has lower endurance-limit fatigue strength because of its low strength but it also detracts from the surface compressive stress shown in Fig. 8.9. It is not unusual for a minute fatigue crack to

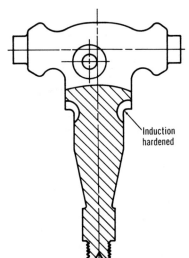

Induction hardened

Fig. 8.8 Induction hardening of a fillet in a forged-steel wheel spindle.

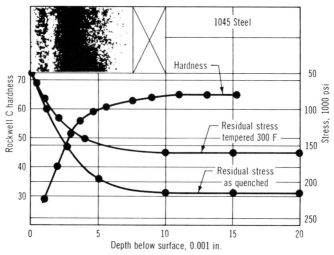

Fig. 8.9 Effect of decarburization on hardness contour and residual stress of water-hardened steel bars.

initiate in this decarburized layer and be arrested when it reaches the part of the steel that is in a sufficiently high state of compression. This sharp-bottomed crack, however, has a tremendously high stress-concentration coefficient and premature failure can result from reduction in both long- and short-life fatigue strength or from impact.

Long-life fatigue strength of 4340 steel is typically reduced by decarburization:

	Tensile Strength (psi)	Endurance Limit (psi)	
		Decar-burized	Not Decar-burized
Pearlitic	157,000	47,000	67,000
Martensitic	275,000	47,000	98,000
Tempered martensitic	178,000	50,000	90,000

Turned and polished or turned, ground, and polished steel bars cannot be assumed to be free of decarburization unles this requirement is negotiated with the supplier. Carbon-restored bars that are essentially decarburization free are available; however, the surface carbon content may vary, for example, from 0.05% below the mean carbon to as much as 0.25% above the mean carbon level. This amount of variation could create problems on the processing and performance of some parts.

Accordingly, all surfaces known to be subject to significant fatigue loading should be carbon-restored and machined. (See Appendix V for recommended stock removal.) Surface imperfections such as seams cause severe stress concentrations and should not be allowed in critically stressed areas. As-rolled bars of basic open-hearth quality can be expected to have 0.001 in. max seam depth per 0.062 in. of diameter. A shallower limit can be negotiated with some producers for a price. Steels are available with various levels of surface and internal quality, as shown in Table 8.4.

9. In applications that require extreme long-life fatigue strength (more than 50×10^6 cycles) it is necessary to use a steel and heat treatment that will provide a microstructure free of ferrite, at least in the critical sections of the part. Typical examples are the connecting rods and crankshafts of high-performance internal combustion engines.

SELECTION FOR TOUGHNESS

Toughness is the ability of a material to deform plastically under conditions of high stress concentration that may result from a notch or crack. This is one of

Table 8.4 Surface Quality Available in Raw Steels

Type of steel	Surface Quality (in order of increasing integrity)
Semifinished steel (carbon)	Rerolling, forging
Hot-rolled carbon bars	Merchant, special, axle shaft, shell steel (C), shell steel (B), restrictive cold working
Carbon steel plates	Regular, cold flanging, firebox, forging, marine boiler
Alloy Steel plates	Regular, flange, firebox, bearing, aircraft
Hot-rolled carbon steel sheets	Commercial, physical, drawing, special soundness steel

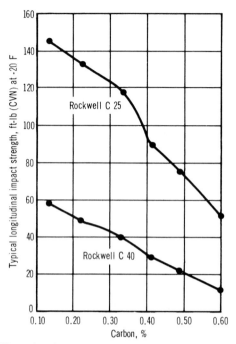

Fig. 8.10 Typical effect of carbon content on Charpy impact strength of the 8600 series steels tempered to hardnesses shown.

the most important characteristics of steels specified for low-temperature service. Many materials, including steel, undergo a decrease in toughness as the temperature is lowered.

Proper quenching and tempering make a steel so tough that, compared with as-rolled steel, composition (except for carbon, sulfur, and phosphorus) is irrelevant for all practical purposes; for example, as-rolled 8620 in a 2.50-in. square has a transverse Charpy V-notch (CVN) impact strength of approximately 10 ft-lb at –20 F: When water quenched and tempered to 25 Rc, this same steel shows an impact strength of more than 50 ft-lb. Toughness is so improved by quenching and tempering that some low-carbon, semikilled steels can be applied to parts requiring good low-temperature toughness.

Maximum toughness for a given grade, however, is provided by thorough deoxidation, preferably by carbon and silicon under vacuum. Aluminum, vanadium, or columbium additions are used to produce a fine-grain size. The positive effect of fine grain on toughness is well documented and is covered in detail in Volume 1 of the ASM Metals Handbook. Steels are now being produced on a laboratory basis with grain sizes as fine as ASTM 15. Some producers are willing to work to a specification of ASTM 7 or finer.

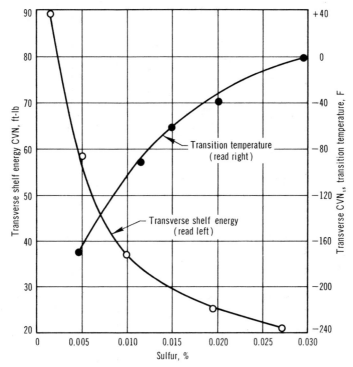

Fig. 8.11 Effect of sulfur content on the impact strength of HSLA steel. (*55th National Open Hearth & Basic Oxygen Proceedings*), Vol. 55, Chicago, IL, April 10–12, 1972, – Sulphur – Some Effects on Steel Processing and Properties. G. J. Roe, P. P. Slimmon, G. F. Melloy (Reported by permission of the American Institute, Met. and Petroleum Engineers).

Treatment with zirconium to change sulfide inclusion shape, and rare-earth additions, will also improve toughness. In addition to the killing technique and grain-size control, an important consideration in choosing a steel for maximum toughness is to keep the carbon, phosphorus, and sulfur content as low as possible. The effect of carbon content on Charpy impact strength is shown in Fig. 8.10. With the basic oxygen process phosphorus content can normally be held below 0.015% without great difficulty. Sulfur is more of a problem, but some producers will accept orders calling for 0.005% max. The strong effect of low sulfur on improving toughness is shown in Fig. 8.11 and 8.12. When maximum toughness is required, the materials engineer should specify a sulfur content that is as low as economically practical. Note, however, that when steel has sulfur content below 0.020% the machinability is adversely affected. When maximum transverse toughness is required, the supplier will cross-roll more than is normally required to minimize directionality. It is not recommended that

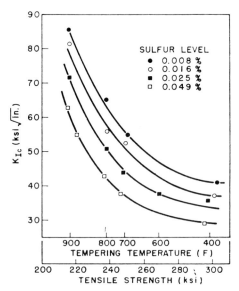

Fig. 8.12 Influence of Sulfur Level on Plane Strain Fracture Toughness of AISI 4345 Steel. ([5] Reprinted by permission of the American Society of Testing and Materials, Copyright).

rolling technique become part of a material specification. Instead, the required properties should be called out and the producer should be permitted to furnish suitable steel by whatever method he determines best.

In heat treating for maximum toughness, the main objective is to obtain tempered martensite or lower bainite (when austempered). Carburizing furnace atmospheres must be avoided at all times. For steels of 0.35% C content or less embrittlement must be avoided by selecting a grade in which the final hardness can be achieved without tempering between 475 and 750 F.

SELECTION FOR WEAR RESISTANCE

There are three principal types of wear: *adhesive, abrasive,* and *corrosive. Adhesive* wear occurs when two surfaces are rubbed together without the presence of an abrasive agent. Good adhesive wear resistance against steel is usually provided by metals that have low melting points, are insoluble in iron, and do not form intermetallic compounds with iron. The adhesive wear resistance of various metals on steel is as follows:

Good	Fair	Poor	Very Poor
Germanium	Carbon	Magnesium	Beryllium
Silver	Selenium	Aluminum	Silicon
Indium	Tellurium	Zinc	Calcium
Tin		Barium	Titanium
Antimony		Tungsten	Chromium
Lead			Iron
			Cobalt
			Nickel
			Platinum
			Gold
			Molybdenum

The adhesive wear of steel running against steel is known as scoring or galling and seizing. This mechanism is a common problem in heat-treated steel parts such as a case-hardened differential spider running against a case-hardened gear. Scoring can be minimized by lubrication and cooling, the latter being extremely important. A small amount of scoring takes place between most steel parts. The result is burnishing of the surfaces from a finish of, say, 50 down to 25 μin. or less, but then the process stops. Destructive scoring will continue to a point at which pitting or actual seizing takes place.

The materials engineer often finds it difficult to prevent scoring if the parts have full hardness and proper microstructure. A common metallurgical solution employs a coarse phosphating treatment such as Parco-Lubrite (a product of Park Chemical Co.). This treatment deposits a manganese phosphate layer on an etched surface. The thickness of this layer is normally about 0.0005 in. Surfaces ground to a very fine finish are difficult to coat properly and a light grit blast before coating is desirable. Beyond this, the design engineer must provide for better lubrication and/or the maintenance of lower lubricant temperatures. With lubricants that do not contain extreme pressure additives the interface oil temperature must be kept below 400 F. Most extreme pressure oils will permit lubricant temperatures 50 to 100 F higher. If the interface temperature exceeds these levels, the lubricant will break down and scoring may occur. A typical formula for determining the tendency to score is that for the interface temperature between two gears:

$$T_t = T_1 + \frac{1.297f(W/\cos\theta)^{3/4}(\sqrt{V_1} - \sqrt{V_2})}{(1 - \text{RMS}/50)\,[P_p P_g/(P_p)]^{1/4}} \qquad (8.5)$$

where T_t = maximum temperature at contact point, F

T_1 = bulk oil temperature, F

W = maximum unit load in pounds per inch of face

V_1 and V_2 = surface velocities at contact point in inches per second

f = coefficient of friction, usually 0.15

RMS = surface finish in microinches after initial run-in

$P_p = [r_0^2 - (r \cos \theta)^2]^{1/2} - (\pi \cos \theta /P)$

r_0 = outside radius of pinion in inches

r = pitch radius of pinion in inches

θ = pressure angle in degrees

P = diametrical pitch

$P_g = [R_0^2 - (R \cos \theta)^2]^{1/2} - (\pi \cos \theta /P)$

R_0 = outside radius of gear in inches

R = pitch radius of gear in inches

From this it can be seen that bulk oil temperature, relative velocities, and contact stress (which is directly related to unit load) are important in the scoring mechanism.

The parameters affecting *abrasive* wear resistance have been quite well established and are essentially as follows:

1   Each set of conditions is unique and test results vary widely.

2   Being a dynamic phenomenon, the wearing of heat-treated steel by an abrasive material is too complex to predict by simple appraisal of two materials of known characteristics rubbing together. Certain chemical and physical reactions take place as a result of the dynamics of the situation and are significant in the results.

3   Generally speaking, abrasive wear resistance increases slightly with carbon content and more strongly with hardness (Fig. 8.13).

4   All else being equal (hardness, carbon content, and microstructure), steels containing chromium, molybdenum, and vanadium, singly or in combination, show somewhat better wear resistance than plain carbon steels. For composition to be substantially effective the carbon and alloy contents must be far above those of the constructional steels; for example, 2% C or higher and 3% Cr or higher. These materials are expensive and very brittle and great care must be exercised in using them in dynamically loaded machine parts.

5   When steels such as 1050 are induction-hardened to 60 R_c min and do not provide adequate abrasive wear resistance, there is little that can be done economically to increase part life other than to provide more pounds of hard steel to wear away. Special high-carbon, high-chromium tool steels provide better wear resistance but at a larger increase in cost and a loss in toughness.

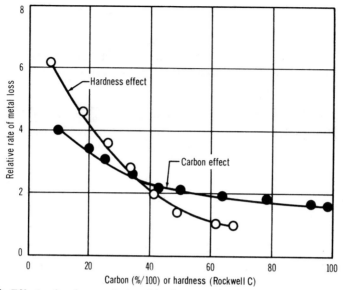

Fig. 8.13 Effects of carbon content and hardness on abrasive wear resistance of constructional steels.

The selection of steels for abrasive wear resistance is covered in greater detail in Chapter 14.

Corrosive wear, as the term implies, is the loss of section thickness by chemical attack on a steel in dynamic contact. A typical example is the life of steel railroad rails in various environments:

$$\text{Wear per year} = K \sqrt{\text{millions of axles traffic/year}}$$

Values of K:

Electric subway	0.004 –0.006
Electric commuter	0.010 –0.018
Rural steam	0.007 –0.022
Industrial steam	0.015 –0.030
Dry tunnel (electric)	0.0015–0.0025
Tunnel	0.050 –0.100
(steam and smoke)	

The amount of alloy in the constructional steels is inadequate to provide any substantial degree of resistance to corrosive wear. Some investigators, however, have offered preliminary evidence to show that the 1% Cr contained in the 5100 series is of some benefit.

REFERENCES

1 R. E. Petersen, *Stress Concentration Factors,* Wiley, New York, 1953.

2 *Fatigue Design Handbook,* Society of Automotive Engineers, New York, 1968.

3 Microscopic Determination of Inclusions in Steels, SAE Recommended Practice J422A, *SAE Handbook,* 1972, p. 99.

4 R. F. Kern, *Metal Progr. Data Book,* 1974, p. 179.

5 R. P. Wei, Fracture Toughness Testing in Alloy Development, *Fracture Toughness Testing,* ASTM STP 381, American Society for Testing and Materials, 1965, p. 286.

6 *Metals Handbook* Vol. 1, Properties and Selection of Metals, American Society for Metals, 1961, p. 222.

CHAPTER NINE

GENERAL PRINCIPLES FOR
THE SELECTION OF STEELS
FOR CASE HARDENING

C ASE HARDENING CAN BE ACHIEVED by selectively heating only a surface layer (e.g., flame or induction) or by changing the composition of a steel surface. The case hardening processes are discussed in Chapter 8 and elsewhere throughout this book, and this chapter is limited to those processes that involve a change in surface composition. Numerous proprietary processes that claim to enrich surfaces with exotic elements ranging from boron to tungsten are known to exist. Although they may be useful in some special applications, processes based on the addition of carbon and/or nitrogen have proved to be reliable over many years and generally are more predictable, less expensive, and more readily available. These processes, generically known as carburizing, carbonitriding, or nitriding, are used to provide the following:

1 Resistance to wear.
2 Resistance to scoring.
3 Improved bending and/or torsional static strength.
4 Improved bending and/or torsional fatigue strength.
5 Improved rolling contact fatigue strength.

A comparison of the characteristics contributed to parts by the case-hardening methods of carburizing, carbonitriding, and nitriding is made in Table 9.1. Any or all of these characteristics can be provided by case hardening with com-

159

Table 9.1　Case-Hardening Processes and Their Characteristics

Carburizing	Hard, highly wear-resistant surface (medium case depths); excellent capacity for contact load; good bending-fatigue strength; good resistance to seizure; excellent freedom from quench cracking; low-to-medium-cost steels required; high capital investment required
Carbonitriding	Hard, highly wear-resistant surface (shallow case depths); fair capacity for contact load; good bending-fatigue strength; good resistance to seizure; good dimensional control possible; excellent freedom from quench cracking; low-cost steels usually satisfactory; medium capital investment required
Nitriding	Hard, highly wear-resistant surface (shallow case depths); fair capacity for contact load; good bending-fatigue strenth; excellent resistance to seizure; excellent dimensional control possible; good freedom from quench cracking (in the pre-treatment); medium-to-high-cost steels required; medium capital investment required

parative ease up to well established limits. Usually, but not always, steels for carburizing or carbonitriding are limited to a carbon content of 0.25% max for one or both of the following reasons:

1 Parts are provided with a hard surface, yet with a high degree of toughness. Lower carbon steels also have superior machinability, especially grades like 1117 and 11L17. Grades like 1010 and 1015 have optimum formability to permit cold heading and cold forging.
2 High residual compressive stresses in the case resulting from large differential carbon contents between surface and core permit higher unit load-carrying capacity; maximum long-life bending and torsional fatigue qualities are also assured.

The toughness of two case-hardened alloy carburizing steels is compared with that of a medium-carbon through-hardening steel (8660) in Fig. 9.1. The compressive stresses typically present in a carburized gear tooth are shown in Fig. 9.2. These stresses add significant cyclic load-carrying capability.

The selection of steels for case hardening requires an even more precise determination of the minimum engineering requirements than is necessary for the through-hardening grades because the case depth and microstructure of the case as well as core properties must be anticipated.

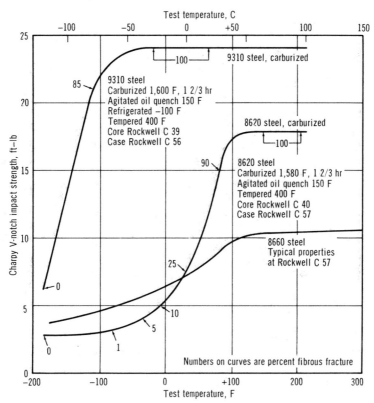

Fig. 9.1 Relationship of Charpy impact strength to test temperature for three steels (from *Selecting Steels and Designing Parts for Heat Treatment,* American Society for Metals, Copyright 1969).

In selecting steels solely for resistance to wear under conditions of low contact stress and/or bending or torsional loading, the materials engineer need only consider hardenability to the extent that he is sure that the case will harden adequately. The hardened layer performs no significant engineering function other than extending the wear life of the part. The only heat-treating requirements would normally be an adequate case depth and hardness for the life required. For wear resistance under conditions of high compressive loading, however, the hardened layer must not only be deep enough for the wear life of the part but must also support heavy contact loads, usually under conditions of sliding or rolling or combinations thereof. For this purpose the case must be fully martensitic except for some allowable retained austenite which will vary with the application; for example, most rolling bearings could not tolerate the normally acceptable limit of 25% austenite because they require excellent

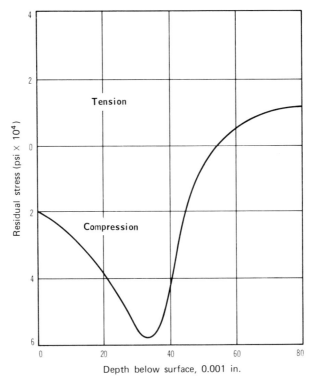

Fig. 9.2 Subsurface residual stresses in a carburized gear (from Pedersen and Rice [2]).

dimensional stability. With high unit loading the combination of case depth and core hardness must also be adequate to prevent case crushing. When case hardening is employed to prevent scoring between two parts under heavy contact loading, it must produce an extremely hard surface. Those processes that add nitrogen to the surface (carbonitriding and nitriding) are exceptional in this respect and provide a "slippery" surface with maximum scoring resistance.

Case hardening is also used to increase bending or torsional strength. As with the direct hardening steels, however, the strength at all points from the surface to the neutral axis must exceed the operating stresses. Because case-hardening steels are usually tempered at 300 to 350 F after quenching, the core tensile strength is also high.

When the materials engineer is called on to select a steel and a process for case hardening, his first consideration should be to meet engineering requirements. As shown in Table 9.1, no process is best for all applications, for each has at least one advantage over the others. The material and process combination, that will produce a satisfactory part regardless of cost and finally the combination that will produce the part at the lowest total cost should be determined. The engineering requirements for parts to be case-hardened can be obtained from experience, from calculation, or from tests. Experience is usually quite reliable; however, it is not uncommon that new designs cannot rely completely on prior knowledge. Calculation is a necessity in most case-hardening applications, but it too requires making assumptions, some of which may be incorrect. In all but the simplest parts calculation plus experience, followed by testing, are necessary to ensure proper performance at lowest cost.

SELECTING STEEL FOR CARBURIZING

The composition of a steel selected for carburizing should have these characteristics in varying degrees, depending on application:

1 It must be capable of being formed (forged, extruded, etc.) with a minimum of difficulty.
2 It must generally be easy to machine.
3 The hypereutectoid portions of the case must have adequate hardenability as measured by hardness and/or microstructure.
4 The steel must be capable of developing the desired microstructure with a quench and maintain distortion at a consistently low level from heat to heat.
5 It must have adequate hardenability so that the desired properties can be obtained in the hypoeutectoid zones of the case and in the core.
6 After carburizing and hardening the part must have sufficient toughness to perform without failing in a brittle manner.

The steel-selection process must take into consideration *case and core* hardenability on one hand and quench-cooling rate on the other; for example, during World War II steels like 4820 were replaced by leaner alloy grades like 8627 on the basis of equivalent core hardenability. The results were often disastrous because the case hardenability was inadequate, the residual compressive stresses in the case were lowered by the substantially higher core-carbon content, and the desirable effect of nickel on short-life fatigue strength was lost.

The milder the quench, the more highly alloyed the steel. Water, brine, or caustic quenching may be preferred to minimize steel cost, but the attendant distortion and the serious deleterious effect of the drastic quenching on furnace fixtures must be considered. Figure 9.3 provides a means of making a quick estimate of case hardenability required for hardening round bars to 60 Rc with 0.85% surface carbon. Note that the case carbon should not exceed 0.90% when water quenching alloy grades to prevent microcracking. When unit loading

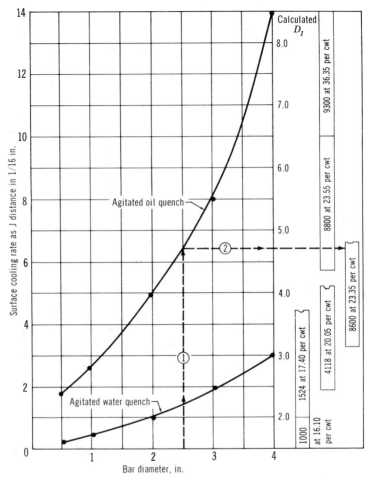

Fig. 9.3 A guide to the selection of low-price steels that will provide a 60 R_c case when carburized to the 0.85% C level, reheated, and quenched in agitated oil or water. A 2 1/2-in.-diameter bar [1] processed by an oil quench would require an alloy level equivalent to that found in 8800 or 8600 steel [2]. Prices are for OH-base hot-rolled bar plus grade (from *Selecting Steels and Designing Parts for Heat Treatment*, American Society for Metals, Copyright 1969).

is very low, especially in pure compression, almost any steel can be used in a carburized part; for example, a ball socket loaded in pure compression is machined from 5-in.-round 1215 steel and carburized and hardened with a water quench. The machinability of the 1215 is desirable to form the 2.75-in.-diameter spherical seat. When high bending, torsion, or tension stresses must be resisted, this type of steel should never be used because of its brittleness. A very impor-

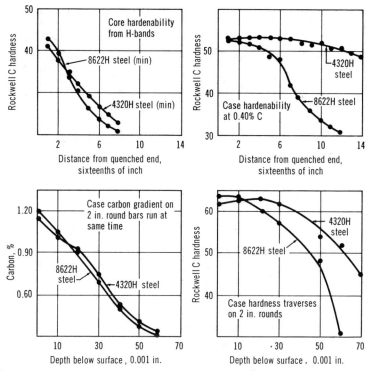

Fig. 9.4 Comparison of core hardenability, case hardenability, case-carbon gradient and case-hardness traverses of 4320H and 8622H steels (from *Selecting Steels and Designing Parts for Heat Treatment*, American Society for Metals, Copyright 1969).

tant matter to keep in mind is that steels of equivalent core or base hardenability can have widely different case hardenabilities, as shown in Figure 9.4.

The proper selection of steels for carburizing requires that the quench-cooling rate be known in significant locations of a part or estimated as suggested in Chapter 6. With this information it is possible to select steels that will produce a fully martensitic (with austenite) microstructure in the carburized case if the minimum DFB (*J* distance from first bainite, determined by the method described in Chapter 6) is known. The calculated distances to 10% or even less transformation products can be determined by using the methods developed by Jatczak and Girardi [1]. It must be remembered that case hardenability varies with the heat treating practice (direct quench or reheat and harden), as shown in Table 10.2. The DFB minima in the table are estimates based on limited test data.

In addition to ensuring adequate case hardenability and determining the core hardenability from published bands, the materials engineer must keep the following points in mind:

1   Grades in which the major contribution to case hardenability is provided
    by carbide-forming metals such as chromium are somewhat sensitive to
    microcracking, particularly when vigorously direct-quenched. This can
    be controlled by restricting the case carbon to 0.90% max. The safe limit
    for the 8600 series at 0.90% case carbon is a quench-cooling rate of J 3.
    Grades like the 8800 series are preferably not direct-quenched.

2   Grades that have substantial percentages of nickel (e.g., as the 4800 and
    9300 series) may form excessive amounts of austenite (more than 30%)
    when direct-quenched unless the case carbon is maintained below 0.75%;
    for example, a four-pitch final-drive pinion made of 8620 failed by pit-
    ting because of excessive bainite in the case microstructure. The heat-
    treating practice used was a direct quench from 1550 F with a case
    carbon content of 1.0 to 1.2%. The material was changed to 4820 to
    eliminate the bainite, but with high case carbon and direct quenching
    the austenite in the microstructure exceeded 40%, the case hardness was
    as low as 53 Rc, and failures continued. The failures were still by pitting
    but resulted from the loss in load-carrying ability engendered by the
    excessive austenite. They were finally solved by a change back to 8620
    with a more vigorous quench which eliminated the bainite. In fine-pitch
    gears problems may occur from excessive austenite when 8620 is direct-
    quenched. Deep chilling to –100 F is sometimes proposed as a method of
    salvage. Unfortunately, this will often produce severe microcracking,
    as shown in Fig. 9.5, and can lead to cracking in grinding and premature
    bending-fatigue failure.

3   Carburizing of sections larger than 3-in. round presents problems of
    obtaining satisfactory case and core hardnesses and microstructures
    because of the low quench-cooling rates, at least with oil quenching.
    Processes such as induction hardening or nitriding should be considered
    for such applications. When this is not possible, highly alloyed steels
    like 9310 are usually required.

Some reasons for using grades other than those given in Fig. 9.3 are the following:

1   Grades like the 4000 and 4600 series are substantially more machinable
    than 1524 and are normally preferred when expensive cutters such as
    hobs are used and superior finish is required. A comparison of the ma-
    chinabilities of numerous carburizing steels is shown in Fig. 9.6.

2   The user who buys his steel from warehouses is usually limited to selec-

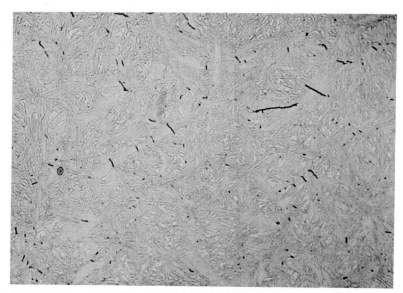

Fig. 9.5 Microcracks in 8617 steel carburized, reheated to 1540 F and quenched. Typical of cracks produced by deep chilling.

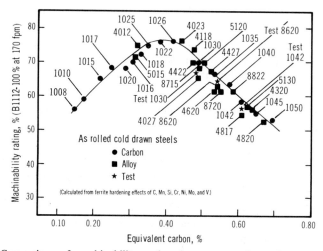

Fig. 9.6 Comparison of machinability and carbon content for various steels (courtesy Bethlehem Steel Co.).

167

tion from 1018, 1019, 1117, 1118, 4620, 8620, and, in a few instances, 3310 or 9310.

3 The grades shown are those suggested for firms whose heat-treating facilities provide precise control of temperature and atmosphere and uniformly vigorous quenches. When this metallurgical control is not available, the use of steels more highly alloyed with nickel is suggested. These steels — 4300, 4600, 4800 and 9300 — have a degree of case-carbon control built in and will fully harden in even sluggish quenches; for example, a production heat treatment used on three-pitch final-drive pinions made of 4820 steel is

(a) Carburize at 1700 F, slow cool to room temperature.
(b) Reheat to 1490 to 1510 F.
(c) Cool in unheated chamber (individually) for 5 to 6 min.
(d) Quench in agitated oil.

In 0.90% case carbon this heat treatment produces file-hard parts (58 to 63 Rc) with no network carbide or upper bainite and a core micro-structure of almost 100% martensite.

4 Some of the standard grades of steel have much narrower core-harden-ability bands than others. In these steels the case hardenability spreads are also narrower; for example, the band spread of 8620H at 4/16 in. (J 4) is 14 points Rockwell C, whereas 9310 has a spread of only eight points. This provides a higher degree of distortion control as well as certain engineering characteristics that are required for some parts. It should be borne in mind that, as previously mentioned, alloy grades can usually be purchased to as much as a 50% restriction in hardenability for an extra of approximately 10%.

The surface of a carburized part is usually subjected to the highest engineering demands that make it important to call for a steel and heat-treatment combina-tion that will ensure adequate surface properties. Beyond this provisions must be made to provide necessary core properties. These include adequate case support (to prevent case crushing), yield strength to transmit torque and/or support bending loads, and toughness for insurance against brittle fracture.

The work of Pederson and Rice [2] on case crushing in carburized gears is helpful toward an understanding of this phenomenon. Although the work was done on gears, it is applicable to other parts like pins, shafts, and bushings which are ground after heat treatment, thus removing substantial percentages of the carburized case. For general use with a core hardness of 30 to 45 Rc the required case depth can be roughly estimated by the following:

$$\text{Case depth to } 50 \, R_C = \frac{7.5 \times 10^{-6} \, (W) - .0175}{F} \tag{9.1}$$

where W = the force in pounds pressing the surfaces together
$\quad F$ = length of line of contact in inches

The strength of the core can usually be determined from the hardness, provided the steel was properly austenitized (an important factor in long-life fatigue applications); it should have adequate hardenability so that, on quenching, the structure will be essentially martensitic. It must be remembered that case crushing occurs by shear. The shear yield strength of heat-treated steels is shown in Fig. 9.7. Pederson and Rice [2] found that the subsurface stress/strength ratio should not exceed 0.55 for long-life cyclic applications.

Proper austenitization and quenching to martensite are also required for maximum core toughness. Any ferrite present, especially of a primary type in

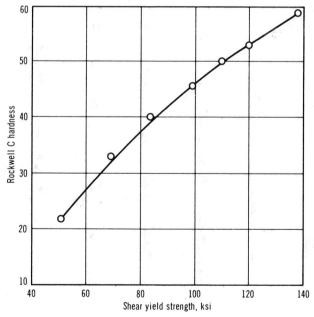

Fig. 9.7 Shear yield strength of carburizing steels as affected by hardness (from Pederson and Rice [2]).

blocky form, will seriously detract from the toughness. Toughness, especially at low temperatures, is improved by the use of nickel-bearing grades, as typically shown by AISI 9310 in Fig. 9.1. DePaul [3] found a marked superiority of nickel-bearing steels in impact fatigue (Fig. 9.8). DeBarbadillo [4] also found that

- (a) nickel-containing carburizing steels with normal amounts of austenite (20%) do not lose their long-life fatigue qualities after impact pre-stressing to the same degree as nickel-free steels (4820 versus 4118 in Fig. 9.8);
- (b) steels such as EX-1 and 4820 were found to have their long-life fatigue qualities actually somewhat improved by impact prestressing (shown in Fig. 9.9).

STEELS FOR CARBONITRIDING

The carbonitriding process adds nitrogen to a steel surface along with carbon. The effect of the nitrogen is to increase the case-hardenability substantially. Typical effects are shown by the results of tests on case-hardened, end-quench specimens in Fig. 9.10. By applying the hardenability factors of Jatczak and Girardi [1] it is possible to conclude that the improvement in hardenability on the 1020 steel in Fig. 9.10 is equivalent to a molybdenum content of more

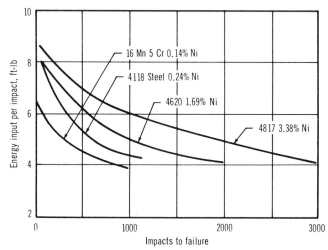

Fig. 9.8 Impact fatigue strength of various grades of carburized alloy steels with case depths of 0.025 to 0.034 in. (from DePaul [3]).

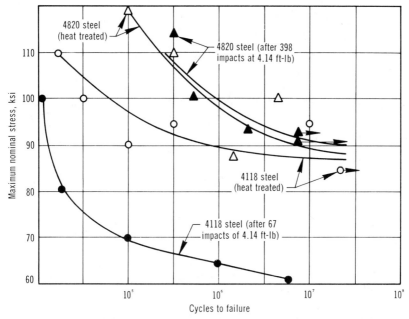

Fig. 9.9 Effect of impact prestressing on long-life fatigue strength of carburized steel (from De Barbadillo [4]).

than 0.50% (hardenability factor of 4). This effect is, of course, shallow (0.005 to 0.025 in.) and also decreases sharply as the alloy content of the base steel is increased. Using the same Jatczak and Girardi technique on the 8822 steel in Fig. 9.10, we find that the nitrogen factor is approximately 1.40 (equivalent to approximately 0.13% Mo).

The strong positive influence of nitrogen at hypereutectoid carbon levels in plain carbon and low-alloy steels often permits the use of oil quenching with attendant low distortion even when full hardness is required. A D_I and quench-cooling rate at J distance can be selected from Fig. 9.11. In using this chart, the quench-cooling rate must be estimated or determined according to the recommendations in Chapter 6. The following points must also be considered:

1 When using resulfurized steels, keep in mind that a portion of the manganese content equaling approximately three times the percentage of sulfur is tied up as manganese sulfide and therefore is not available to enhance hardenability.

2 Figure 9.10 is for 3% ammonia atmospheres in the heat-treating cycle. Higher percentages will increase the percentage retained austenite (and

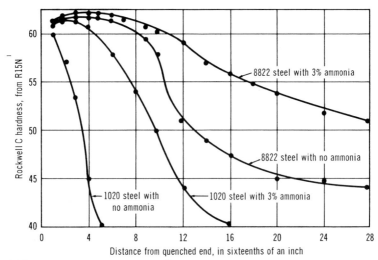

Fig. 9.10 Results of carbonitriding of case-hardened end-quench specimens of 8822 and 1020 steels (from *Selecting Steels and Designing Parts for Heat Treatment,* American Society for Metals, Copyright 1969).

lower the hardness), whereas lower percentages of ammonia will reduce case hardenability.

3 It should also be remembered that if penetration hardness is required steels containing more than 0.70% Ni or 1.65% Mn have strong austenite-forming tendencies that may create problems for example, in meeting, a 58 Rc min surface-hardness requirement.

4 The steels to which the D_I applies in Fig. 9.11 will contain less than 0.45% C.

The good hardening capability of the carbonitrided case permits the material selection for many parts on the basis of lowest factory cost up to the heat-treating operation. Consequently, when only a small amount of machining is required, plain carbon steels such as 1018, 1019, 1020, and 1022 are used. If machining is more extensive, grades like 1116, 1117, and 1119 are used. Because of their manganese content, these grades, along with 1118, also have substantially higher minimum case hardenability than those with only 0.30 to 0.60% or 0.60 to 0.90% Mn. This makes them a good choice when section thickness exceeds 0.75 in. When machinability is a major consideration, grades such as 1212, 1213, and 1215 can be selected. It must be remembered that the high phosphorus content of the 1200 series make it extremely brittle in the case-hardened condition. These steels are also usually coarse-grained, which further increases brittleness. A typical guide for selecting steels for carbonitriding is given in Table 9.2.

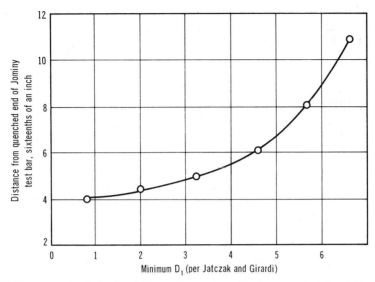

Fig. 9.11 Approximate hardenability and quench-cooling rate required to obtain 58 R_c min in carbonitriding.

Table 9.2   Steel Grades Versus Size for Hardening by Oil Quenching to R_{15N} 88 Minimum Surface Hardness

Grades	Manganese (%)	Maximum Diameter (in.)
1010, 1015, 1020	0.30–0.60	0.50
1016, 1018, 1019	0.60–1.00	0.75
1117, 1119	1.00–1.30	1.00
1118, 1524	1.30–1.65	1.50

When higher core properties are required than can be provided by the plain low-carbon steels or the screw stocks, alloy steels can be used. For critical applications coarse-grained steels should be avoided, especially for heavy cases (0.010 to 0.030 in.), because deleterious microcracking can occur; 5135, 5140, 8635, and 8640 are popular carbonitriding steels for gears and hydraulic valve components. Core hardnesses of small-diameter bars for popular oil-quenched carbonitriding grades of steel are given in Table 9.3.

The discussion of steel selection for carbonitriding up to this point is based on those grades that are used primarily in bars for which oil or hot oil quenching is to be employed. Also, the type of carbonitriding involved is carried on at

Table 9.3 Minimum Core Hardnesses for Small-Diameter Oil-Quenched Car-bonitrided Rounds (H-Value of Quench 0.55, Hardness Values at 10% Radius Depth)[a,b]

Diameter of Bar (in)	Hardness (Rc) as Quenched							
	1018	1022	1026	1117	1118	1132	1524	1527
0.25	41	43	45	41	41	48	44	46
0.50	39	43	45	41	41	48	44	45
0.75	22	25	26	25	31	37	36	34
1.00	18	19	21	21	22	28	27	26

[a] All steels contain 0.15 to 0.30% Si. All are coarse-grained except 1524 and 1527, which are fine-grained.
[b] Courtesy Keystone Steel and Wire Corp.

1550 ± 50 F in an atmosphere of 1 to 5% ammonia, 1 to 3% enriching gas, and the balance in endothermic carrier gas. Another type of carbonitriding is carried out at a temperature of 1425 to 1475 F in an atmosphere with an ammonia content of 5 to 10%. This process is usually on thin parts, such as thrust washers, stamped from commercial-quality steel sheet or strip. The lower temperature results in less distortion, especially if parts are properly racked and gas-quenched in a cool atmosphere. Many such parts require a case depth of only 0.005 to 0.010 in. The case hardness is only 50+ Rc converted from Knoop, but the high-nitrogen surface is resistant to scoring.

Case depths in carbonitriding are time-temperature-dependent, as shown in Fig. 9.12. The strong influence of temperature is evident and must be kept in mind when case depth is selected for parts to be carbonitrided at the lower ranges.

SELECTION OF STEELS FOR NITRIDING

The nitriding process is used for one or a combination of the following reasons:

1 To provide a thin, but very hard, case for wear resistance.
2 To provide a surface that will be highly resistant to scoring.
3 To improve long-life fatigue qualities by the development of high resid-ual-surface compressive stresses.
4 To improve corrosion resistance (except stainless steel).
5 To provide a hard case that is resistant to tempering up to the nitriding temperature.
6 To case harden with the lowest and most predictable distortion of any heat-treating process.

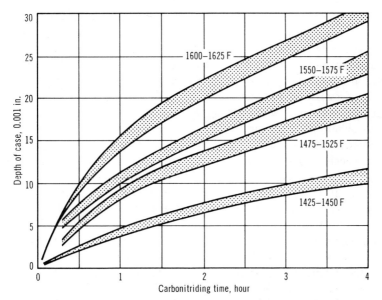

Fig. 9.12 Comparison of case depth of steels carbonitrided with varying times and temperatures (from *Selecting Steels and Designing Parts for Heat Treatment*, American Society for Metals, Copyright 1969).

Nitriding can be done in molten salt or in a partly dissociated ammonia gas atmosphere almost always at a temperature between 950 and 1050 F. A conventional nitriding treatment (without quench) is not considered suitable for use on carbon steels. The nitrided case is very brittle on carbon steel and tends to spall. The constructional grades of steel suitable for nitriding must contain one or a combination of the following alloying elements: aluminum, chromium, molybdenum, titanium, and vanadium. Accordingly, they must be purchased at an alloy base price. Besides carbon content, selection is usually based on the level of case hardness required, as shown in the tabulation below:

	Working Hardness	
Steel	Knoop (min)	Rc (approx)
4330, 4340, 8640	400	40
4130, 4140	475	48
4140 with 1% Mo (VCM)	550	53
Nitralloy 135	1000	70
H11 tool steel	1000	70
Type 430 or 466 stainless	1200	—

In most instances the hardness gradient through the nitrided case is important to the life of the part. The type of alloy base that affects this is shown in Fig. 9.13. As in carburizing, the case depth and its supporting core must be selected to withstand the contact stress. Also, the hardenability of the steel for the section involved must be adequate to provide a tempered martensite microstructure, which is necessary for optimum nitriding at the surface as well as at a depth sufficient to withstand shear, bending, and torsional stresses.

Because nitrided cases are shallow, core hardness is usually high. It is not uncommon that the contact stresses are such that the required tempering temperature associated with the preheat treatment is nearly the same as the nitriding temperature. In all instances, however, the tempering temperature should be no less than 50 F above the nitriding temperature. Core hardnesses above 35 Rc can create serious problems in machining. One solution calls for 4140 to have an addition of 0.15 to 0.35% Pb. For production shop personnel who have never had to machine Nitralloy, having this grade thrust on them is a difficult experience. The presence of aluminum compounds make the material abrasive to cutting tools. Whenever possible, the EZ grade of Nitralloy, which contains 0.20% Se, should be used for improved machinability.

The scoring resistance of nitrided steels is exceptionally high; the nitride constituents present contribute to making a slippery surface as well as one that is extremely hard. An eminently useful application of nitrided steel is in hydraulic valves and control mechanisms. In fact, it is nearly a standard for this industry. Parts such as timing gears for internal-combustion engines do not carry high operating loads, but their speed and high operating temperature can cause severe scoring that often requires the use of nitrided 4140. The gear blanks are

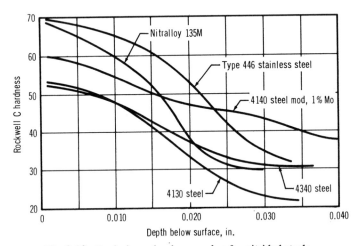

Fig. 9.13 Typical case-hardness probes for nitrided steels.

quenched and tempered to a machinable hardness before cutting the teeth and nitriding.

The nitriding process induces residual compressive stresses in the surface of a part. These stresses are so pronounced in small parts that they will elongate or enlarge in other directions to try to accommodate the nitriding effect. Thin sections of 135 Nitralloy, unrestrained, will grow approximately 0.020 in./in., whereas 4340 will grow about 0.006 in./in. The restraining effect of section thickness increases the compressive stress and thereby the long-life fatigue strength. Boegehold [5] reported that nitrided 4340 test specimens 0.300 in. in diameter showed a fatigue strength of 95,000 psi, whereas a full-sized crankshaft with sections up to 2 in. thick of the same material showed a fatigue strength of 120,000 to 130,000 psi. The effect of nitriding crankshafts of VCM steel (equivalent to 4140 with 1% Mo) is shown in Fig. 8.6. Insofar as steel selection is concerned, the higher the hardness attainable in nitriding alloy steels contain- ing 0.40 to 0.50% C, the higher the compressive stress and the long-life fatigue strength. Also, the deeper the case, the greater the size change in a given part, as shown typically in Fig. 9.14. In specifying a steel for nitriding, it is imperative to select a stock size and processing such that the part will be free of decarbur- ization when it is ready for nitriding. If this is not accomplished, cracking and spalling of the case are likely to occur.

Although nitriding is not widely used as the only means of improving the corrosion resistance of constructional steels, this characteristic is of profound importance in many parts; for example, nitrided 4140 is far superior in service for oil-field pump sleeves to steels that are hard-faced with an alloy composed of 16% Cr, 67.5% Ni, and 3% Mo.

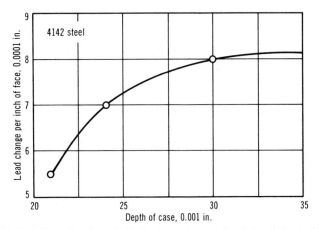

Fig. 9.14 Typical distortion in a nitrided gear versus case depth. Lead change is the out-of-parallel between the teeth of the gear and the centerline through its bore.

Stainless steels are beyond the composition ranges of the common constructional steels, but occasionally the materials engineer will encounter a part that requires a substantial level of corrosion resistance, yet must be hard enough to resist wear. Typical are pump impellers, pump shafts, and face-type seals exposed to water with mildly corrosive chemical content. In these instances stainless types such as 302, 321, 420, 430, and 446, nitrided, may be the answer. The attainable case hardnesses with these steels are above 1000 Knoop (70 + Rc).

Nitriding is used to provide a surface that will retain a large percentage of its hardness at temperatures as high as the nitriding temperature. One of the all-time major uses of type 135M Nitralloy was for cylinder liners in air-cooled aircraft engines during World War II. These liners had to be hard to resist abrasive wear, but they also had to remain hard at operating temperatures above 550 F. The hardness of a nitrided surface is the same at 750 F as at room temperature. A comparison of the resistance to tempering of carburized 8620 versus nitrided 4140 is shown in Fig. 9.15.

The most significant factors affecting distortion in nitriding are the following:

1 The design of the part to compensate for induced compressive stresses.
2 The quality and uniformity of the prenitriding heat treatment.
3 The uniformity of heating and gas circulation around the parts during nitriding.

Because nitriding is done at such a low temperature and no quench is employed, the distortion is not only low but consistent. A hydraulic valve spool, 0.750 in. in diameter by 11 in. in length, distorted as follows during heat treatment (500 pieces from three heats of steel per test):

| | | TIR Bow on Centers | |
| | | $\bar{X}$ (in.) | Standard Deviation (in.) |
Steel	Heat Treatment		
1042	Induction hardened	0.014	0.0008
8620	Carburized (hot-oil quenched)	0.011	0.0006
4140	Nitrided	0.002	0.00012
Nitralloy EZ	Nitrided	0.003	0.00013

In this test the 1042 was normalized before maching and the 8620 was also normalized from a temperature 50 F above the carburizing temperature. The

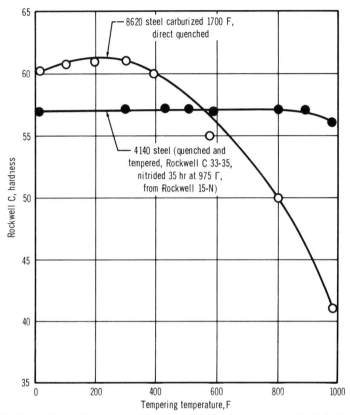

Fig. 9.15 Comparison of tempering effects on 8620 carburized and 4140 nitrided steels (surface hardness, 1.50-in.-diameter round).

4140 and Nitralloy EZ were heat-treated by the user because it was found that the mill heat-treated bars distorted excessively, probably as a result of cold straightening. Numerous other examples of nitriding distortion are given in Volume 2 of *Metals Handbook* [6] .

REFERENCES

1 C. F. Jatczak and D. J. Girardi, *Trans. ASM,* **51,** 335 (1959); C. F. Jatczak, *Met. Trans.,* **4,** 2267 (1973).

2 R. Pederson and S. L. Rice, *Case Crushing of Carburized and Hardened Gears,* SAE Farm, Construction, and Industrial Equipment Meeting, Milwaukee, 1960.

3 R. dePaul, *Metals Eng Quart,* **10,** No. 4, 25 (1970).

4 J. J. deBarbadillo, *The Effect of Impact Fatigue Prestressing on the High Cycle Fatigue Resistance of Carburized Gear Steels*, SAE International Automotive Engineering Congress, Detroit, January 8–12, 1973.

5 A. L. Boegehold, *Metal Progr.*, 57, No. 3, 349 (1950).

6 Gas Nitriding, Heat Treating, Cleaning and Finishing, *Metals Handbook*, Vol. 2, American Society for Metals, Metals Park, Ohio, 1964, p. 149–163.

CHAPTER TEN

SELECTION OF STEELS
FOR CARBURIZED GEARS

GEARS ARE MADE OF A NUMBER
of different materials and heat-treated by a variety of processes, but in current experience those made of alloy steel heat-treated by carburizing have the greatest load-carrying capacity. Depending to a considerable extent on the life required, these gears are now operating successfully at contact stresses of more than 300,000 psi and bending stresses of more than 100,000 psi. As the demand for these stress levels becomes more widespread a thorough knowledge of the factors surrounding the selection of steel for gears and their heat-treating characteristics becomes ever more important.

RELEVANT VERSUS IRRELEVANT ASPECTS

A considerable portion of the attention given to the manufacture of carburized gears is often irrelevant to their performance. Whether the maximum case-carbon content at the surface is 1.0 or 1.15% or the carburizing temperature is 1650 or 1700 F does not significantly affect the engineering qualities or the total cost. On the other hand, machinability, case microstructure, hardened depth, and whether expensive alloys and/or steel qualities are required are relevant.

A discussion of Fig. 10.1, which is a chart relating gear pitch to several steel types, illustrates the importance of these engineering requirements. The grades

181

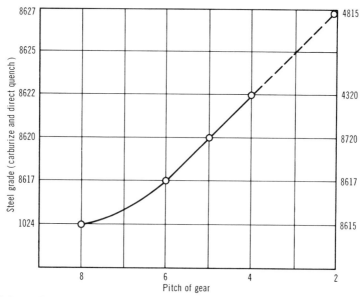

Fig. 10.1 Typical steel selection chart for carburized gears based on pitch of gear. Steels on left may prematurely fail because of inadequate case hardenability.

selected on the left of the chart (i.e., the 8600 series) increase in core hardenability with increasing coarseness of pitch. With the same alloy content, however, the case hardenability remains the same even though the core hardenability increases. Therefore the active surfaces of the heavier teeth will contain increasing percentages of transformation products (bainite and pearlite). The grades on the right in Fig. 10.1 contain increasing percentages of alloy (and therefore higher case hardenability) as the pitch coarsens; thus transformation products are prevented from forming. The presence of bainite decreases the contact load-carrying capability approximately as shown in Fig. 10.2. If steels are selected from the left of Fig. 10.1, this decreased load-carrying ability may occur in the worst possible place, namely, the coarse-pitch final-drive gears that have the highest contact stresses. The following is a case in point:

A manufacturer of heavy machinery specified the 8600 series of steels for all pitches of carburized gears; carbon was increased to provide a suitable level of core hardness for coarse pitches. Six-pitch transmission gears were made from 8620 steel and a final-drive, two-pitch pinion weighing 78 lb was made from 8627. Some of these final-drive pinions failed in less than 100 hr because of inadequate strength in the carburized case caused by the presence of more than 15% bainite at the lowest point of single-tooth contact. Even where the equipment was used less severely failure still occurred after several hundred hours of

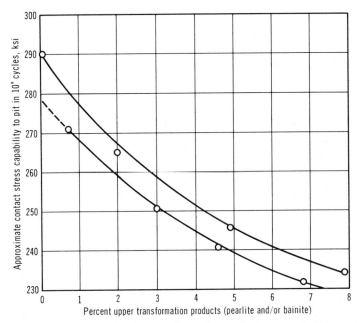

Fig. 10.2 Estimated effect of upper transformation products on contact-stress capability of 8620, 8720, and 8822 carburized steels.

operation. This was the result of pitting because of low contact stress capability of the high bainite carburized case. A change was made to 4320 steel, which improved the case microstructure, although not enough, and it was necessary to go to 4817 to provide the necessary metallurigcal quality in the pinion.

It should be noted that the grades suggested on the right in Fig. 10.1 are by no means 100% reliable for all designs and heat-treatment variations.

DEMANDS ON A GEAR TOOTH

In order to manufacture high-quality gears at the lowest overall cost, the demands on a gear tooth must be recognized. A gear tooth must be considered as a structural member of a mechanical device subjected to the following forces and interactions:

1 The gear profile surfaces are subjected to high contact loads in rolling and sliding, the direction of slide changing abruptly at the pitch line.
2 The loads imposed on a gear tooth cause it to act as a cantilever beam,

where the contact stress produces a bending moment concentrated at the root fillet. Most gears are expected to operate for many thousands or even millions of revolutions; therefore a long-life bending-fatigue situation exists in this fillet.

3 The relative velocity at the interface of two mating gears can be very high, thereby generating heat and destroying the protective lubricant film which leads to scoring.

4 The high contact stresses and the tendency to score, plus the usual presence of abrasive materials, require that the surfaces be very hard to preserve their contour and dimensions.

5 The operation of any gear set is required to withstand severe overloads, at least occasionally, without failing in a brittle manner. This is an outstanding characteristic of the carburized gear.

6 In some applications like passenger-car gearing distortion must be held to a low and consistent level to minimize noise.

The selection of steel should also include consideration of availability, formability, and, of course, cost.

A close examination of these engineering requirements indicates that they are related primarily to the properties of the surface and of the immediate subsurface zones of the gear tooth. It is here, therefore, that steel selection should begin. Mechanical properties at the radial centerline in the core of the tooth are significant only in a relative sense, somewhat similar to those at the neutral axis of a beam.

It is perhaps in order to admit that selection of steel for gears is not always done by careful materials engineering. Standardization for procurement and stocking and familiarity with processing peculiarities are often good reasons for using a certain steel or alloy series. As shown in Fig. 10.1 and its attendant example, this can be overdone, and the time usually comes when failures become epidemic or it is recognized that even slight increases in performance requirements would be disastrous. Likewise, it is not possible to attain the lowest cost without the optimum combination of design, materials, heat-treatment, and processing.

Before the steel is selected the complete gear design should include the following determinations:

1 The contact stress at the lowest point of single-tooth contact on the pinion.

2 The maximum bending stress in the root fillet.

3 The case-crushing load at the lowest point of single-tooth contact.

4 The interface temperature of the gear set at maximum anticipated speeds.

5 The level and rate of application of overloads and the lowest temperature at which they will occur.

Except for the overloading, which can usually be estimated from experience or from tests in simulated operation, the methods of calculating these requirements are readily available in the literature. Arbitrary selection of a steel for a gear from lists of those that have been used in passenger cars or farm equipment, for example, is dangerous. These lists are interesting and in some cases their investigation is justified; a quality gear manufacturer, however, will determine the specific requirements for his applications and select steel accordingly.

LIMITS OF DESIGN

If the designer is not limited by space, weight, or cost, he can rely on over-design to bypass the need for the highest metallurgical quality. Unfortunately this is not always expedient because either the cost or one or more of the design requirements will exceed the following limits for this approach (carburized gears):

Contact stress	220,000 psi
Bending stress	95,000 psi
Subsurface shear stress	55% of shear yield strength
Interface temperature	500 F max with bulk oil temperature of 200 F

These values are based on shaved gears made from basic open-hearth or basic-oxygen–quality alloy steel in which at least the chromium, nickel, or molybdenum contents, either singly or in combination, are above residual quantities normal for carbon steel. Plain carbon steels are capable of carrying 80% of the stresses listed when made by conventional heat-treating practices, although the reason is not fully understood.

THE SELECTION PROCESS

The selection process should automatically exclude steels obviously unsuited to high-quality gears:

1 Steels known to contain large nonmetallic inclusions; for example, semi-killed and/or resulfurized grades (above 0.05% S).
2 Grades exhibiting erratic and poor machinability (e.g., 1320).

3 Coarse-grained or aluminum-killed steel (silicon-killed, fine-grained is required).

4 Grades containing poor balances between carbide formers and matrix strengtheners (e.g., 5120 and 6120). Grades such as 4118 are being used with success; the case carbon content, however, should preferably be lower than 1%.

The next step in steel selection is to estimate or determine the quench-cooling rate as close as possible to the root fillet surface. Figure 10.3 can be helpful in making this determination for gears of simple shapes. Note the large difference in quench-cooling rate between conventional free-oil and impingement quenches. Actual cooling-rate determinations are suggested in Chapter 6.

Next, a decision must be made regarding the microstructural quality desired. Here are some guidelines:

1 If the contact and bending stresses are substantially below (i.e., up to 50% lower than) the maximum listed, microstructure is not of major importance. Selection can then be made on the basis of case hardenability in terms of hardness by using Table 10.1 as a guide.

2 As the stresses approach the suggested maximums case microstructure becomes increasingly important.

3 If the suggested maximums are exceeded, microstructure control should be considered a requirement unless experience has shown otherwise or unless the required life is less than 10^5 load applications. It is advisable not to exceed a contact stress of 275,000 psi or a bending stress of 120,000 psi if the life requirements are at least 10^7 load applications.

When microstructure control is necessary, the microstructural capabilities of

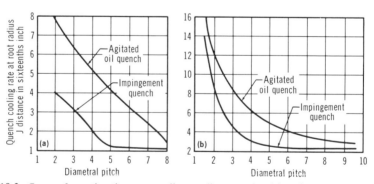

Fig. 10.3 Curves for estimating root radius cooling rate for (a) web-type gears and (b) solid pinion gears.

Table 10.1 Hardness Criteria for Gear-Steel Selection

Cooling Rate, J Distance $(1/16$ in.$)^a$	Direct-Quenched Gears			Reheated Gears		
	Case Carbon					
	1.10%	0.90%	0.80%	1.10%	0.90%	0.80%
1	1018	1018	1018	1018	1018	1018
2	1524	1524	1524	1524	1524	1524
3	4026	4026	4026	8620	8620	4118
4	4118	4118	4118	8620	8620	8620
5	4118	4118	4118	8620	8620	8620
6	8620	8620	8620	8720	8620	8620
7	8620	8620	8620	8720	8720	8720
8	8720	8720	8720	8822^b	8822	8822
9	8822^b	8822	8822	8822	8822	8822
10	8822	8822	8822	8822	8822	8822
11	8822	8822	8822	4320	8822	8822
12	8822	8822	8822	4320	4320	4320
13	8822	8822	8822	4320	4320	4320
14	8822	8822	8822	4320	4320	4320
15	8822	8822	8822	4820	4820	4820
16	8822	8822	8822	4820	4820	4820
18	4320	4320	4320	4820	4820	4820
20	4320	4320	4320	4820	4820	4820
More than 20	4320	4820	4820	9310	9310	9310

^{a} Cooling rate required for 60 Rc min, as quenched.
^{b} 8822 may need a lower carbon content if used for gears requiring machining after hardening. Also, 8822 is susceptible to microcracking at more than 0.90% C when direct-quenched.

candidate steels must be known or the materials engineer must have sample gears to section and examine microscopically. For maximum load-carrying ability upper bainite, pearlite, and network carbide must be absent from the case microstructure.

As pointed out in Chapter 9, the J distance to 1% transformation can be calculated by using the Jatczak and Girardi method or determined by the DFB method suggested by Kern (both described in greater detail in Chapter 6 and Appendix VII). Although exact minimum values of DFB for carburizing grades of alloy steel are not known, some typical results are shown in Table 10.2.

Table 10.2 DFB for Carburizing Steels[a]

Composition (%)	Direct-Quench[b]		Reheat-harden[b]		Cost[c] ($/cwt)
	typical	est min	typical	est min	
10B16 (1.00 Mn, 0.17 Cr, 0.07 Mo)	0.138	0.075	0.122	0.062	15.55
1018	0.075	0.050	0.055	0.030	13.75
10B22 (0.84 Mn)	...	0.075	0.105	0.062	15.55
15B24 (1.40 Mn)	0.122	0.100	0.116	0.100	16.20
1117 (1.27 Mn, 0.06 Cr)	0.122	0.062	0.116	0.062	14.55
1118	...	0.062	...	0.075	14.70
1213 (0.98 Mn, 0.06 P, 0.30 S, 0.18 Ni, 0.03 Mo)	0.122	0.062	0.118	0.062	14.80
1524	...	0.100	...	0.100	15.05
3310	2.000+	2.000	2.000+	2.000	29.05
4118	...	0.062	0.085	0.075	15.25
4120 (0.80 Mn, 1.00 Cr, 0.05 Ni, 0.25 No)	...	0.075	0.114	0.100	15.25
41B16	...	0.100	0.186	0.125	16.35
4320	0.960	0.875	...	0.875	21.25
4620 (with 0.40 Mo)	...	1.250	0.250	0.200	21.50
4620	...	0.750	0.272	0.250	20.85
4817	2.000+	2.000	2.000+	2.000	26.80
5120	...	0.050	0.080	0.062	15.05
8620.	0.232	0.200	0.108	0.100	17.50
8720	...	0.300	0.132	0.100	17.75
8822 (low side)	....	0.750	0.189	0.185	18.20
8822 (medium composition)	1.270	1.000	0.300	0.250	18.20

X9115	...	0.075	0.104	0.075	NR
9120	...	0.075	0.084	0.075	NR
94B17	...	0.500	0.173	0.150	19.00
EX–15 (1.00 Mn, 0.50 Cr, 0.16 Mo)	...	0.200	0.116	0.100	15.80
EX–24 (0.87 Mn, 0.55 Cr, 0.25 Mo)	0.385[d]	0.300	...	0.200	15.95
EX–29 (0.87 Mn, 0.55 Cr, 0.35 Mo, 0.55 Ni)	0.760[d]	0.750	...	0.300	18.40
EX–31 (0.80 Mn, 0.55 Cr, 0.35Mo, 0.85 Ni)	2.000[d]	2.000	...	2.000	20.95
20 Mn Cr 4	0.375[d]	0.285	0.188[d]	0.100	16.30
16 Mn Cr 5	0.375[d]	0.250	0.188[d]	0.150	15.80

[a] In inches, at 1.00% C.

[b] Heat treatments: Direct-quench consists of carburizing at 1700 F, cooling to 1550 F and quenching. Reheat hardening consisted of carburizing at 1700 F, slow cooling to room temperature, reheating to 1550 F, and quenching. No tempers.

[c] Prices are for as-rolled, hot-rolled, special-quality carbon bars and as-rolled alloy steel bars.

[d] Data courtesy of Climax Molybdenum Co.

189

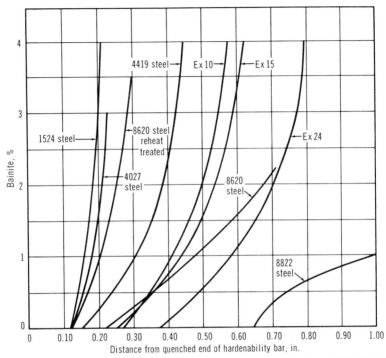

Fig. 10.4 Typical bainite profiles of carburizing steels, carburized to 1.0% C, direct quenched, except as shown (curves constructed from data provided by Climax Molybdenum Co. of Michigan [1]).

An example of the use of such data on a five-pitch T-section gear to be direct-quenched is as follows:

1 The quench-cooling rate at the root fillet with a well agitated oil quench, as taken from Fig. 10.3, is approximately $J = 4$ (4/16 in.).

2 From Table 10.2 it appears that the lowest cost steels having a DFB value of 4/16 in. min will be Ex-24 and 16 Mn Cr 5.

It will be noted from the tabulation of DFB data that there is need for additional steels for heavy gearing with microstructural capabilities of resisting bainite formation at distances slightly less than those for 4320 and 4820. Ex–29 and Ex–31 have such capabilities, but their high molybdenum content may pose microcracking problems if these steels are direct-quenched.

Although the load-carrying capacity of a gear decreases rather sharply with bainite content at the surface (see Fig. 10.2), the steel cost to eliminate bainite is high. Another factor to consider, however, is the rate at which bainite forms beyond the DFB. This interesting phenomenon was discovered by the Climax Molybdenum Co. of Michigan [1] and designated as the bainite profile of steels. Some of these bainite profiles are shown in Fig. 10.4 and are of great practical significance. It can be seen that 1524, for example, has a distance to first bainite (DFB) of 0.125 in. ($J = 2$). The distance to 4% bainite, however, is only 0.210 in. ($J = 3.4$). In other words, if for reasons of design or irregularity in heat treatment the quench-cooling rate is reduced by the equivalent of only 0.09 in. Jominy distance, the microstructure will contain 4% bainite and will

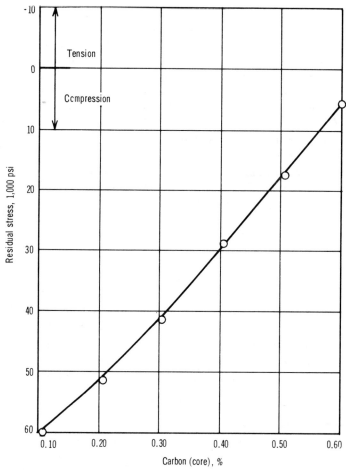

Fig. 10.5 Surface residual compressive stresses improve long-cycle bending-fatigue strength. Curve shows effect of core carbon content on these stresses (curve is for 1.0-in.-round 8600 series bars, carburized to 0.050-in. case depth and oil quenched).

exhibit substantially reduced pitting resistance. The benefits of more highly alloyed steels such as 8620 and 8822 in achieving fully martensitic microstructures are apparent. Figure 10.4 also shows very vividly the loss in bainite profile of 8620 by changing from direct quenching to reheat hardening.

Steels whose microstructures do not contain upper transformation products will provide maximum load-carrying capacity in contact stress and bending fatigue (other characteristics being equal). Long-life bending-fatigue strength of a gear tooth can be optimized by using steel with the lowest possible carbon content (see Fig. 10.5). A low carbon content and vigorous quenching result in the maximum possible surface compressive stress, thereby enhancing bending fatigue resistance.

The usual mode of failure caused by high contact stress is pitting. Excessively high contact stress can be the result of an error in calculation or, more often, of misalignment (e.g., due to deflection in the mounting). The pitting phenomenon is characterized by a starter crack normal to the surface in the dedendum of the driving pinion. This crack progresses a short distance into the carburized case, turns parallel to the pinion surface (toward the tip), and back out to the surface. Particles then break out. The mechanism of pitting appears to be a fatigue failure that starts at the point of the maximum shear component of the Hertzian stress. Its progress is also enhanced by the stress caused by friction and heat. Pitting can also start at a nonmetallic stringer located at the surface, but unless the inclusion is very large it will usually "heal" in operation. Pitting problems are extremely difficult to solve metalurgically and usually require a design change if a properly carburized, clean steel at maximum hardness fails to perform satisfactorily.

CASE CRUSHING

After a steel that will provide a surface with proper hardness or both hardness and microstructure has been selected, provision must be made for sufficient hardened depth to resist case crushing. Specification of case depth by simple rules of thumb, such as one-sixth of the chordal thickness, are inadequate for highly loaded gears. Insufficient strength through the case, particularly at the junction of case and core, will result in case crushing, as shown in Fig. 10.6. This phenomenon is considered to be a subsurface failure in shear. Typically, a crack starts in the case and progresses into, or close to, the soft core. The crack progresses in the transition zone with offshoots at right angles to the surface. This transition zone is particularly vulnerable to fatigue failure because of the abrupt change in chemical composition and/or hardness and the residual tensile stresses that are usually present. Maximum case crushing occurs at the lowest point of single tooth contact on the pinion. Resistance to case crushing is

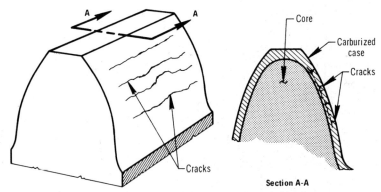

Fig. 10.6   Typical case crushing failure of a carburized gear.

provided by increasing the subsurface strength. The method suggested by Pedersen and Rice and referred to in Chapter 9 should be used to determine the strength required.

The required level of subsurface strength can be provided by one of the following methods or combinations thereof:

1   The subsurface strength that can be expected with a given quench of a carburized steel depends primarily on the carbon content; for instance, steels with nominal carbon content of 0.20 to 0.25%, carburized deeply, can provide the required case depth. Steels like 8640 (with a thin carburized case) can be used for gears in which shock loading is not an important factor and nominal bending stress is low.

2   The second method of developing high subsurface strength consists of using highly alloyed steels with high hardenability (e.g., 4820). In these steels even with mild quenches, all of the carburized case down to the core will have hardened to essentially 100% martensite.

3   The third method consists of using vigorous quenches such as the impingement-quench arrangement shown in Fig. 10.7. The effectiveness of this type of quench is shown in Fig. 10.3.

All of these methods of increasing subsurface strength have both advantages and shortcomings. Deep carburizing raises heat-treating costs and increases the danger of distortion. High-core carbon reduces the residual compressive stresses in the surfaces of the gear tooth and the toughness. Highly alloyed steels increase costs of both raw material and machining. Quenching should be as vigorous as possible, although individual quenching in impingement fixtures is sometimes impractical.

A compromise approach is generally used. This consists of holding the nominal core carbon to 0.25% max, quenching as vigorously as is practical, and vary-

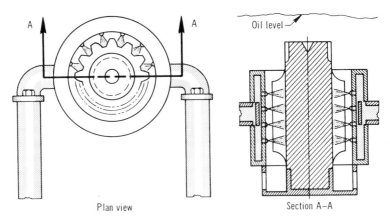

Fig. 10.7 Typical fixture for impingement quenching of a gear.

ing alloy content and carburized case depth to obtain the required effective depth of hardening. To calculate the depth required, the following technique may be used:

1 Determine the quench-cooling rate in the core at the location in Fig. 10.8 (e.g., it may be $J = 4$).

2 From published hardenability bands (see Appendix VI) select a grade of steel that has a hardness range at the quench-cooling-rate distance determined that will approximate 30 to 45 Rc (e.g., 4320H = 32–44 Rc, 4621H = 30–44 Rc, 4815H = 30–43 Rc, 8622H = 29–44 Rc, 8720H = 30–42 Rc, 8822H = 34–46 Rc, and 94B17 = 36–44 Rc, all at Jominy distance $J = 4$). Unless the gear is used in an abusive application, 8622H is probably the preferred choice because it is lowest in cost.

3 Case depth to 50 Rc at the lowest point of single-tooth contact is determined as follows:

$$\text{minimum case depth (inches)} = \frac{11.25 \times 10^{-6} \ W_t \ - .0175}{F \cos \theta} \qquad (10.1)$$

where W_T = tangential tooth load in pounds,
    F = face width in inches,
    θ = pressure angle.

Note. This formula contains a 1.7 misalignment factor. If good alignment can be assured, this factor can be reduced to 1.50 or even less and the required case depth will, accordingly, be shallower.

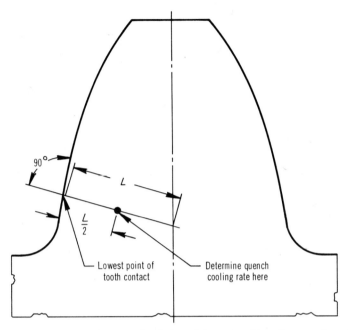

Fig. 10.8 To determine proper case depth to resist case crushing, the quench-cooling rate in the core should be determined at the location shown.

Case depths to a hardness of 50 Rc can be increased by three methods. Tests run by Kern, for example, indicate that for each 0.10% increase in carbon at a point 0.06 in. below the surface of a gear tooth an increase in hardness of 1.5 to 2.0 points Rockwell C can be expected if the quench-cooling rate corresponds from $J = 3$ to $J = 9$, inclusive. For a properly diffused case this means an increase of only 0.010 to 0.020 in. in carburized case depth. Other tests conducted to evaluate the effect of higher alloy content showed that the hardness increase in Rockwell C points per unit of actual D_I increase is three plus two-thirds times quench-cooling rate in sixteenths of an inch. For example, at the hypoeutectoid levels of the carburized case that cooled at a rate corresponding to $J = 9$, an increase of 1 in the actual D_I of the steel would increase the hardness approximately $3 + 2/3(9) = 9$ points Rockwell C. As the D_I approaches six, however, the hardness increase per unit of D_I increase drops off sharply.

Vigorous quenches should always be an objective because they are usually a low-cost way to get the most out of money spent for steel. As we can see, deciding which course to follow in providing adequate subsurface strength is not simple.

BENDING FATIGUE (SHORT LIFE)

With the field of candidate steels for a gear narrowed to provide proper case hardness and microstructure and adequate subsurface strength to resist case crushing consideration must be given to the importance of bending fatigue.

As in direct-hardening steels, case-hardened steels have different parameters for short-life than for long-life fatigue strength. For maximum short-life fatigue strength toughness is a very important factor. The toughness of carburized steel can be increased by the following methods:

1 Tempering from an as-quenched case hardness of 58 Rc or higher down to 55 Rc or lower. (Avoid tempering temperatures of 475 to 700 F because of embrittlement of the core.)
2 Quenching to produce 15 to 30% retained austenite in the case microstructure.
3 Applying steels with contain nickel as a major (more than 1%) alloying element.

When using tempering to improve toughness, the materials engineer must make sure that the problem is *solely* short-life fatigue. The reason for this is that as the part is tempered its contact stress capability is significantly reduced and so is the scoring resistance. Also, because the residual compressive stresses in the case will be reduced, both the long-life fatigue in bending and the resistance to case crushing will be lowered. In some designs in which nothing else works manufacturers have found tempering to be an acceptable practice. Its use should always be justified by tests.

Troiano and his associates [2,3], Razim at the Institut für Harterei-Technik [4], and Brugger and Kraus [5] have shown the desirable effects of retained austenite in carburized gears. It was also found by Troiano et al. that there are differences in the fracture resistance of steels of different base compositions containing like amounts of retained austenite in the carburized case.

The beneficial effect of nickel in improving the toughness of carburized steel is indicated by both tensile and bend test results, as shown in Figs. 10.9, 10.10, 10.11, and 10.12 [6]. As demonstrated by dePaul and deBarbadillo (see Refs. 3 and 4, Chapter 9) in both impact-fatigue tests and combinations of impact prestressing and bending fatigue, the toughness imparted by nickel results in improved short-life fatigue strength.

Most of the tests we have referred to were on laboratory specimens of various sizes and shapes. Many gear manufacturers can report actual field tests in which changing from a low-nickel steel (e.g., 8620) to one with substantially more nickel (e.g., 4820) solved a short-life tooth-breakage problem. Some typical examples are the following:

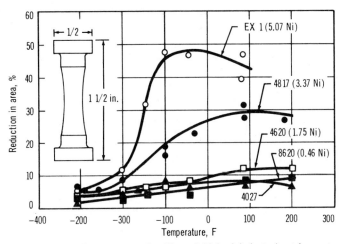

Fig. 10.9 Tensile tests show greater ductility of high-nickel steels at lower temperatures (from Johnson and Shelton [6]).

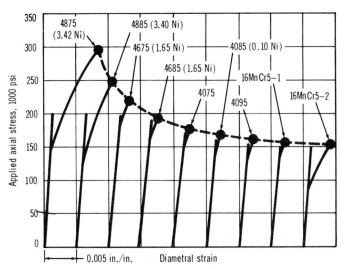

Fig. 10.10 Tensile tests show greater strength and ductility of nickel steels in comparison with nickel-free steels (from Johnson and Shelton [6]).

197

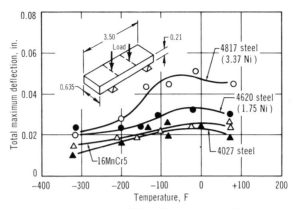

Fig. 10.11 Bend tests show greater ductility of nickel steels (from Johnson and Shelton [6]).

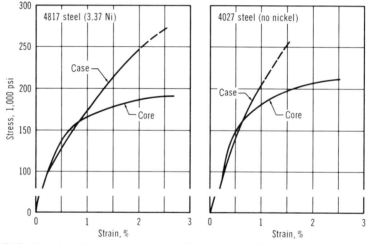

Fig. 10.12 Nickel steels have more case ductility than nickel-free steels, as shown in tensile tests (from Johnson and Shelton [6]).

1   A solid four-pitch, final-drive pinion used in a crawler tractor was made of 8622 steel. Failures by tooth breakage were occurring in 100 to 300 hr of operation. Impingement quenching was producing excellent microstructures but even shot peening did not cause significant improvement. A change to 4815 with impingement quenching solved the problem.

2   A solid two-pitch hoist pinion on a crane made of 8625 steel was failing by tooth breakage in 50 to 200 hr of operation. (Fixture-quenched car-

burized 8625 had already replaced 4340 quenched and tempered to 269 to 321 Bhn, which had a history of failure by pitting.) When the material was changed to 4320 with the same processing, the problem was solved.

3   A six-pitch sun gear made of 8620 in the final drive of an industrial trac- tor was failing by tooth breakage in only 25 to 100 hr of operation. A change to 4320 solved this problem.

4   A six-pitch, final-drive pinion made of 8617 (0.55% Ni) for a fork-lift truck was failing through the 1.50-in.-diameter stem in 25 to 200 hr. A change to 4320 (1.8% Ni) increased life to only 500 hr min, which was still inadequate. A change to 9310 (3.3% Ni) was made, and no further failures occurred.

As previously pointed out, when changes are made in gear steels, (e.g., from 8617 to 9310), machining becomes difficult even with the best postforging treatment. The materials engineer should alert his manufacturing department to this problem.

BENDING FATIGUE (LONG LIFE)

The long-life bending-fatigue strength of carburized gear teeth is optimized by the following conditions:

1   Freedom from bainite, pearlite, and network carbide in the case micro- structure.

2   Absence of microcracks, especially near the surface in the root fillet.

3   Presence of maximum residual compressive stress in the case (by using a steel with the lowest possible carbon content).

4   Freedom from defects in the root-fillet surface, whether they take the form of nonmetallic inclusions or poor finish from machining.

5   Use of nickel-bearing steels when shock loading is likely to occur.

Upper transformation products in the case microstructure decrease long-life fatigue strength simply because they are soft, weak constituents and there- fore act as nuclei for failure.

Microcracks usually, but not always, occur through long martensite needles. The mechanism is not fully understood except that it is at its worst when case carbon intent exceeds 0.90%, and parts are direct-quenched, and when the har- denability of the steel results primarily from carbide-forming elements such as chromium and molybdenum (more than 0.90% Cr and 0.30% Mo). Subzero treatment of parts to transform retained austenite often results in microcracks that can reduce the long-life fatigue strength. Figure 10.13, from the work of

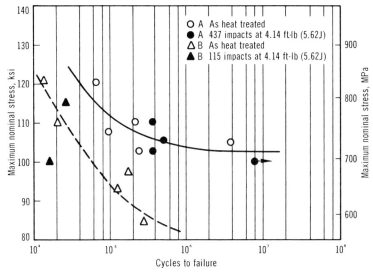

Fig. 10.13 Bending-fatigue properties of 4820 steel. Sample A received standard heat treatment, while B was refrigerated at –73 C (–100 F). (from De Barbadillo [4] Chapter 9).

deBarbadillo (Ref. 4, Chapter 9), shows the great loss in long-life fatigue strength of carburized 4820 steel that had been cold-treated.

The relation of residual compressive stress in the surface of a part to its long-life fatigue strength was discussed in Chapter 8 and Figure 10.5 shows how this stress varies with core carbon content. Somewhat surprisingly, it has been found that the level of core hardness to which a given steel has been quenched has little to do with the level of surface compressive stress developed; for example, the surface compressive stress in a three-pitch pinion tooth of carburized 8822 steel was 37,000 psi when quenched in well agitated oil. The core hardness was 35 Rc. When quenched in an impingement fixture, the core hardness increased to 44 to 49 Rc, but the residual compressive stress was still 36,000 psi (which is within the limits of accuracy of x-ray residual-stress measuring equipment). It is common practice today to temper carburized gears at 300 to 350 F after quenching. This reduces the compressive stresses in a carburized case as much as 15%. When there is no clashing or shock loading and especially no grinding, elimination of the tempering operation can provide useful improvement in long-life fatigue strength.

Shot peening is usually considered as a means of increasing the compressive stress in the teeth of a carburized gear. With conventional shot (which is 40 to 50 Rc), however, x-ray diffraction studies indicate that the level of compressive stress may actually be reduced. Peening with shot of a hardness of 60 + Rc will increase the compressive stress. Any improvement in long-life fatigue strength by peening with conventional shot is probably the result of one or a combination of the following:

1 An improvement in finish, at least in the root fillet.
2 An improvement in lubrication qualities of the gear surface at the lowest point of single-tooth contact.
3 Peening will improve the long-life fatigue properties of any areas having upper transformation products.
4 Peening will tend to wear away surface areas that are intergranularly oxidized, thus improving the overall quality of the surface.
5 Razim [4] found that the maximum long-life fatigue strength in rotating-beam specimens of carburized steel could be achieved by heat treating to a high percentage of retained austenite (e.g., 80%) and then shot peening or rolling to cold-work the surface. However, we know of no gears made in production in this manner that were successful. The four-pitch, final-drive pinion cited in Chapter 9 as failing by pitting resulting from excessive retained austenite was made in this way. This gear was shot-peened with 60 + Rc shot.

Carburizing grades of steel inherently tend to have more inclusions than higher carbon steels made by the same melting practice because of their low carbon content. Some producers, knowing that these steels will be used in critical parts, melt as clean as possible and then either vacuum degas or carbon deoxidize and vacuum degas. When high-quality carburized gears are made, steps should be taken to ensure that maximum steel cleanliness consistent with the required cost is provided. Melting practices that use minimum amounts of aluminum should be encouraged. Also, for very highly loaded gears magnetic-particle inspection for the detection of large nonmetallic inclusions should precede final shot blasting.

When the surface finish produced by hobbing in the root fillet is erratic, it is sometimes necessary to grind this area *before carburizing* to provide the necessary long-life fatigue strength. Several manufacturers are using this practice on heavily loaded, final-drive pinions made of 4320 and 4815 steels. Grinding after carburizing usually reduces fatigue strength and is not recommended for highly loaded gearing.

SCORING

Although scoring is not a parameter of steel selection, the materials engineer is often confronted with scoring problems that designers are prone to assign to metallurgical causes. When a gear tooth is harder than 58 or 59 Rc, little can be done to eliminate scoring. This failure results from high interface temperatures that cause lubricant breakdown; it is a function of bulk oil temperature, the relative velocities of the tooth faces, their finish, the coefficient of friction of the gear steels, and the load per inch of tooth width. Formula 8–3 can be used to determine interface temperature, which should be calculated for several anticipated operating speeds and loads to ensure that the 500 F limit is not exceeded.

The reason that scoring is sometimes blamed on metallurigcal deficiencies is that failure starts by scoring, then pitting, then more scoring, followed by more pitting. A proper investigation of a gear failure requires careful calculation of all loads and of the interface temperature.

OVERLOADS

The overloads that most gears must be able to withstand without premature failure are very high, often as much as two or more times the normal load and even more in shock loading. Failures can be a result of normal operation of a device, such as "shift shock" in heavy-duty transmissions, or of misuse by the operator. In designing a new machine, it is difficult to anticipate the level and frequency of this type of loading; hence the need for prototypes and stress studies in actual operation. Sometimes a situation that appears to involve extreme shock loads is found to be much less drastic; for example, tests on final-drive gears of a crawler-tractor bulldozer by ramming it into a large tree stump showed shock loads of only one-half those absorbed by the timing gears in a diesel engine in normal operation. Friction losses in the backlash of mechanisms ahead of the gears can often absorb much of the imposed shock loads.

Shock loading is an extreme condition of short-life fatigue and suggestions made for its control can apply. In some instances, such as swing pinions on large drag lines and power shovels and in some steel-mill applications, carburizing cannot be used. Design contact stresses must be lowered to a point at which induction- or flame-hardened 4330 or similar steels can be used. In other applications, however, carburizing grades such as those containing 9% Ni can be economically justified.

HEAT TREATMENT

With proper steel selection the carburizing heat treatment of gears is relatively simple. With adequate case and core hardenability and a composition selected for suitable short- and long-life fatigue qualities the major day-to-day problem is to control distortion as it affects tooth loading and noise. Here, though, are some important points to remember in the heat treatment of carburized gears:

1 If forgings are used, normalize or anneal, as required by grade, from a temperature at least 50F above the carburizing temperature. If bar stock is used, heat-treat in a similar manner after cutting to length.

2 Make sure the gears are machined to the dimensions shown on the drawing before heat treatment.

3 When carburizing, it is preferable to bring the entire gear uniformly to the carburizing temperature in a nearly neutral atmosphere and then expose it to the carburizing gas, which should be adequately circulated in the furnace.

4 Adjust the carburizing atmosphere and time such that for deep cases (more than 0.060 in.); the carbon penetration follws a smooth curve downward from surface to core. A decrease of 0.15 to 0.20%/0.010-in. increment of depth is close to ideal. For shallower cases the gradient can be steeper.

5 For steels on the 4300, 4600, 8600, 8800, and 9400 series, the surface-case carbon recommended is 0.90 to 1.10%. Grades such as 4800 and 9300 are preferably reheated for hardening, in which case 0.90 to 1.10% is also satisfactory. If high nickel steels (e.g., 4800 series) are to be direct-quenched, the recommended surface-case carbon is 0.65 to 0.85%.

6 Use direct quenching wherever possible because it minimizes cost and distortion.

7 Make sure that quenching dies and plugs are properly maintained to ensure dimensional accuracy.

8 Quench as vigorously but uniformly as practical. Use impingement-quench fixtures at least on large solid pinions four-pitch and coarser.

9 For fine-pitch gears, for which dimensional accuracy is very important, use hot-oil quenching if possible.

10 Case hardness of the finished gear should preferably be 60 Rc or higher, although 58 and 59 Rc are commonly permitted.

11 If practical, file-test each gear to detect partial decarburization and/or upper transformation products.

Case depths used to be checked by measuring carbon penetration on carburized unhardened pins. The trend is toward a hardened depth to 50 Rc. Usually it is not economical to provide extra gears to be cut up for control purposes, and a relation between the hardened depth on a test pin and the gears run with it must be established. Some manufacturers of high quality gears go to the expense of sectioning good gears, some even weighing several hundred pounds, to make sure that the specified depth to $50 R_C$ is met. By comparing Fig. 6.11, which shows the surface-cooling rates of rounds quenched in well agitated oil, and Fig. 10.3, which shows the quench-cooling rate by pitch of gears, the relationship shown in Fig. 10.14 can be established. It could reasonably be expected, for example, that the case depth to 50 Rc at the root radius of a four-pitch, web-type gear quenched in well agitated oil would be approximately the same as that found on a 2-in. round of the same heat of steel, heat-treated in a similar manner. These curves are eminently more useful if they are constructed from data reflecting plant conditions.

Another important factor in the control of distortion of carburized gears, particularly as it affects noise, is uniformity of surface carbon content from gear to gear and heat to heat. Snyder [7] found this to be very important in the manufacture of passenger car gearing.

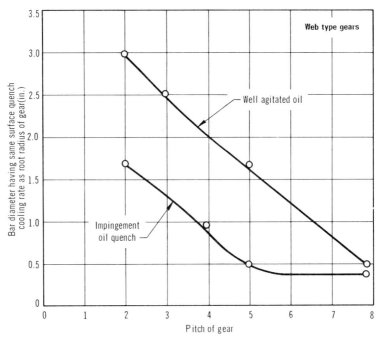

Fig. 10.14 Curve comparing quench-cooling rate in root radius of a web-type gear and surface of round bars quenched and agitated oil.

REFERENCES

1 J. W. Hallock, *Reproducibility of the Measurement of Distance to First Bainite,* unpublished report of Climax Molybdenum Co. of Michigan Report L-193-84, May 22, 1972.

2 L. J. Ebert, F. T. Krotine, and A. R. Troiano, A Behavior Model for the Fracture of Surface-Hardened Components, *Trans. ASME,* Ser. D, Basic Eng., 87, 871 (1965).

3 F. T. Krotine, M. F. McGuire, L. J. Ebert, and A. R. Troiano, The Influence of Case Properties and Retained Austenite on the Behavior of Carburized Components, *Trans. ASM,* 62, 829 (1969).

4 C. Razim, Influence of Retained Austenite on the Strength of Case Hardened Specimens under Conditions of Fluctuating Stress, *Harterei-Technic Mitt,* 23, No. 2, 1 (1968).

5 H. Brugger, G. Kraus, Influence of Ductility on the Behavior of Carburized Steel During Static and Dynamic Bend Testing, *Eisenhutten,* 32, 529 (1961).

6 J. Johnson and C. H. Shelton, Progress Report on Nickel Carburizing Steels, *Metal Progr.* 104, No. 7, 37 (1973).

7 W. Snyder, *Metal Progr.* 88, No. 4, 121 (1965).

CHAPTER ELEVEN

SELECTION OF STEELS
FOR SHAFTS

WEBSTER [1] DEFINES A SHAFT as a bar used to support rotating pieces or to transmit power. Although perfectly adequate for most this definition is an oversimplification in an engineering sense. Many shafts perform not only one but both functions and also act as a main support structure. It is common for shafts to support significant loads as well as to transmit power. Shafts can be further classified as "live" or "dead." A live shaft is one that rotates, whereas a dead shaft does not turn or transmit torque and is therefore subjected only to bending loads. Because live shafts are subjected to both torque and bending, they are more complicated in configuration and contain splines, flanges, and keyways to attach components. Therefore live shafts normally require a more careful stress analyses and selection of steel and processing than most dead shafts.

Whether a shaft is live or dead does not exclude it from cyclic loading and the effect of fatigue. Accordingly, anything that can be done in the design to maintain the stresses at the lowest possible level consistent with weight and cost limitations is effort well spent. Few shafts fail because of lack of static strength because most designs are based on the well known static strength factors (usually yield strength) and failures nearly always result from fatigue because of inadequate long-life fatigue strength.

Heat-treated shafts are typically made by one of the following methods:

Type 1. Use prehardened raw steel at a machinable hardness and finish machine.

Type 2. Same as type 1, shot-peened, surface-rolled, nitrided, or selectively induction-hardened after machining.

Type 3. Hot-rolled or forged-stock-finish-machined, except for grinding, then induction-hardened all over, and finish-ground.

Type 4. Hot-rolled or forged-stock-finish-machined, except for grinding, then shell-hardened,* and finish-ground.

Type 5. Hot-rolled or forged-stock-finish-machined, except for grinding, then carburized, hardened, and finish-ground.

Generally speaking, the materials and processing of types 1 through 4 provide a progressively increasing long-life fatigue capability. Type 5 is in general, but not always, used for combination shafts with gears or cams.

The first step in steel selection is to make sure that the minimum engineering requirements for yield strength are sufficient to resist permanent set in twisting or bending. Long- and short-life fatigue requirements must also be known so that the fatigue limit and toughness requirements can be determined. The designer must know or estimate the life required in terms of normal stress cycles as well as peak or unusual stresses.

TYPE 1 SHAFTS

An example of a Type 1 shaft is shown in Fig. 11.1. Suppose the designer reported that considering proper stress concentration factors, the required strength at the 2-in. diameter is 100,000 psi. The normal fully reversed cyclic stress is 45,000 psi and a life of 10^6 cycles is required at that level. Under severe conditions, peak stress of 90,000 psi ($\Delta\epsilon_p \approx 0.004$) may be encountered up to 10^3 times in the life of the part. How should the material and heat treatment be selected?

---

*Shell hardening constitutes making a part from shallow hardening carbon or carbon-boron steel (e.g., 1035 or 10B35) by heating it above the Ac_3 point and vigorously quenching it in water, brine, caustic, or water-polymer quenchants. Tempering is usually at 475 F max. Depending on the section size and hardenability of the steel, a hardened layer 0.1 to 0.2 in. thick is produced over the entire surface of the part; hence the term shell hardening.

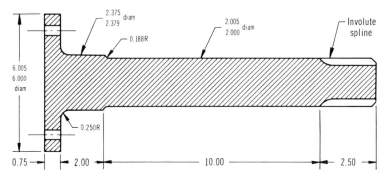

Fig. 11.1 A typical heat-treated shaft, type 1. Part is forged, quenched, and tempered to a machinable hardness and is then finish machined.

From Fig. 11.2 it can be seen that for a 100,000-psi yield point a minimum tempered hardness of 27 Rc is required. This will provide an ultimate strength of 130,000 psi, which is more than adequate to meet the long-life bending fatigue requirement of 45,000 psi. To estimate the short-life fatigue strength the reduction in area must be known or estimated in order to use the equation (8.4) discussed in Chapter 8.

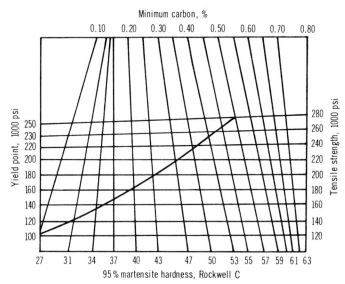

Fig. 11.2 When the required yield point or tensile strength is known, the minimum carbon content and necessary hardness can be determined from this chart.

$$2N_f = \left(\frac{\Delta \epsilon_p}{2\epsilon_f} \right) 1/c$$

Approximate relationships between hardness carbon content and reduction in area for alloy steels are shown in Fig. 11.3. Assuming that the part is oil quenched, a nominal 0.40% carbon alloy steel would normally be used. With a hardness of 27 to 33 Rc a 56% reduction in area could be expected. This would provide a life of approximately 26,000 cycles at 90,000 psi, well exceeding the 10^3 requirement. A more rigorous analysis would include consideration of cyclic stress-strain behavior and cumulative fatigue damage at the two stress levels; in this instance, however, these are not necessary.

To attain the expected fatigue qualities a microstructure of at least 90% martensite should be provided and should extend to 30% of the distance from the surface to the center. For a 0.40% carbon steel this means an as-quenched hardness of at least 48 (preferably 50) Rc. The surface and 30% depth quench-cooling rates in well agitated oil on a 2.25-in. (as-forged-size) round are $J = 5$ and $J = 9$, respectively. Therefore a 0.40% carbon steel of 50 Rc min at $J = 9$

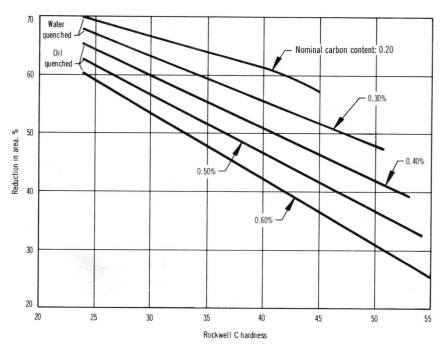

Fig. 11.3 Approximate relationships between hardness and reduction in area for heat-treated alloy steels of different carbon contents.

is required. An examination of the H bands (Appendix VI) shows that standard 4140H steel will not have adequate hardenability. Grades such as 4340H and 9840H should be used unless 4140H is obtained to a restricted (high-side) hardenability band.

TYPE 2 SHAFTS

It is not uncommon that Type 1 shafts fail to perform satisfactorily in the field. Prototype parts are precisely made on engine lathes and other precision machines, but production equipment introduces dimensional-tolerance stackups that can result in misalignment and accompanying operating stresses beyond those expected. The users may also overload and otherwise abuse the machinery. Another common source of difficulty is the failure to recognize inertia loads; for example, the swing shaft of a hydraulic excavator may experience significantly higher torque during certain digging operations than during full-swing motor power.

If metallurgical examination of a failed part shows that the material and heat treatment were as specified, it is important to resist the temptation to "make it harder" without proper investigation. Identifying the mode of failure is extremely important, for if it was caused by stress of the short-life type additional toughness would be needed and 4330 or even 4320 may be an appropriate choice even at the expense of some hardness. If a long-life failure occurred, a change to a higher hardness would be the proper course. All too often the steel will be arbitrarily changed from, say, 4140 to 4340 and the hardness increased to the maximum that the machining facility can handle. At best this causes unnecessary expense and at worst it only delays proper solution to the problem. Volumes 9 and 10 of the *Metals Handbook* [2] can be useful in establishing the exact mode of failure.

If the failure is known to be of the long-life type, the change in material and hardness will extend the life only slightly. If redesigning to increase the highly stressed section is out of the question, the following procedures are often effective:

1 Shot peening.
2 Nitriding.
3 Induction hardening.
4 Surface rolling.

Which of these processes to use depends on the availability of equipment and many other considerations that equate to cost. All processing must be performed by properly trained personnel if it is to be effective and good quality control must be exercised.

Shot Peening

Among the advantages of shot peening are (a) it does not require a material change; (b) its effect is obtained over the entire surface of the part (except in areas in which grinding may be required after peening); (c) it can be used to salvage inventory; and (d) it is comparatively low in cost. A disadvantage is that some areas, such as those that have been ground, could become the origins of failures after peening unless they are carefully masked. In extreme cases failure can also occur at recesses that cannot be properly impacted by the peening shot. The proper shot size and peening intensity depend on the configuration and hardness of the part. For details in setting up shot-peening specifications the materials engineer should call in experts who do this work. The article on shot peening in Vol. 2 of the *ASM Metals Handbook* [3] is helpful. The cost of shot peening varies widely with the size of the part, the quantity required, and the equipment available, but generally is between 1 and 10¢/lb.

On direct-hardened parts the depth affected by shot peening will be 0.005 to 0.030 in., depending on the hardness of the part, shot size, and peening intensity. Shot-peened parts should not be straightened because this operation will remove the compressive stress induced by peening often where it is needed the most (e.g., at the bottom of a snap ring groove). If parts distort excessively as a result of shot peening, it is sometimes possible to avoid this problem by selectively peening only the areas that require it.

Nitriding

If the shaft can be made from a suitable steel like 4140 or 4340, nitriding is an effective way of increasing the long-life fatigue strength. In addition to the examples of fatigue-strength improvement cited in preceding chapters, Spencer reported that this process increased the bending-fatigue strength of diesel locomotive crankshafts by 40 to 100% [4]. He also reported on the effect of case depth and core hardness on long-life fatigue strength (see Fig. 11.4). For optimum improvement of shafts approximately 1 in. in diameter the case depth should be at least 0.020 in.

Nitriding for long-life fatigue improvement has a number of important advantages:

1 The part can usually be entirely finish-machined before heat treatment, except possibly for some light polishing in bearing areas primarily to remove the white layer (which is about 0.0001 to 0.0003 in. thick when the Floe process is used for nitriding). Distortion on parts such as shafts is minimal and growth is predictable.

2 The nitriding operation is effective over the entire surface of a part, including the insides of oil holes and the bottoms of deep keyways.

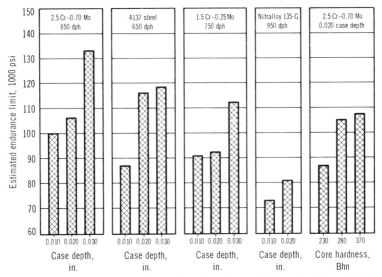

Fig. 11.4 Bar-graph summary of the estimated endurance limits of various nitrided steels and various combinations of case depth and core hardness (from R. M. Spencer, National Forge Co.).

3 The residual compressive stresses induced by nitriding are resistant to tempering that might occur, for example, on the shaft of a blower for moving hot gases.

The disadvantages of nitriding are high cost and somewhat limited availability, especially when the weight of each piece is low and fractional furnace loads must be run. Comparative costs, from a large commercial heat-treating firm, along with those for other heat treatments, are as follows (based on a 10-lb gear): harden and temper, $0.13/lb; carburize to 0.030 to 0.040-in. depth and harden, $0.18/lb; and nitride to 0.020 to 0.025-in. depth, $0.34/lb.

As mentioned in Chapter 8, heat treatment and processing prior to nitriding must ensure freedom from decarburization and a microstructure of well-tempered martensite. When a great deal of steel must be machined away, for example, from a deeply splined shaft, it would be well to have a thermal stress relief treatment at 1000 F min between rough and finish hobbing to minimize distortion from machining stresses.

Induction Hardening

Localized induction hardening is a popular method of improving the long-life fatigue strength in critical areas of shafts. Radii that must, of necessity, be quite sharp (e.g., to accommodate the inner race of a rolling bearing) can be substantially strengthened by this process. The wheel spindle shown in Fig. 8.8 is a typical example. The fillet hardening of crankshafts shown in Fig. 8.6 is another example of the effectiveness of selective induction hardening. The hob runout areas of splined shafts, snap-ring grooves, and threads are also trouble spots in maintaining long-life fatigue qualities. These areas can be locally induction-hardened, but the irregularities present require that some precautions be observed:

1 The steel should be as low in carbon content as possible to minimize the tendency to quench-crack (preferably 0.35% max).
2 The heating rate should be as low as possible to maximize uniformity and prevent overheating of sharp corners.
3 A somewhat mild quench such as 10 to 15% Ucon* or 5% soluble oil may be required in extreme cases to minimize quench cracking and still obtain a microstructure of 100% martensite.
4 Magnetic particle inspection should be regularly employed to ensure freedom from quench cracks.

When using induction hardening for localized improvement of long-life fatigue strength, it should be kept in mind that hardened depths under 0.075 in. are usually residually stressed in tension and may exhibit poor fatigue strength. It is necessary to harden deeper than about 0.105 in. to ensure development of compressive stresses. In general, deeper case depth is better, even though distortion may be slightly increased. In addition to favorable residual stresses, the higher strength of the hard steel supplies a higher long-life fatigue strength. The hardness for optimum long-life fatigue strength in a 0.40% carbon steel is approximately 40 Rc. For a steel of 0.60% carbon content the maximum fatigue strength occurs at approximately 50 Rc. In many cases it is advisable to use higher than optimum theoretical hardnesses to avoid reducing the favorable residual stress pattern in the tempering cycle.

Induction heating can also be used to improve short-life fatigue strength in local areas by providing a superior microstructure on quenching and attendant improved toughness after tempering. This technique is useful in shafts larger than 4 in. in diameter when the preliminary heat treatment does not produce a proper microstructure to obtain maximum toughness. Shafts of 4140 steel up to 12 in. in diameter are being made in this way.

Surface Rolling

The effect of surface rolling on local improvement of long-life fatigue strength is shown vividly in Figs 1.8 and 8.6. A 1954 supplement to the ASM *Metals*

*A product of Union Carbide Corp.

Handbook is an excellent reference on the subject [5]. With the proper tooling and technique, rolling is an effective, low-cost method of improving long-life fatigue qualities, particularly in fillets and threads. When higher fatigue strength is required in threads, rolling them into the bar after heat treating provides the best fatigue properties known at this time.

The design of rollers, the application pressures, and details such as the required finish before rolling constitute a highly specialized science. Manufacturers of this equipment should be consulted for advice on specific problems.

TYPE 3 SHAFTS

Type 3 shafts, which are induction-hardened all over to a high hardness, are standard for the passenger-car industry. This hardening procedure is also widely used in farm tractors and some trucks. Formerly most of these shafts were scan-hardened, a process that required relatively long hardening cycles. The production rate was later improved by installing scanning units with four or more work stations. Today, scan hardening has largely been replaced by "one-shot" heating, developed by the Tocco Division of Park-Ohio Industries. With this method the entire surface of a complex shaft can be heated in 5 to 10 sec. The depth of the hardened layer is about 0.130 in., the heated depth, about 0.3 in. Typically, power supply for this type of installation is solid state, about 600-kW capacity, with an operating frequency of 3 kHz. Tempering is also done by induction. Figure 11.5 is closeup of a shaft and an inductor and Fig. 11.6 a schematic of the arrangement. To justify this type of equipment the number of pieces processed per set-up must be very large — at least 1000.

The reliable operation of Type 3 shafts in passenger cars has resulted in the installation of large scanning units for smaller lot sizes and for much larger shafts (up to 6 in. in diameter). Shafts of the larger sizes normally require deeper heating for deeper hardening, and some firms use frequencies as low as 1 kHz. Typical sections from a shaft approximately 3 in. in diameter are shown in Fig. 11.7. To provide the necessary torsional strength for this shaft the heating and hardened pattern extend through most of the section. Also, as the depth of heating approaches the center, and provided a drastic quench is used on the surface, the residual compressive stresses developed in the surface layer become the same as those in a shell-hardened (Type 4) shaft. This would provide the magnitude of compressive stress and fatigue strength shown for the 1045 specimens in Fig. 1.8.

The selection of steel for Type 3 shafts does not differ greatly from this choice for other hardening applications. The strength throughout the cross section must exceed the stresses at any point, and hardness and microstructure must provide the necessary fatigue limit. These additional considerations should

Fig. 11.5 Single-shot induction heating of a shaft Inductor moves down to rotating shaft to bring it to hardening temperature (courtesy *Manufacturing Engineering,* a publication of the Society of Manufacturing Engineers, July 1970).

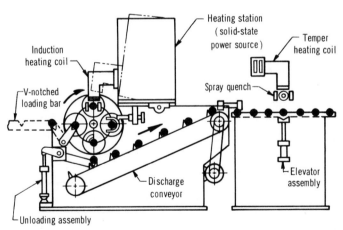

Fig. 11.6 Four-stage machine (shown in Fig. 11.5) heats, quenches, and transfers part to tempering unit (courtesy *Manufacturing Engineering,* a publication of the Society of Manufacturing Engineers, July 1970).

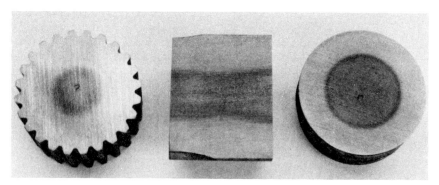

Fig. 11.7 Sections from an induction hardened shaft having diameters up to 3.75 in.

also be kept in mind:

1 Drastic quenches are usually employed; therefore plain carbon steel can be used for shafts up to 2 in. in diameter. Carbon-manganese boron steel is suitable up to approximately 3 in. in diameter.

2 Because the quenches are so drastic and absolute uniformity in quenching is impossible to achieve, the carbon content should preferably be held under 0.38% (ladle analysis) to eliminate quench cracking.

3 The hardenability of plain carbon steels can vary widely because of residual alloy content. To achieve adequate subsurface strength, some measure of hardenability control must be used.

4 Because of the surface heating effect of high-frequency electric power, surface defects and inclusions may initiate cracks and therefore must be minimized.

5 Because the heating rate is so high and the time at temperature is short, some alloy steels do not austenitize and harden properly.

Shafts made by this method are not ordinarily subjected to loading close to the yield point. In other words, they are used where long-life fatigue strength is a major engineering requirement. Therefore the final hardness can and, for reasons of residual stress, often must remain high. This means that they are tempered at relatively low temperatures (300 to 600 F) and final bending and torsional static strength is approximately equivalent to as-quenched strength. The factors for estimating the hardness at a 3/16-in. depth on the SAC traverse can be used to determine the approximate chemical composition (other than carbon content) required. It is good practice to work closely with two or three suppliers of plain carbon steel for induction-hardened shafts, and it is suggested that all be high-residual or low-residual mills, as required by the part. Mixing of the two kinds of carbon steel will lead to difficulties. If strength is critical or cracking tendencies are high, arrangements should be made to purchase to hardenability or restricted carbon limits. For shafts 2 to 3 in. in diameter the C-Mn-B steels can be purchased to mutually aggreable hardenability limits at no extra cost. Standard grades 15B28H, 15B30H, 15B35H, and 15B37H are available. Nonstandard grades are also available if ordered in heat lots.

Steel Type	Time Above Ac_3 (sec)	Hardness Deficiency* (Rc)
1050	25–30	9
1144	25–40	2
2340	25–30	3
4150	25–30	27
4150	300	17
4150	900	7
4340	25–30	20
6150	25–30	17

For shafts greater than 3 in. in diameter alloy steel is almost always used. The short austenitizing times with induction heating can result in less than full hardening when carbide formers are the major alloy constituents. For the most uniform induction hardening alloy steels with manganese content up to 1.40%, chromium to 0.90%, and boron are suggested.

In critical shafts the cleanest possible steel should be used. Seams, laps, and nonmetallic inclusions (especially alumina) detract greatly from fatigue strength. Money spent on carbon deoxidation and vacuum degassing is often a good investment.

TYPE 4 SHAFTS

When properly made, this type of shaft is the ultimate insofar as load-carrying ability is concerned. Through-heating in a furnace (or by induction) and vigorous quenching in a roller-die machine result in the maximum residual compressive stress in the surface. Liss, Massieon, and McKloskey (Ref. 9, Chapter 1), have found that more than 50% of this stress is created by thermal effects in quenching and that carbon contents as low as 0.30% can be used. Even as-quenched (Rkw C50) 0.30% carbon steel has the excellent toughness required for short-life fatigue strength.

*Hardness deficiency is defined as the difference in quenched hardness between a furnace-heated and induction-heated part held above Ac_3 for the times shown.

An etched longitudinal cross section of a shaft made in this way shows in Fig. 3.3 how the desired martensitic layer follows the contour of the shaft. The flange, however, is purposely shielded from the quench to facilitate drilling after heat treatment. Shielding or even prequenching a flange or other section in oil before transfer to the main roller quench is an effective means of keeping this area soft.

The success of the shell-hardening process is heavily dependent on the drastic quench on the outside surface and the thermal effects of the core. The thermal conditions must not be disturbed, particularly in the important areas of a shaft. A change in thermal conditions on a shell-hardened transmission shaft similar to the one in Fig. 11.1 brought disastrous field failures. The shaft was being made of a high-hardenability carbon steel, roller quenched in water, and tempered at 425 F. A field failure had never been reported, even though thousands of shafts had operated for thousands of hours. The designer then decided that it would be convenient to carry transmission lubricant through the shaft and changed the part to call for a 0.375-in.-diameter hole to be drilled on its centerline. An epidemic of failures started immediately; service lives were only 50 to 250 hrs. Investigation revealed that although the surface compressive stress on the solid shaft was greater than 140,000 psi in the drilled shafts it was as low as 40,000 psi. Drilling the hole after heat treatment retained the high compressive stress and solved the failure problem.

Because Type 4 shafts are usually finish-machined before heat treatment, except for operations like grinding bearing diameters, it is important that decarburization be prevented. Scale should also be avoided because it may adhere and interfere with the quench, thereby creating nonuniformities in the desired residual compressive stress at the surface. Decarburization reduces both the level of compressive stress obtained and the final hardness of the surface; both factors seriously detract from long-life fatigue strength. The superiority of this type of shaft in bending fatigue is shown by the results of the 1045 specimens in Fig. 1.8.

The carbon level of steel for Type 4 shafts can vary from 0.30 to 0.95%, depending on the complexity of the part as affected by machinability and the tendency to quench crack. For the most intricate shaft designs which contain snap-ring grooves, keyways, splines, and radial oil holes steel of nominal ladle carbon content between 0.30 and 0.35% should be specified. Parts with variations in diameter up to a ratio of 1.5:1.0 and shallow splines without snap-ring grooves can usually tolerate nominal ladle carbon content up to 0.40%. Axles for highway trucks, especially when roller-die-quenched in hot 5% caustic soda, can tolerate low-residual steels of 0.45% nominal carbon content. Plain rounds, hexagons, and octagons can tolerate eutectoid and over-eutectoid carbon contents, provided the ends are shielded from the quench or machined with smooth generous radii.

The hardened depth must be adequate to support the bending and/or torsional loads. Because shafts made in this manner are outstanding in long-life fatigue qualities, designers as well as materials engineers are prone to assume that they should also be outstanding in bending and torsional strength. This is not necessarily so because the shell-hardening effect is quite shallow (e.g., 0.125 in.). Beneath this very hard layer, especially in plain carbon steels, the hardness drops off rapidly to less than 35 Rc—making the shaft subject to permanent deformation under conditions of extreme overload.

To provide for uniform hardening and adequate hardened depth the minimum manganese content of plain carbon steels should be 0.70%. Standard grades like 1037 and 1039 can be specified, but because a grade extra of $0.15/cwt is charged for 1.10% maximum manganese nonstandard grades with 0.80 to 1.10% Mn can be used if needed in sufficient tonnages. The factors in Chapter 6 for predicting the hardness on the SAC traverse can determine the manganese level required. It is recommended that the maximum ladle manganese content not exceed 1.40% for reasons of machinability and uniformity. Suggestions similar to those for induction-hardened shafts are the following:

1 Plain carbon steel can ordinarily be used up to a maximum diameter of 2 in.
2 Some control of hardenability is essential. As a minimum control steel sources should be restricted to those supplying low residual alloy content (preferred) or to those supplying high residuals. In either case evidence should be given that the residual alloy content is consistent from heat to heat. The next step up in control would be to a chemical factor range, and the best control is to specify the hardness at some location on the SAC traverse.

The limitations on carbon steel, of course, are strictly a function of design stress. We are aware of a 7-in.-diameter shaft made of 1035 steel and shell-hardened to a fully martensitic microstructure on the surface. When the manganese is above 1.10% max and the shaft has irregularities such as splines machined in it, the specified carbon content should be reduced to prevent cracking; for example, if 0.80 to 1.10% Mn is inadequate to provide the proper hardened depth with 0.35 to 0.42% C, then 1.10 to 1.40% Mn can be used, but the carbon range should be reduced to 0.32 to 0.39%.

For shafts 2 to 4 in. in diameter C-Mn-B steels are recommended. Standard grades like 15B30H, 15B35H, and 15B37H are suggested. (See Appendix VI for hardenability bands.) The 1.50% Mn permitted by 15B37H should be avoided, however, particularly in closed-die-forged shafts in which transverse properties at the flash line may be erratic.

Shafts more than 4 in. in diameter should preferably be of alloy steel.

If possible, however, grades that have a rather sharp knee in the maximum hardenability curve to avoid through hardening (e.g., 50B40) should be selected. Another alternative is to purchase steel to a restricted hardenability requirement. It should be remembered that vigorous water quenching on the outside surface will develop high-residual compressive stresses only if through hardening is avoided.

A major objective of the shell-hardening process is to develop high residual compressive stresses on the outside surfaces of a part. Tempering therefore should be at a temperature as low as possible to prevent excessive relief of these stresses. Straightening should also be avoided, particularly if yielding occurs at a critical area. Yielding will remove the compressive stresses induced by the heat treatment.

TYPE 5 SHAFTS

Most shafts that are case-hardened are heat-treated by carburizing or carbonitriding because the unique qualities so developed in the steel are required for an attached gear or cam. The case hardening not only satisfies the requirements of the gear or cam but it also enhances the torsional and bending strength of the shaft. The hard cases are almost always residually stressed in compression, which is desirable for long-life fatigue strength. In many instances, however, the shaft portions require straightening and the good fatigue qualities are due primarily to the inherently high long-life fatigue strength of the hard surface.

Usually the steel required in the gear has adequate hardenability for the shaft. The selection of steel for gears, as discussed in Chapter 10, is applicable to most shafts that are required to have significant load-carrying capacity.

There are times when case hardening a plain shaft is necessary. A typical example is one in which, for economic reasons, a shaft must act as the inner race of a roller bearing in addition to transmitting torque. When this situation confronts the materials engineer, he must call not only for a steel that will have adequate case hardenability to provide the necessary high-quality case microstructure but also for a quality that will be essentially free of significant nonmetallic inclusions in the raceways. A good policy is to specify bearing quality steel and a 100% magnetic particle check of the parts in the raceways.

In small diameters (1.25 in. and under) case hardening can also sometimes provide outstanding static and long-life fatigue strength beyond that provided by shell and induction hardening. An interesting example is the timing-gear idler shaft for a diesel engine shown in Fig. 11.8. This shaft was originally made of 8640 cold-finished bar stock 2.25 in. in diameter. After rough turning on a turret lathe it was heat-treated to 262-321 Bhn and then finish-ground. The test shafts failed in the 0.25 in. radius which was 0.188 in. in approximately 500 hr.

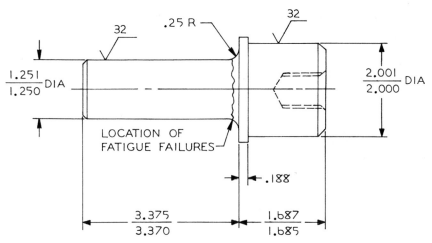

Fig. 11.8 Design for a timing-gear idler shaft for a diesel engine that posed a severe fatigue-failure problem.

After the radius was increased to 0.25 in. the life improved to approximately 750 hr, which was still grossly inadequate. Shot peening did not solve the problem. A serious situation existed because the shaft could not be made larger due to interference in the engine-block casting. Inasmuch as induction hardening worked so well on the front-wheel spindle (Fig. 8.8) it was decided to try this process on the 0.25-in.-radius fillet. It, too, failed to give satisfactory life. Then, because shell-hardened 1035 was successful in axles, it was also tried but again satisfactory life was not achieved. Finally, a number of shafts were made of 8620, carburized 0.065 to 0.090 in. deep, and lightly shot-peened after finish-grinding. This solved the problem, but the reasons are still obscure.

No doubt many similar case histories exist for carbonitriding, especially in small diameters or thin sections. The relationship between residual compressive stress and long-life fatigue strength holds. However, when section sizes are less than 1 in., the compressive stresses that can be developed by shell or induction hardening decrease as does the improvement in long-life fatigue strength.

REFERENCES

1 *Webster's New Collegiate Dictionary*, G. & C. Merriam Co., Springfield, Massachusetts, 1960.

2 *Metals Handbook*, 8th ed., American Society for Metals, Metals Park, Ohio, 1974–1975.

3 Shot Peening, Heat Treating, Cleaning and Finishing, *Metals Handbook*, Vol. 2, American Society for Metals, Metals Park, Ohio, 1964, pp. 398–405.

4 R. M. Spencer, Gas Nitriding of Large Components, William Hunt Eisenman Conference on Heat Treating, American Society for Metals, Metals Park, Ohio, 1968.

5 *Metals Handbook Supplement,* American Society for Metals, Metals Park, Ohio, 1954, pp. 106–107.

6 *Nickel Alloy Steels Data Book,* Sec. 2, Bull. A, The International Nickel Co., New York, 1965.

7 *Modern Steels and Their Properties,* 7th ed., Bethlehem Steel Corp., Bethlehem, Pennsylvania, 1972.

8 T. D. Kelso, Single Shot Induction: A New Approach to Hardening, *Manufacturing Engineering and Management,* July 1970.

CHAPTER TWELVE

SELECTION OF STEELS FOR
HEAT-TREATED SPRINGS

A MECHANICAL SPRING MAY BE defined as an elastic body with the primary function of deflecting or distorting under applied load and returning to its original shape when the load is removed. In this chapter the effect on performance of important processing operations, as well as selection factors, is covered. Also, in order that the materials engineer may be reasonably conversant in spring discussions, some of the nomenclature of the spring industry is explained. Fundamentals of spring design are discussed, and the following basic types of springs are covered:

1 Multiple leaf.
2 Helical compression and tension.
3 Torsion bar.

The uninformed may wonder why it is necessary to have so many different kinds of spring. Surely one must be best; but just as in structural sections each has its place in engineering design. Often the spring is expected to perform more than a pure spring function; for example, the rear leaf springs in an automobile act structurally to maintain rear-axle alignment. In other applications there may not be room to accommodate a leaf spring and a helical spring or a torsion bar must be used. Because a spring's efficiency is dependent on its specific capability to store energy per unit weight, strength levels are above 200,000 psi. The absorption and release of energy involves cyclic loading and a long-life fatigue situation exists. The comparative efficiencies of different springs are typically as follows:

Type of Spring	Energy per Pound of Spring (in.–lb)
Single leaf or all leaves, full length	173
Properly stepped, multileaf	380
Single leaf ($H = 0.16, J_c = 0.40$)	488
Helical, round bar (160,000 psi stress)	2050
Torsion bar, round (140,000 psi stress)	1570

$$H = \frac{c}{1} \qquad (12.1)$$

where 1 = cantilever length
c = length of constant cross section at end

$$J_c = \frac{T_e}{T_o} \qquad (12.2)$$

where T_e = thickness at load application
T_o = thickness at encasement

LEAF SPRINGS

Leaf springs are used mainly in suspensions, the characteristics of which are affected essentially by the spring rate and the static deflection. The spring rate is the change of load per inch of deflection. This is not the same at all positions of the spring and is different for a spring alone and for a spring installed on a piece of machinery. Static deflection of a spring equals the static load divided by the spring rate at a static load. Spring leaves are usually round-edged flat bars of uniform width; the edges are circular arcs of a radius equal to at least 75% of the thickness of the bars.

The design of leaf springs is a complicated science, but the parameters of greatest concern to the materials engineer are the operating stress and the life required. Figure 12.1 shows some typical spring designs with formulas for calculating the approximate stresses. An outstanding reference for all details of leaf-spring design is the SAE Leaf Spring Manual [1]. When the types of leaf section, ends, clamping, and so on, are selected, the approximate spring dimensions can be determined, but final design is preferably worked out with

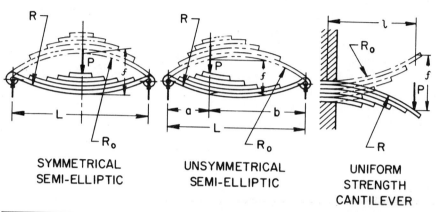

SYMMETRICAL
SEMI-ELLIPTIC

UNSYMMETRICAL
SEMI-ELLIPTIC

UNIFORM
STRENGTH
CANTILEVER

		SYMMETRICAL SEMI-ELLIPTIC	UNSYMMETRICAL SEMI − ELLIPTIC	UNIFORM STRENGTH CANTILEVER
I	LOAD RATE	$k = \dfrac{wN}{12} \cdot \left(\dfrac{1000t}{L}\right)^3$	$k = \dfrac{wN}{192} \cdot \left(\dfrac{1000t}{L}\right)^3 \cdot \dfrac{(Y+1)^4}{Y^2}$	$k = \dfrac{wN}{192} \cdot \left(\dfrac{1000t}{l}\right)^3$
II	LEAF THICKNESS FROM DEFLECTION	$t = \dfrac{l}{1000} \cdot \dfrac{L^2}{f} \cdot \dfrac{s}{125000}$	$t = \dfrac{l}{1000} \cdot \dfrac{L^2}{f} \cdot \dfrac{s}{31000} \cdot \dfrac{Y}{(Y+1)^2}$	$t = \dfrac{l}{1000} \cdot \dfrac{l^2}{f} \cdot \dfrac{s}{31000}$
III	STRESS, FROM DEFLECTION	$S = 125000 \cdot \dfrac{1000tf}{L^2}$	$S = 31000 \cdot \dfrac{1000tf}{L^2} \cdot \dfrac{(Y+1)^2}{Y}$	$S = 31000 \cdot \dfrac{1000tf}{l^2}$
IV	NECESSARY WEIGHT OF ACTIVE SPRING STEEL, POUNDS (APPROXIMATE)	WEIGHT $= \left(\dfrac{5000f}{S}\right)^2 k$		

L = ACTIVE LENGTH OF THE
SEMI-ELLIPTIC SPRING   (INCHES)

l = ACTIVE LENGTH OF THE
CANTILEVER SPRING   (INCHES)

w = WIDTH OF THE LEAVES   (INCHES)

t = THICKNESS OF THE LEAVES   (INCHES)
FOR PRELIMINARY CALCULATIONS ALL
LEAVES ARE ASSUMED EQUALLY THICK

k = LOAD RATE   (POUNDS PER INCH)
(LOAD INCREASE / DEFLECTION INCREASE)

f = DEFLECTION   (INCHES)

S = STRESS   (POUNDS PER SQUARE INCH)

N = NUMBER OF LEAVES

Y = CANTILEVER LENGTH RATIO

FORMULAE ARE BASED ON A MODULUS OF $29 \cdot 10^6$ psi AND A STIFFENING FACTOR OF 1.08

Fig. 12.1   Design formulas for typical steel leaf springs [1].

the manufacturer. These details include the following:

1 **Leaf Thickness.** In automotive practice springs are usually composed of leaves of two or three different thicknesses. The main leaf and those immediately adjacent are generally the thickest for the following reasons:

 (a) to give the main leaf more strength to resist I forces;
 (b) to allow more tolerance of radius on the short leaves;
 (c) because the designed spring rates can be approached more closely by combination of gages than with all leaves of a single gage.

2 **Leaf Radii.** The curvature of the unassembled leaves of a spring is not the same as that of the main leaf. A typical design will provide clearance such that the leaves will fit tightly together when all are clamped in assembly. In the main leaf the assembly stresses are opposed to operating loads, whereas in the short leaves they are additive to the load stress. Leaf radii are usually hot-formed in conjunction with the hardening operation.

Operating Stress and Fatigue Life

To keep the weight of a spring as low as possible it is necessary to use the highest achievable stress. This is limited by three parameters:

1 Settling under load.
2 Fatigue life.
3 Quality of processing.

The settling of a spring of a given hardness is a function of the maximum stress. All springs settle somewhat during the first few load cycles, but this can be minimized by proper presetting. Presetting consists in stressing a spring slightly beyond its yield point in the direction of the operating loads. The result is the development of residual compressive stresses (which are desirable for fatigue strength) but the main purpose is to minimize settling in service.

For automotive suspensions the design load stress is usually 100,000 psi max and for industrial machinery, 75,000 psi max. All operating stresses combined should never exceed the yield strength of properly quenched and tempered spring steel, which should be 180,000 psi min.

Accelerated fatigue testing of springs is sometimes a rather unsound practice from a materials engineering standpoint; for example, some fatigue test devices are of constant displacement rather than constant stress. In such testing a spring of poor metallurgical quality will settle down and look better in fatigue than a properly made spring. Another problem that often confronts the materials engineer is that designs are usually made for long life (more than 10^6 cycles), yet evaluations of fatigue qualities are often made at lives of 50,000 to 75,000 cycles. This sort of testing may lead to erroneous conclusions, as illustrated in Fig. 12.2. The parameters discussed in Chapter 8 for maximum short-life fatigue strength are different from those for long life in springs, just as they are for shafts and other parts. Accelerated testing can mislead a designer into believing that he has made an improvement when, in fact, the opposite may be true. Consequently, fatigue tests with failures occurring at less than 200,000 cycles are often invalid for designs based on millions of reversals.

Steels for Leaf Springs

Although springs can be made from any steel, steels of at least 0.55% C heat-treated to approximately 40 to 45 Rc are generally used. As shown in Fig. 1.2, attainable fatigue strength increases with carbon content. Although steel with carbon content above eutectoid (e.g., 1095) is used, most high-quality heat-treated springs are made of alloy steel with a nominal 0.60% C. Exceptions are 6150 and 4068. The most popular leaf-spring steels are 4161H, 5160H, 50B60H, 8660H, and 9260H.

The 9260 grade, and modifications thereof with chromium and/or molybdenum, is perhaps one of the oldest alloy steels in use for springs today. Its

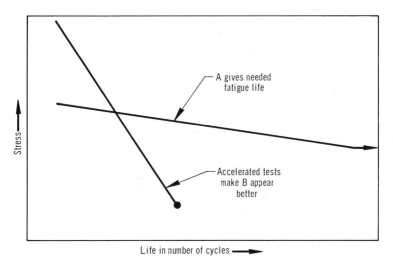

Fig. 12.2 A common mistake in fatigue testing, as shown by S-N curves for two combinations, A and B, of material and heat treatment. B appears to be the best combination because of better life at very high stress levels which is not the engineering requirement.

development enabled passenger-car manufacturers to guarantee springs for the life of the car. It was a difficult steel to make, however, because its 1.70 to 2.20% Si content usually required reladling, especially when made in an open-hearth furnace. The chilling effect of large ladle additions of silicon, and the reladling, made the attainment of good internal cleanliness and surface quality difficult. For high quality springs internal cleanliness and surface integrity must be carefully controlled. Today, with basic-oxygen or electric-furnace practice which provides hotter metal, the grade has regained some of its former popularity. The hardenability of 9260 and 9262 is low; therefore these steels are suitable only for leaf springs up to 0.500 in. thick.

The greatest tonnage of 5160 H steel is used for leaf springs, attesting to its acceptance as a high-quality spring material. It can be melted relatively easily to a high degree of cleanliness and rolled with a good surface. It is suitable for leaf springs up to 0.625 in. thick. Its shortcomings are the following:

1 Low hardenability, which makes it suitable only for relatively thin leaves
2 Chromium banding, which leads to somewhat erratic properties
3 Reliance on supplies of chromium, which may not be dependable

The grade extra for 5160H is $1.55/cwt; for 50B60 H, $2.40. Even at the $0.85 premium, however, 50B60 H is a bargain. First, the strong effect of boron in inhibiting the formation of proeutectoid ferrite is an important characteristic of high-quality spring steel. Second, the 99.9% martensite hardness for 50B60 occurs at J 5 min, whereas for 5160 it is only J 3. Third, the fracture toughness and notch tensile strength of 50B60 H are substantially better than for 5160 H. Fourth, the use of 50B60 H conserves supplies of often-scarce chromium.

The substantially higher hardenability of 50B60 H makes it suitable for use in leaf springs up to 1.375 in. thick. The boron present will provide superior microstructural capability at the same hardenability as a boron-free steel which should enhance its load-carrying ability. It can be rolled with a good surface and internal cleanliness and the potential hardenability of the grade permits adding less than the optimum amount of boron (see Chapter 15).

The 8660 H grade is perhaps the highest quality spring steel available. Its 0.5% Ni gives it superior load-carrying capability, but it tends to form a tight scale in rolling, and when hardening from direct-fired heating furaces this scale interferes with the quenching process and causes the surface to be somewhat irregular. For a superior steel of this composition the addition of boron at $1.10/cwt will significantly increase hardenability and microstructural capability. For maximum load-carrying ability in leaf springs electric-furnace-quality 8660 H, surface-ground before heat treating, is suggested. This steel is suitable for use in leaf springs up to 1.375 in. thick without boron and 2 in. thick with boron (55 Rc min at $J = 20$).

The 4161 H grade has high hardenability that makes it suitable for spring leaves up to 2 in. thick. The tendency of the high chromium to "band" sometimes results in poor transverse properties. High-side heats have also given trouble in quench cracking.

Although alloy steel of basic-open-hearth or basic-oxygen-process quality permits 0.035% P max, heats of spring steel should contain less than 0.025% and preferably less than 0.020% P. Sulfur is permitted up to 0.040%, but high-quality spring steel should never contain more than 0.025% S and in extreme applications sulfur should be restricted to 0.010% max.

Selection of steel grades for leaf springs other than the five grades we have noted can usually be made on the basis of hardenability. A good rule of thumb is to have a completely martensitic surface with at least 80% martensite in the center of the section. The steel should be one that can be readily melted to a high degree of cleanliness and rolled to a surface that is essentially free of seams and nonmetallic stringers. When maximum fatigue properties are required, electric-furnace-quality steel, carbon deoxidized and vacuum degassed, should be specified. In extreme cases aircraft or magnaflux quality may be necessary. When fatigue-strength requirements necessitate steel with a higher quality than the basic open-hearth or basic oxygen process provides, at least the highly stressed areas of the main leaves should be given a magnetic particle inpection with no defects permitted.

Heat Treatment of Leaf Springs. The objectives of heat treatment are to increase strength and produce maximum resistance to long-life fatigue failure. Therefore steels with adequate hardenability measured by hardness and microstructure must be selected. The heat-treating practice must preserve or improve the surface qualities of the steel. Austenitization must be complete, but grain coarsening from excessively high temperatures, avoided. Improperly controlled austenitization atmospheres can decarburize a steel much faster than an air atmosphere, and control of this factor must be absolute. If hardening and forming are performed in the same heating cycle, adequate mechanization must ensure that the parts are at the proper temperature when quenching begins.

Quenching should be as vigorous as practical to ensure maximum transformation to martensite; slack quenching is to be avoided. Quenching should continue until the part is completely below the martensite-finish temperature (M_f). As-quenched surface hardness below any decarburization or surface irregularities present should correspond to 99.9% martensite for the carbon content of the heat in question. As a guide, use Fig. 12.3.

Tempering temperatures should be set to meet the Brinell hardness range specified on the drawing, which should correspond to a spread of 0.2 Bmm with

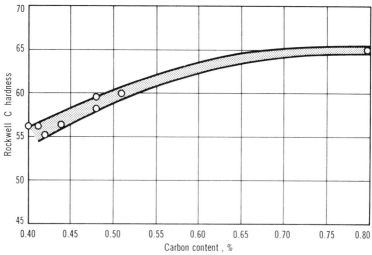

Fig. 12.3 Hardness of 99.9% martensite as a function of carbon content (from Hodge and Orehoski [3] Chapter 1).

a 10-mm-diameter ball and 3000-kg load. The typical total hardness range for all thicknesses is 2.75 to 3.10 Bmm. Usually, but not always, the thinner leaves are specified with a high-side hardness range of 2.75 to 2.95 Bmm and heavier leaves (more than 1 in. thick), with a range of 2.90 to 3.10 Bmm. Typical mechanical properties of heat-treated leaf-spring steels are as follows:

Hardness	3.10 to 2.85 Bmm
	42 to 49 Rc
Tensile strength	190,000 to 245,000 psi
Yield strength (0.2% offset)	170,000 to 225,000 psi
Reduction in area	25% min
Elongation in 2 in.	7% min

Properly made and heat-treated spring steel should exhibit a V-notch impact strength at –20F of 6 ft-lb min at 42 Rc.

Serious problems of breakage, even before the steel is put in service, can occur as a result of inadequate tempering time. It must be remembered that positive segregation, particularly of chromium and molybdenum, can occur, and this causes the core to be more resistant to tempering than the surface. A single test on the quality of tempering is to retemper a full cross section for 0.5 hr/in. of section at the original tempering temperature. A reduction in hardness greater than two points Rockwell C anywhere along a surface-to-center traverse indicates inadequate tempering time.

Decarburization on a leaf spring from the rolling of the steel and/or its heat treatment can significantly reduce its fatigue qualities, especially if the decarburization exceeds the following limits (as heat treated):

Carbon-free depth   0.003 in.
Total affected depth   0.015 in.

If a spring is to be shot-peened, an additional 0.001 in. on the carbon-free depth and 0.005 in. on the total affected depth can be tolerated. Carburization beyond 0.10% of nominal ladle carbon must also be avoided.

Mechanical Prestressing. Presetting, shot peening, and/or strain peening are effective means of improving the fatigue durability of a spring leaf. Actually these processes are likely to be more effective than a change in material. Control, however, is a problem because the only reliable nondestructive test for the measurement of residual compressive stress uses x-ray diffraction equipment that does not lend itself very well to production and to large leaf-spring assemblies. The mechanism of prestressing in improving the fatigue strength consists in creating residual compressive stresses in the surface of a leaf. Because residual stresses are algebraically additive to load stresses, the introduction of residual compressive stresses in the tension surface of a leaf by prestressing reduces the mean stress level and thereby increases fatigue life. Shot peening also has the capability of smoothing out even minor defects in the surface of a spring and this has a positive effect on fatigue life.

Presetting has already been discussed in connection with minimizing settling under load. The details of presetting require the experience of high-quality spring manufacturers.

Shot peening is an effective and controllable process for improving the fatigue qualities of leaf springs. Cut wire shot, size CW–23 to CW–41, and cast steel shot, S230 to S390, are generally used. Peening of leaf springs is done on the concave or tension side of a part. The intensity is usually in the range of 0.010 to 0.020 in. Almen A for springs up to 0.75 in. thick and 0.008 to 0.017 in. Almen C for heavier leaves, single-leaf springs, and strain peened parts. (See SAE J441, J442, J443, and J444 for a description of Alemn strips and specifications for shot.) Coverage of at least 90% is required. Ranges for Almen strip specifications are usually 0.004 in. when using A or C strips. Typical improvements in fatigue life of leaf springs that can be expected from shot peening are shown in Fig. 12.4.

Strain peening is shot peening done to the tension side of a leaf spring while it is loaded in the direction of subsequent service loading. This induces higher compressive stresses in the surface of the leaf than are possible by conventional peening and attendant increase in long-life fatigue strength.

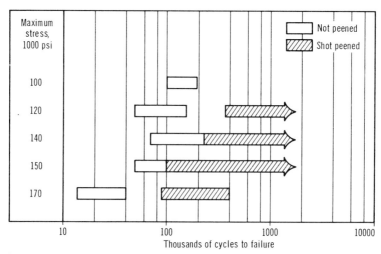

Fig. 12.4 Effect of shot peening on fatigue life of automotive leaf springs (from R. Schilling [6].

Protective Coatings. It is always desirable, and sometimes necessary, to protect leaf springs from general corrosion, stress-corrosion cracking, and/or the loss of the surface metal that has been residually stressed in compression. Any protective material used in a leaf-spring assembly should completely envelop the leaves and have suitable flexibility as well as sufficient adhesion for the required protective period. Finally, it is very important that the effect of the coating on fatigue life be known. When protection beyond that provided by an oil, paint, or plastic film is required, conventional electroplating can be used; however, a hydrogen relief bake must be specified. Optimum plating (e.g., with cadmium) is best done in vacuum followed by a plastic coating. Although we know of no actual use in this application, 17–4 PH stainless steel seems to have the necessary properties to be used in leaf springs for some corrosive environments.

HELICAL COMPRESSION AND TENSION SPRINGS

The helical spring is perhaps more widely used, in terms of numbers of pieces, than any other design. Its simplicity of manufacture and its energy-storing efficiency make it an outstanding engineering component. It is used in mechanisms many times smaller than mousetraps and in mountings for gigantic equipment such as large gyratory crushers and 100-ton freight cars. As in leaf springs, there

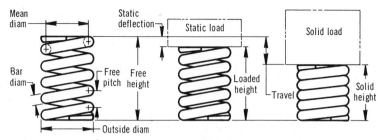

Fig. 12.5 A typical helical compression spring design, with ends closed and squared.

are many modifications of the simple helical tension or compression spring, the differences usually being adopted to achieve a desired load rate or means of attachment.

A typical compression-spring design and the most widely used terminology for helical springs appear in Fig. 12.5.

Helical springs are made in two general ways:

1 Wire of suitable mechanical properties is wound into an essentially finished spring.
2 Soft wire of suitable composition and hardenability is wound into a spring (sometimes hot); the spring is then quenched, tempered, and further processed into a finished part.

The highest quality springs are made by using preheat-treated wire; however, this practice is limited to wire from 0.25 to 0.625 in. in diameter, depending on the type of wire. Springs of larger wire must be wound before heat treatment.

Excellent references for helical springs are SAE Information Report HS-J795A [7] or the ASM *Metals Handbook* [8].

As for leaf springs, to select material and processing for a helical spring, the materials engineer must know (*a*) the stress level, (*b*) the minimum number of load cycles required from the part, and, (*c*) the temperature and other parameters of the environment in which the spring will operate. For helical compression springs the spring wire is subjected principally to torsional stress when compressed.

Similar formulas apply for tension springs except that initial tension must be taken into account. Also, in tension springs the stress concentration in the attachment hooks may be the limiting design factor. From (12.3) and (12.4) it is immediately obvious that the most effective means of reducing stress is by increasing the wire size. This will, of course, require additional changes to maintain the same spring rate.

Round wire	Square and Rectangular Wire

$$S \text{ (psi)} = \frac{8PD}{d^3} (K_w) \qquad (12.3) \qquad\qquad S \text{ (psi)} = \frac{PD}{K_2 bt^2} \qquad (12.4)$$

where S = maximum tensile stress at surface of wire,
P = load in pounds,
D = mean coil diameter in inches,
d = wire diameter in inches,
K_w = Wahl correction factor (see Table 12.1),
K_2 = stress constant (see Fig. 12.6),
b = width (long side) in inches,
t = thickness (short side) in inches.

Table 12.1   Tabulated Values for Wahl Stress Correction Factor K_W

c^a	0.0	0.1	0.2	0.3	0.4	0.5	0.6	0.7	0.8	0.9
2	2.058	1.975	1.905	1.844	1.792	1.746	1.705	1.669	1.636	1.607
3	1.580	1.556	1.533	1.512	1.493	1.476	1.459	1.444	1.430	1.416
4	1.404	1.392	1.381	1.370	1.360	1.351	1.342	1.334	1.325	1.318
5	1.311	1.304	1.297	1.290	1.284	1.278	1.273	1.267	1.262	1.257
6	1.253	1.248	1.243	1.239	1.235	1.231	1.227	1.223	1.220	1.216
7	1.213	1.210	1.206	1.203	1.200	1.197	1.195	1.192	1.189	1.187
8	1.184	1.182	1.179	1.177	1.175	1.172	1.170	1.168	1.166	1.164
9	1.162	1.160	1.158	1.156	1.155	1.153	1.151	1.150	1.148	1.146
10	1.145	1.143	1.142	1.140	1.139	1.138	1.136	1.135	1.133	1.132
11	1.131	1.130	1.128	1.127	1.126	1.125	1.124	1.123	1.122	1.120
12	1.119	1.118	1.117	1.116	1.115	1.114	1.113	1.113	1.112	1.111
13	1.110	1.109	1.108	1.107	1.106	1.106	1.105	1.104	1.103	1.102
14	1.102	1.101	1.100	1.099	1.099	1.098	1.097	1.097	1.096	1.095
15	1.095	1.094	1.093	1.093	1.092	1.091	1.091	1.090	1.090	1.089
16	1.088	1.088	1.087	1.087	1.086	1.085	1.085	1.084	1.084	1.084
17	1.083	1.083	1.082	1.082	1.081	1.081	1.080	1.080	1.079	1.079
18	1.078	1.078	1.077	1.077	1.077	1.076	1.076	1.075	1.075	1.074

(a) $c = \dfrac{D}{d}$

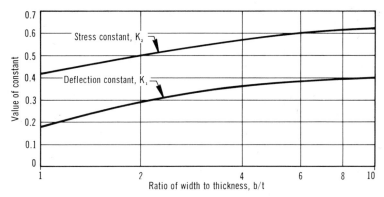

Fig. 12.6 Stress constants for springs of rectangular cross section.

Steels for Helical Springs

Types of spring wire suitable for the manufacture of helical springs without further heat treatment are shown in Table 12.2. The wire is produced from hot-rolled rods by cold drawing through carbide dies to obtain the required size, surface finish, dimensional accuracy, and mechanical properties. By varying the composition the amount of cold reduction and other mill processes, including heat treatment, a wide diversity of mechanical properties and finishes are available, as described in the following paragraphs.

Hard-Drawn Carbon-Steel Spring Wire. This wire is usually made from silicon-killed steels containing 0.45 to 0.75% C and 0.60 to 1.20% Mn. Its properties are developed by cold drawing after patenting in the hot-rolled condition. Patenting might be called high-temperature austempering because transformation is carried out between 850 and 1050F, which is substantially above the temperatures normally used for the austempering heat treatment. The same bainitic microstructure can be produced in the hot-rolled rods by a suitable combination of composition and cooling rate. Cold drawing increases the strength without a significant loss in ductility. The toughness of the cold-worked bainite make this wire suitable for more severe forming than oil-tempered wire. Because of the necessary isothermal transformation, the maximum size is usually 0.5 in. The main application is in springs subject to static load of low stress with infrequent stress cycles. The tensile strength is shown in Fig. 12.7. To avoid settling of springs during static loading, the maximum design stress should not exceed the torsional elastic limit, as shown in Fig. 12.8.

Table 12.2 Properties of Common Spring Materials

Name	Specification	Modulus of Elasticity in Tension (E)		Modulus of Elasticity in Shear (G)		Available wire sizes (in.)	Maximum Temperature	
		10^6 psi	GPa	10^6 psi	GPa		°F	°C
High-carbon steel								
Hard-drawn	SAE J1113	30	206.8	11.5	79.3	0.020–0.531	225	107
Hard-drawn valve	SAE J172	30	206.8	11.5	79.3	0.092–0.250	225	107
Music	SAE J178	30	206.8	11.5	79.3	0.004–0.250	275	135
Oil-tempered	SAE J316	30	206.8	11.5	79.3	0.020–0.650	325	163
Oil-tempered valve	SAE J351	30	206.8	11.5	79.3	0.062–0.250	325	163
Annealed	AISI 1065/AISI 1066	30	206.8	11.5	79.3	Less than 0.1875 / More than 0.1875	⋯	⋯
Alloy steel								
Chromium-vanadium	ASTM A 231	30	206.8	11.5	79.3	0.020–0.500	450	232
Chromium-vanadium valve	SAE J132	30	206.8	11.5	79.3	0.020–0.500	450	232
Chromium-silicon	SAE J157	30	206.8	11.5	79.3	0.032–0.438	475	246
Stainless Steel								
Austenitic (18–8)	SAE 30302/30304	28	193.1	10	68.9	0.005–0.375	550	288
17–7 PH	SAE J217	29.5	203.4	11	75.8	0.030–0.500	650	343

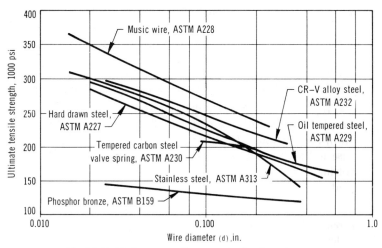

Fig. 12.7 Minimum tensile strength of spring wires.

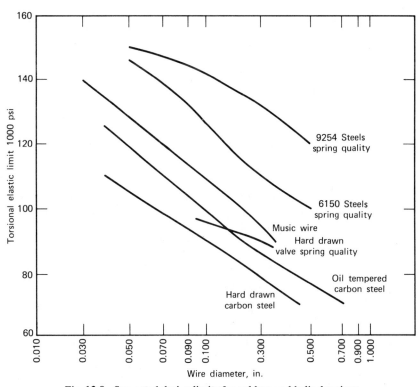

Fig. 12.8 Suggested design limits for cold wound helical springs.

237

Music Spring Wire. This grade represents the highest quality of hard-drawn steel spring wire. It is the least subject to the embrittling effects of plating of any spring wire. Steel for this grade is melted and/or selected for maximum cleanliness. It is then carefully rolled to prevent laps and other rolling defects. It is patented several times between drawing operations as it is reduced in size. The speed of drawing is slower than for hard-drawn carbon steel wire in order to prevent rise in temperature. Music wire is used in more severe spring applications involving dynamic loading in which fatigue strength is a major engineering requirement.

Oil-Tempered Carbon-Steel Spring Wire. The mechanical properties of this grade are developed by heat treating the finished wire. Treatment consists of austenitizing, oil quenching, and tempering by passing the wire through a molten lead bath at a temperature that will produce the required mechanical properties. The tempered martensite microstructure results in a wire more resistant to relaxation under cyclic loading than hard drawn wire. Because hardness is determined by tempering temperature rather than cold work (as with hard-drawn wire), the tensile strength can be controlled more closely, permitting more precise coiling. The tempered martensitic microstructure of oil-tempered wire makes it more subject to embrittlement by plating than hard-drawn wire.

Oil-Tempered Carbon-Steel Valve-Spring Wire. Valve-spring wire is the highest quality oil-tempered wire. It is intended to be used in applications such as valve springs of internal-combustion engines in which maximum life expectancy is required. Because valve-spring quality requires internal cleanliness and freedom from surface defects deleterious to fatigue life, it is also available in hard-drawn carbon-steel wire (e.g., SAE J172) [9]. For reasons of relaxation, however, valve-spring wire is used in the oil-tempered condition. The surface of this wire must have no areas of carbon-free decarburization or even of substantially reduced carbon. It must also be essentially free of seams, scratches, die marks, pits, or other defects that would detract from the fatigue life of the spring.

Alloy Steel Spring Wires. Alloy steels are used in the manufacture of springs because of their superiority over carbon grades for one or a combination of the following reasons:

1 Resistance to relaxation at temperatures up to 450 F.
2 Higher tensile strength.
3 Higher torsional elastic limits.

The compositions used are a chromium-vanadium type similar to 6155 (SAE J132) [10] or a chromium-silicon steel similar to 9254 (SAE J157) [11]. These wires are oil-tempered when purchased to these SAE specifications; however, they may also be purchased hard-drawn or annealed and in commercial or valve-spring qualities. For maximum fatigue life oil-tempered valve-spring quality is recommended. At one time the 9254 composition was thought to be the ultimate in constructional grades of steel, especially for use in springs operating at temperatures up to 450 F and requiring fatigue life of more than 10^8 cycles. The high silicon content of this grade makes it somewhat difficult to melt clean; however, its severest shortcoming is a tendency to crack transversely when drawn or coiled. Accordingly, we have a strong preference for the chromium-vanadium steel, especially when maximum fatigue life is required.

Stainless Steels. Although these grades are alloyed far beyond the limits of constructional steels and are therefore beyond the scope of this book, it is desirable to review briefly their use in springs because it is not uncommon for the materials engineer to have to select a spring to operate in a corrosive and/or high-temperature environment. Corrosive conditions may be beyond the resistance capability of the coatings usually applied to carbon or alloy steel wire. For small diameter wire, up to about 0.020 in., stainless steel is sometimes less costly than plating a hard-drawn or oil-tempered carbon steel wire.

Types 302, 303, and 304 stainless steel (SAE J230 [12] have excellent corrosion resistance and also resistance to heat relaxation up to 500 F. Type 17–7PH (SAE J217) [13] is age hardened by heating to 900 F for one hour after coiling. This grade has corrosion resistance equal to type 302 and in addition greater relaxation resistance up to 650 F. Springs made from these wires must be passivated after heat treatment to ensure maximum corrosion resistance. All of these grades of stainless are magnetic at spring hardnesses.

Large, Hot-Coiled Helical Springs

Most helical springs that have a mean coil diameter of more than 2 in. and that use wire more than 3/8 in. in diameter are made by selecting a bar of steel of sufficient length to make one spring, heating it to a temperature of 1575 to 1650 F, and hot coiling. Springs made in this manner are used in the suspension of railroad cars.

Most helical springs are subjected in service to a static load if for no other reason than to keep them in place. Large, hot-wound helical compression springs like those in railroad cars, however, carry substantial static loads and are commonly compressed to solid height (see Fig. 12.5) by shock loading. The stress rate q in psi per inch of deflection for such springs is defined as follows:

$$q = s \; \frac{8DK_z}{\pi d^3} \tag{12.5}$$

where

$$s \text{ (load rate)} = \frac{Gd^4}{8nD^3} \tag{12.6}$$

D = mean diameter = OD - d, in.,
d = bar diameter, in.,
K = correction factor for curvature
$$= \frac{4c - 1}{4c - 4} + \frac{0.615}{c},$$
c = spring index = $\dfrac{D}{d}$,
G = modulus of elasticity in shear for steel
  = 11,440,000 psi,
z = correction factor for eccentricity
$$= 1 + \frac{0.5}{n/d} + \frac{0.16}{(n/d)^2} + \frac{2}{(n/d)^3},$$
h = solid height, in.,
n = number of active turns = h/d - 1.2.

Suggested typical limits on proportions are

outside diameter	4 to 8 times bar diameter,
face height	1 to 4 times outside diameter,
minimum height	5 times bar diameter.

Before heating for coiling the ends are given the proper configuration by heating and forging. After hot coiling, which is usually done from direct-fired furnaces, some manufacturers direct-quench in oil. Others allow the springs to cool to room temperature and then reheat for hardening. Because the scale breaks off the bar as it is being coiled, the opportunity for maximum quenching effectiveness exists immediately after coiling; and direct quenching in oil is therefore considered the preferred practice.

The steel grades used for hot-wound springs depend on (a) the design stress, (b) the section size, and (c) the required life. The recommended maximum solid design stresses for preset springs are as follows:

Loading	Carbon Steel (psi)	Alloy Steel (psi)
Moderately severe dynamic (a)	140,000	150,000
Severe dynamic (a)	100,000	120,000
Constant pressure	170,000	180,000

[a] Static stress should not exceed 65% of solid stress. When springs are not preset, maximum design stresses should be 20% less.

The key to selection of steel for high-quality, large, hot-wound helical springs is to use a composition with a nominal carbon content of at least 0.60% and hardenability such that, with the quench employed, the minimum hardness in the center of the bar will be 50 Rc. The schedule of steels used by one firm is as follows:

Carbon steel 1095: moderately severe dynamic loading
to 1 in. diameter

Alloy steel Severe dynamic loading
5160H: to 1.50 in. diameter,
50B60H: more than 1.50 to 2.75 in. diameter,
4161H: more than 2.75 to 4 in. diameter

The microstructure should be free of proeutectoid ferrite (which has a fatigue strength of only 10,000 to 15,000 psi) for at least on-third of the distance from the surface to the center.

Steel Quality

Most steel used in helical springs is of basic-open-hearth or basic oxygen-process quality. Knowing the obvious application of these grades, the manufacturer gives special attention to melting as cleanly as possible. The extremely deleterious effect of dirty steel on fatigue life, both internal and surface, is well known. An excellent example is given in Fig. 12.9. Spring dimensions were outside diameter, approximately 1.00 in.; wire size, 0.100 in.; free height, 3.75 in.; 10 coils. The proprietary superfine-grained, extremely clean Cr-V spring clearly outperformed the standard Cr-Si material, even though the former was approximately 2 to 3 points R_c lower in hardness.

Although various specifications indicate that careful melting, selection of heats, and surface conditioning will be employed, it is a good idea, when highest quality is required, to make sure that carbon-deoxidized, vacuum-degassed steel

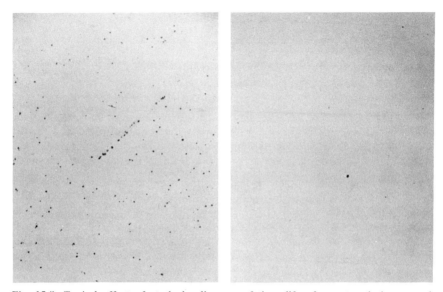

Fig. 12.9 Typical effect of steel cleanliness on fatigue life of preset and shot-peened springs. Steel at left is 9254 valve quality; at right, modified 6150 (unetched; 100X). Fatigue life at 0 to 122 000 psi was 4.4 to 11.5 $\times$ 10^6 cycles for the dirty steel at left; no failures had occurred after 10.2 to 14.8 $\times$ 10^6 cycles for the clean steel at right (courtesy Koehring Co.).

is used. Electric-furnace steel of bearing or aircraft quality and, in extreme cases, vacuum-arc-remelted steel are justifiable. Sulfur is a direct measure of steel cleanliness and therefore should preferably be less than 0.025% (ideally, less than 0.010%). Also, because of its embrittling effect, phosphorus in steels of spring temper should likewise be held below 0.025% (ideally, below 0.010%). Alumina inclusions are perhaps the most damaging because they are so hard and also can be large. Accordingly, steel should be made with minimum aluminum additions. Because most machinery manufacturers purchase their springs from a spring maker, the matter of control of steel cleanliness and surface quality is left to the supplier. Some high-quality spring manufacturers have their own inspectors and observers in the steel producer's plant to ensure the proper quality of raw material.

Processing. The processing of helical springs after coiling is perhaps of equal importance to the inherent metal quality. Processing to reduce relaxation in service and extend fatigue life consists of the following:

   1 Stress relief after coiling of hard-drawn or oil-tempered wire.

2 Heat setting.
3 Shot peening.

When hard-drawn or oil-tempered wire is coiled into a helical spring, the steel is stressed above the yield point. The effect of coiling on residual stress is different on the inside and outside diameters of the spring, the former having residual tensile stresses in the surface of the wire. This decreases its resistance to relaxation and its fatigue strength. Accordingly, springs made of these kinds of wire should be stress-relieved after coiling. The temperature should be as high as possible without lowering the strength of the wire. Springs made of hard-drawn carbon-steel wire and music wire are usually stress-relieved at 450 to 500 F. Oil-tempered wire springs, except for the silicon-chromium grades, should be stress-relieved at 600 to 650 F; silicon-chromium steel springs can be stress-relieved at 700 to 750 F.

Just as with leaf springs, helical springs stressed beyond the yield point in the direction of operating loads will cause the active surface of a spring to be residually stressed in compression when the load is removed. This residual compressive stress increases the long-life fatigue strength. The spring must, of course, be coiled longer than the desired finished length of the spring so that the part will be the correct length after setting. Cold presetting practice for highly stressed springs with indexes* between 5 and 10 is to preset to a height or load at which a corrected stress of 165,000 psi occurs.

Heat setting is used to increase resistance to relaxation at elevated operating temperatures. In this process springs are loaded and clamped while being heated to a temperature beyond the anticipated operating temperature and then cooled before being unclamped. This processing not only minimizes the tendency to relax under load at elevated temperatures but also induces compressive stresses in the outer fibers of the wire, thereby extending fatigue life. The compressive stresses, however, are usually not so high as on cold setting.

A less effective process is hot pressing, a procedure by which the spring is heated in its free position to a temperature beyond the anticipated operating temperature, and, while hot, compressing it to a height below the installed or operating height before releasing the load.

The desirable effects of shot peening on the long-life fatigue strength of helical springs are similar to those described for leaf springs and other parts. Improvements in fatigue life of the order of 4 to 1 are common. To be effective the peening must reach the inside diameter of the coil. As in any process used to induce residual compressive stress, it is useful only when operating stresses are substantially below the yield point (except when used as a means to remove decarburization and improve surface finish). As operating stresses ap-

*Spring index $= D/d$, where D = mean coil diameter and d = diameter of wire.

proach the yield point, localized yielding, and thereby stress relief, will occur, which removes the beneficial effects of the shot-peening treatment. Also, the beneficial effects of shot peening begin to disappear at a spring-operating temperature of 500 F and are gone entirely at 800 F.

Helical springs in extremely dynamic loading conditions, such as valve springs in an internal combustion engine or in a diesel fuel-injection mechanism, must be carefully designed. In addition to simple static and dynamic stress, such factors as environmental conditions, coordination of surge time to compression time, harmonic analysis, and specialized surface treatments should be considered.

Surface Protection

The surface protection of a helical spring depends on six factors:

1 Corrosive severity of the spring environment.
2 The degree of corrosion protection required to give the desired spring life expectancy.
3 Effect of coating on the mechanical properties of the spring (such as the effect of hydrogen embrittlement in plating).
4 Cost of applying the coating.
5 Effect of coating on associated members and environment (e.g., cadmium, lead, and zinc should not be used in contact with food or beverages).
6 Availability of protective materials and equipment necessary for the application of the coating.

Table 12.3 can be used as a general guide for selecting protective finishes.

Quality Control

The quality control of helical springs, especially in highly stressed dynamic applications, is very important. Regular audits of steel composition, steel cleanliness, microstructure, hardness, decarburization, and fatigue qualities should be made. Manufacturers of high-quality springs also make complete examinations of load-carrying capability. When an application requires steel comparable to valve-spring quality, parts are commonly totally inspected by a magnetic-particle method for seams, cracks, and surface inclusions.

The most common causes of premature failure of helical springs (other than steel cleanliness) are the following:

1 Surface defects such as scratches, nicks, seams, and die marks.

Table 12.3 Corrosion Protection for Springs

Criteria for Determining Protection to be Used	Recommended Protection in Order of Preference	Dimensional Buildup (min)/Surface [in. (mm)]	Methods of Applying Protection	Special Advantages, Precautions, and Limitations
Directly exposed to weather, not making close fits	Phosphate coat and paint or paint alone	0.001–0.002 (0.025–0.051)	Dip, spray, or brush	Close fits will peel paint
Directly exposed to weather, making close fits	Cadmium or zinc plate	0.0005 (0.013)	Electroplate	Corrosion protection good; relief of hydrogen embrittlement essential after electroplating
Not directly exposed to weather, making close fits	Cadmium or zinc plate	0.00015 (0.0038)	Electroplate	Same as above
Not directly exposed to weather, making close fits	Phosphate coat plus oil	0.0003–0.0008 (0.008–0.020)	Dip	Corrosion protection poorer than phosphate coat plus paint

Continued

245

Table 12.3 Continued

Criteria for Determining Protection to be Used	Recommended Protection in Order of Preference	Dimensional Buildup (min)/ Surface [in. (mm)]	Methods of Applying Protection	Special Advantages, Precautions, and Limitations
Relatively mild corrosive conditions	Precoated spring wire	Thickness is variable and dependent on size	Hot dip or electroplate	Fair all-purpose coating applied to spring wire; suitable for mildly corrosive conditions
Where corrosive conditions are very mild	Oxide black	Less than 0.0002 (0.005)	Alkaline oxidizing solution, 30–60 min at 300 F (139 C)	
Mildest corrosive conditions	Slushing oil or rust-proofing compounds	Dependent on method of application and/or viscosity of material	Dip or spray	Good for protection before installation and for short period after installation

2 Corrosion, which produces pits and is usually accompanied by hydrogen embrittlement.

3 Improper heat treatment; for example, one that results in grain coarsening because of excessive temperatures and results in improper microstructure (ferrite patches or large undissolved and nonuniformly dispersed carbides).

4 Decarburization, especially when it constitutes a carbon-free ring around the entire circumference of the wire.

TORSION BARS

A torsion bar is a spring mechanism in which torsional resiliency is used to achieve a spring effect. One end is anchored and the other, twisted by applied load via a lever arm. A torsion bar is a simple spring device second to a helical spring in energy-storage efficiency. This type of spring became popular in World War II when it was found that German tanks captured in the Libyan desert were equipped with torsion bars; subsequently, the design of most American tanks was changed to incorporate this type of suspension. Since then torsion bars have found many civilian uses (e.g., in passenger cars).

The stress in a torsion bar can be calculated from the following:

$$S = \frac{16M}{\pi d^3}$$

where M = torsional moment in inch pounds,
   d = bar diameter in inches.

The maximum torsional shear stress [5] for a life of 10^5 load cycles should be 120,000 psi.

A torsion bar, compared with a helical spring, is relatively simple to manufacture. It is essentially a straight shaft with a spline on each end. As might be expected, however, any part operating at such high strength levels requires careful manufacturing practices. Perhaps the most critical area in the design is at the junction of the bar body and the spline. The angle from the root of the spline to the outside diameter of the bar body preferably should not exceed 15°. The root diameter of the spline should be at least $d/7$ larger than the bar-body diameter. The spline should also have full root radius.

The steel selected must have adequate hardenability to ensure a ferrite-free microstructure and adequate hardness (typically 55 Rc) at least to the midradius of the bar body. A typical final-hardness specification is 47 to 51

Rc from the half radius to the outside of the bar body. A hardness of 45 Rc is acceptable from the half radius to the center, provided the total variation across the diameter does not exceed five points Rockwell C. For maximum long-life fatigue strength without brittleness alloy steels with a nominal 0.60% C content are used. As in any highly stressed spring, the steel must be as free of nometallic inclusions as possible. In tank torsion bars this was difficult because the basic spring steel at the time was 9262 (which is difficult to melt clean). It also often had insufficient hardenability so that it was necessary to modify the basic 9262 composition by adding 0.40 to 0.60% Mo or boron, the latter usually in conjunction with 0.01 to 0.05% V. All steel was electric-furnace quality and magnetic-particle testing was required. Present steels are 5160 H, 50B60 H, and 8660 H. We suggest 86B60 H for large diameters to ensure maximum freedom from ferrite in the microstructure.

A typical procedure for manufacturing torsion bars is to upset the ends of a piece of bar stock of suitable length, mill and center, rough turn, hob splines, quench and temper, magnetic-particle inspect, straighten hot if necessary (600 F min), grind, shot-peen, preset, and protective coat.

Because straightening is such an expensive operation (and dangerous to the operator because the steel is hard and brittle), quenching in a roller die machines with a vigorously agitated oil quench is recommended. The hardening-furnace atmosphere should have a 0.60% potential, and the amount of stock removed by grinding should ensure that there is no decarburization on the finished part. For optimum performance shot peening should be intense on the bar body; for example, 0.010 Almen C with 330 steel shot and 0.007 Almen C on the splines. Typical presetting on a heavy-duty torsion bar 60 in. long by 1.463 in. in body diameter is specified as three cycles of $102°$ with a permanent-set allowance of $37°$ max. Fatigue-test requirements on this torsion bar are 45,000 cycles from a loaded torsional shear stress of 20,000 psi to a maximum torsional shear stress of 140,000 psi.

Surface protection of the torsion-bar body is important and usually consists of cleaning, priming, and taping (the latter to prevent nicks and scratches in storage, handling, or use). It is important that the preset direction somehow be indicated on the end of a torsion bar, for if by accident it is installed with the preset direction against the operating load it will probably fail. An even better idea is to design the attachments so that the bar cannot be installed incorrectly.

REFERENCES

1 Manual on Design and Application of Leaf Springs, SAE Information Report HS–H788a, Society of Automotive Engineers, New York, 1974.

2 Cut Steel Wire Shot — SAE Recommended Practice J441, *SAE Handbook,* Society of Automotive Engineers, New York, 1974, p. 191.

3 Test Strip, Holder and Gage for Shot Peening — SAE Standard J442, *SAE Handbook,* 1974, p. 191.

4 Procedures for Using Standard Shot Peening Test Strip — SAE Recommended Practice J443, *SAE Handbook,* 1974, p. 193.

5 Cast Shot and Grit for Peening and Cleaning — SAE Recommended Practice J444, *SAE Handbook,* 1974, p. 195.

6 R. I. Schilling, Effect of Shot Peening on Fatigue Life of Automotive Leaf Spring, *SAE Trans.,* **54,** 366 (1946).

7 Manual on Design and Application of Helical and Spiral Springs, SAE Information Report HS J795a, Society of Automotive Engineers, New York, 1974.

8 Steel Springs, Properties and Selection of Metals, *Metals Handbook,* Vol. 1, American Society for Metals, Metals Park, Ohio, 1961, pp. 160–174.

9 Hard Drawn Carbon Steel Valve Spring Quality Wire and Springs — SAE Recommended Practice J172, *SAE Handbook,* 1974, p. 156.

10 Oil Tempered Chromium-Vanadium Valve Spring Quality Wire and Springs — SAE Recommended Practice J132, *SAE Handbook,* 1974, p. 153.

11 Oil Tempered Chromium-Silicon Alloy Steel Wire and Springs — SAE Recommended Practice J157, *SAE Handbook,* 1974, p. 155.

12 Stainless Steel SAE 30302 Spring Wire and Springs — SAE Recommended Practice J230, *SAE Handbook,* 1974, p. 158.

13 Stainless Steel 17–7 PH Spring Wire and Springs — SAE Recommended Practice J217, *SAE Handbook,* 1974, p. 157.

CHAPTER THIRTEEN

SELECTION OF STEELS
FOR FASTENERS

THE SELECTION OF STEEL FOR
fasteners and their heat treatment is an important, yet small, part of successful
fastening. The design for bolting must be precise and assembly is extremely
important. The total environment must be carefully determined to prevent
premature failure due to hydrogen embrittlement or stress-corrosion cracking,
especially at tensile strengths of more than 130,000 psi.

According to the American Standards Association, a fastener is a "mechanical device for holding two or more bodies in definite positions with respect
to each other." This definition, of course, includes numerous types of bolt,
stud, screw, nut, rivet, pin, and several combinations thereof. Because of the
large number of different fasteners many of which are not ordinarily heat-
treated, the scope of this chapter is limited to bolts, cap screws, and studs made
from grades of steel intended for heat treatment.

A bolt is an externally threaded fastener. Studs may be of many kinds, but
we are mainly concerned in this chapter with the bolt stud, which is essentially
a bolt threaded at both ends. A cap screw is a bolt whose surfaces are machined,
or equivalently finished, and which has a closely controlled body diameter, a
flat chamfered point, and a slotted, recessed or socket head of proportions
and tolerances designed to ensure full and proper loading when wrenched or
driven into a tapped hole. Cap screws ordinarily have a finished washer face to
facilitate solid abutment against the clamped parts. A typical bolt, cap screw,
and stud are shown in Fig. 13.1.

250

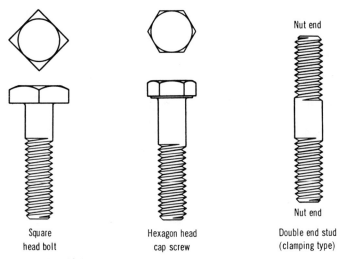

Square
head bolt

Hexagon head
cap screw

Nut end

Nut end

Double end stud
(clamping type)

Fig. 13.1 Typical design of a bolt, cap screw, and stud.

Most of these fasteners exert clamping stresses between the parts and/or perform a location function in shear. Successful mechanical fastening usually relies on adequate clamping force. Fasteners discussed here are seldom of inadequate shear strength and more often fail as a result of inadequate clamping strength. According to one large manufacturer of fasteners,

$$RF + RH + RT = MCP$$

(right fastener + right hole + right torque = maximum clamping performance)

This chapter discusses all of these factors with emphasis placed on the selection of steel and heat treatment.

Most heat-treated fasteners are produced to meet grades 5 through 8.2 of SAE Standard J429g [1], and ASTM Specifications A325, A354BD, A449, or A490 (see Table 13.1). The composition check analyses requirements for these bolts are listed in Table 13.2.

With the exception of Type 3 of ASTM-A325, details of chemical composition are subordinated to meeting the required mechanical properties, which are essentially those shown in Table 13.3. SAE and ASTM specifications also have inspection requirements for defects that affect serviceability. By and large, fasteners produced to these specifications are eminently successful; some users, however, find it necessary to supplement the SAE or ASTM requirements.

Table 13.1 Widely Used Fastener Steels

Specification	Grade or Type	Diameter Range (in.)	Steels	Quench
SAE J429g	5[a]	0.25 through 0.38	1038 coarse-grain	Oil
	5[a]	0.44 through 0.50	1038 Mod (0.80–1.10% Mn) coarse-grain	Oil
	5[a]	0.56 through 0.75	1038 coarse-grain	Water
	5[a]	0.87 through 1.50	1039 or 1040 coarse-grain	Water
SAE J429g	5.1	#8 and smaller	1021 Mod (0.80–1.10% Mn) fine-grain or 1022 coarse-grain with 0.15–0.30 Si	Water
	5.1	#10 through 0.25	1022 Mod (0.80–1.10% Mn) coarse-grain	Water
	5.1	0.31 through 0.38	1522, 10B18 Q-Temp[c], or 10B21 Q Temp[c]	Water
SAE J429g	7	0.25 through 0.44	4037, 4042 1335H	Oil
	7	0.50 through 0.75	8637H, 4137H	Oil
	7	0.87 through 1.00	4137H	Oil
	7	1.00 through 1.50	4140H	Oil
SAE J429g	8[b]	0.25 through 0.44	1335H, 4037, 4042	Oil
	8[b]	0.50 through 0.75	8637H, 4137H	Oil
	8[b]	0.87 through 1.00	4137H	Oil
	8[b]	More than 1.00 through 1.50	4140H	Oil
SAE J429g	8.1	All	1541	None
SAE J429g	8.2	All	10B21 Q Temp[c]	Water
ASTM A325	1	All	10B30 (0.30–0.34% C, 0.80–1.10% Mn)	Water

[a] Also ASTM A325 and A449.

[b] Also ASTM A354BD and A490.

[c] Trade name of U.S. Steel Corp.

252

Table 13.2   Compositions of SAE and ASTM Fastener Steels

Grade		Carbon	Manga-nese (min)	Phos-phorus (max)	Sulfur, (max)	Boron
SAE 5[a]		0.28–0.55	. . .	0.048	0.058	. . .
SAE 5.1		0.15–0.30	. . .	0.048	0.058	. . .
SAE 5.2		0.15–0.25	. . .	0.048	0.058	0.0005
SAE 7[b]		0.28–0.55	. . .	0.043	0.048	. . .
SAE 8 [b,c]		0.28–0.55	. . .	0.043	0.048	. . .
SAE 8.2		0.15–0.25	0.74	0.048	0.058	0.0005
ASTM A325, Type 1		0.27 min	0.47	0.048	0.058	. . .
	Type 2	0.13–0.25	0.67	0.048	0.058	0.0005
	Type 3	Alloy steel to match corrosion-resistant properties of structure in which used				
ASTM 490[d]		0.28–0.55	. . .	0.045	0.045	. . .

[a] ASTM A449 has same composition.
[b] Required to be alloy steel unless fastener manufacturer and purchaser can mutually agree on use of carbon steel.
[c] ASTM A354BD has similar composition.
[d] Must be alloy steel as defined by the American Iron and Steel Institute.

Table 13.3 Mechanical Property Requirements of Heat-Treated Bolts

Specification	Diameter Size (in)	Tests on Bolts		Tests on Specimens Machined from Bolts				
		Minimum Proof Strength (psi)	Minimum Tensile Strength (psi)	Minimum Proof Strength (psi)	Minimum Tensile Strength (psi)	Minimum elongation (%)	Minimum Reduction in Area (%)	Rc
SAE 5	¼–1	85,000	120,000	92,000	120,000	14	35	25–34
SAE 5	1–1.5	74,000	105,000	81,000	105,000	14	35	19–30
SAE 5.1	#6–0.38	85,000	120,000	...	...	...	...	23–40
SAE 5.2	¼–1	85,000	120,000	92,000	120,000	14	35	26–36
SAE 7	¼–1.5	105,000	133,000	115,000	133,000	12	35	28–34
SAE 8	¼–1.5	120,000	150,000	130,000	150,000	12	35	32–38
SAE 8.2	¼–1.5	120,000	150,000	130,000	150,000	12	35	34–42
ASTM:								
A325	½–1.0	85,000[a]	120,000	...	...	...	...	23–35
	1.12–1.5	74,000[a]	105,000	...	...	...	...	19–31
A354BD	¼–1.5	120,000	150,000	...	...	...	...	32–38
A449	¼–1	85,000	120,000	...	...	...	...	25–34
A449	1.0–1.5	74,000	105,000	...	...	...	...	19–30
A490	½–1.5	120,000	...	130,000[b]	150,000 to 170,000	14	40	32–36

[a] Length measurement method.
[b] 0.2% offset method.

254

DESIGN FOR FASTENERS

A designer fastens two or more parts together usually because they cannot be efficiently made in one piece or because one or more of the parts must be removable for inspection, repair, or replacement; for example, if the crankshaft and connecting-rod assembly in an internal-combustion engine could be installed and serviced with a one-piece block casting, there would be no need for separately attached oil pans. Similarly, if structural beams could be rolled, fabricated, and transported in 200-ft lengths, there would be less need for bolted joints in a bridge. The designer selects the type and number of fasteners on the basis of adequate clamping and/or shear strength in the mechanical joint to resist the applied stress. If the fasteners are subjected to cyclic stresses in tension, the total clamping force must always exceed the applied load to prevent loosening and possible failure. In shear the designer makes sure that the combined shear strength of all fasteners exceeds the shear load by a suitable safety factor. The frictional force between two clamped surfaces can seldom be high enough to prevent movement equivalent to bolt-hole clearances. When no movement can be tolerated, dowel pins are usually required.

The length of thread engagement should equal or slightly exceed the diameter of the externally threaded fastener. If nuts are used, the thread stick-through should be 1½ to 2 threads. The depth of engagement into threads should preferably exceed 68% of thread height.

For economic reasons in manufacture, procurement, and stocking the use of standard-size fasteners is encouraged. These can be found in the catalogs of quality fastener manufacturers. It will be noted that fine and coarse series of threads are available. The advantages of fine threads are as follows:

1 Because of the greater cross-sectional areas, fine-threaded fasteners are about 11% stronger on the average than coarse-threaded fasteners.
2 Fine-threaded fasteners have slightly greater fatigue strength than coarse-threaded fasteners.
3 Fine-threaded fasteners are superior in vibrating applications because of their lower thread helix angle.
4 In hard materials fine threads are easier to tap.
5 When the length of engagement or the wall thickness of the mating part is limited, fine-threaded fasteners are best for proper clamping.
6 When the fastener is used as an adjustable stop, fine threads adjust more precisely.

The following are the advantages of coarse threads:

1 Coarse-threaded fasteners normally are easier and faster to assemble than fine-threaded fasteners, with less chance of cross threading.

2 Nicks and burrs from handling are less likely to affect assembly.

3 Coarse-threaded taps are more readily available.

4 Coarse-threaded bolts are less prone to strip when inserted into soft materials.

5 Coarse threads are less likely to seize in high temperature and corrosive applications.

6 It is easier to tap coarse threads in brittle or friable materials (gray iron, etc.)

7 With more clearance between threads, coarse threads usually provide a better plated fastener.

IN-HOUSE SPECIFICATIONS AND METALLURGICAL PROCESSING

Most manufacturers of machinery and structures do not make their own bolts and leave to the supplier the details of selection of steel and metallurgical processing. Because the fastener industry is highly competitive and the trade specifications do not always call out some of the important engineering characteristics, the steels used, the level of quality control employed, and the true engineering capability of the fastener vary widely from one supplier to another. For these reasons most firms requiring high-quality fasteners have specifications of their own; for example, standard specifications either omit or have improper ranges for the following:

1 **Silicon Content.** This should preferably be 0.15 to 0.30% for carbon steels (when permitted at all) and 0.20 to 0.35% for alloy steels, except that when the latter are carbon-deoxidized and vacuum-degassed then 0.10% max is acceptable.

2 **Phosphorus Content.** The phosphorus ranges permitted in trade specifications are excessively high for parts as hard as heat treated bolts and should be kept below 0.020% at hardnesses above Rc 35.

3 **Steel Grade.** The supplier should notify the customer of the grade of steel he will use and not change it unless by mutual agreement.

4 **Heat Treatment.** Detailed austenitization requirements are not covered. Also, permissible tempering temperatures of less than 700 F could produce brittle bolts. Hardness ranges of only four points Rockwell C are too narrow for production heat treatment but, on the other hand, some ranges permitted seem excessively wide.

5 **Quality of Microstructure.** High-quality fasteners should call for 95% min martensite, throughout the threaded section. Some trade specifications, however, permit 0.50% carbon steel with a minimum as-quenched hardness of only 47 Rc, which can be attained with only 50 to 60% martensite. If tempered at 800 F, this bolt could meet the tensile requirements and pass the wedge test (see ASTM Standard A370), but it could be a brittle bolt.

6 **Thread Forming.** The thread-forming technique is often not specified. Rolling the threads after heat treatment gives the highest quality but is very expensive. Rolling before heat treatment is next best, provided it is properly done, and cut threads are the least desirable from an engineering standpoint.

7 **Hardness Testing.** Location of the hardness test on finished parts is sometimes not specified.

8 **Surface Defects.** In some specifications there are no requirements for decarburization and maximum seam depth permissible.

9 **Thread Defects.** In some specifications there are no requirements for laps and other irregularities or defects in the threads.

10 **Plating.** There may be no provisions for plating techniques and/or thermal processing to prevent hydrogen embrittlement or stress-corrosion cracking.

11 **Fatigue and Impact Testing.** There may be no dynamic test requirements for fatigue and impact loading.

Assuming that the designer has correctly determined the clamping force he needs and has selected fine or coarse threads and the strength required, the choice of material and processing for a bolt or cap screw can proceed. The designer should also specify rolled or cut threads, the former being preferred even if rolling is done before heat treatment because of the desirable grain flow following the thread contour. Rolling of threads before heat treatment is almost always lower in cost because this operation can be performed in the boltmaking machine. Bolts and cap screws up to 1.125 in. diameter by 6 in. long are usually made from coils of wire on this type of machine. The steps involved are shown in Fig. 13.2. In the original process (Fig. 13.2a) a flash was formed and trimmed off in the fourth step of the cycle. Figure 13.2b illustrates an advanced process known as "scrapless" cold heading, which dispenses with

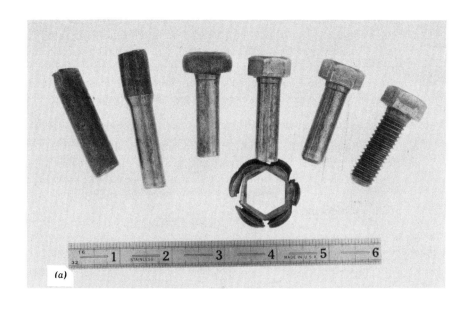

(a)

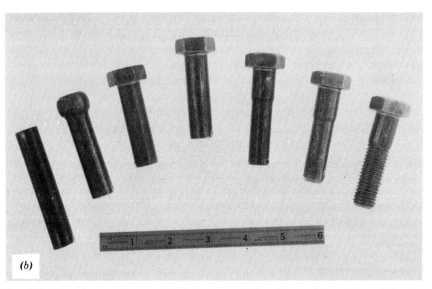

(b)

Fig. 13.2 Steps in manufacture of a bolt on a cold heading machine, using coiled wire: (a) standard process; (b) "scrapless" cold heading.

258

flash trimming. Boltmakers operate at very high speeds; for example, 2300 pieces/hr for a 3/4-in. by 2-in. hexagonal-head cap screw. Accordingly, if the grade of steel, its microstructure, and its surface are of inadequate quality, a great many defective parts may be run and the boltmaking machine may even be damaged. On the basis of manufacturing cost, therefore, the cold-forming qualities of a steel must be given careful consideration.

SELECTION OF STEELS

Use of steels of minimum carbon content is perhaps more important for fasteners than for any other direct-hardened part. Except for phosphorus, carbon is by far the most effective strengthening element in steel, as shown in the following equation developed by Hurley, Kitchin and Crawford [2] for flow stress at .2% strain:

$$\sigma_{0.2} = 47.8 + 70.0(\% \text{ C}) + 30.5(\% \text{ Si}) + 16.2(\% \text{ Mn}) + 11.2(\% \text{ Mo})$$
$$+ 8.1(\% \text{ Cr}) + 6.5(\% \text{ Ni}) \tag{13.1}$$

The relationship between hardenability and cold formability is important to the fastener manufacturer. He must have a material that will heat-treat to the correct microstructure and yet be easily formable. These relationships between alloying elements are shown in Fig. 13.3 [3]. This is by no means the only criterion for selecting a steel for cold forming, but it is an important parameter.

Abutting surfaces for the bolt and nut are seldom perfectly flat and parallel. Consequently, the bolt is expected to bend to conform. When other elements of a bolt composition and the quality of heat treatment are equal, a lower carbon content will provide superior conformability without cracking.

Steel of low carbon content are usually simpler to anneal to a satisfactory microstructure for cold heading. As a matter of fact, some steels, such as the "Q Temp" grades made by U.S. Steel Corp., can be cold-headed in the as-rolled condition, sometimes with the savings shown in Table 13.4 [4]. A typical analysis of mechanical properties is provided by comparing 10B18 and 1038:

	10B18 As-Rolled	1038 Spheroidize Annealed
Tensile strength	70,400 psi	68,450 psi
Yield point	38,670 psi	38,720 psi
Elongation in 8 in.	25%	24%
Reduction in area	62.7%	60.7%
Hardness	72 Rb	73 Rb

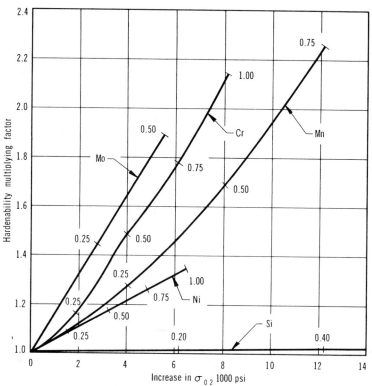

Fig. 13.3 Multiplying factor versus increase in flow stress, $\sigma_{0.2}$, for various alloying elements (from G. Creeger, Y. Smith, and D. V. Doane, unpublished report of Climax Molybdenum Co. of Michigan, Feb. 17, 1972).

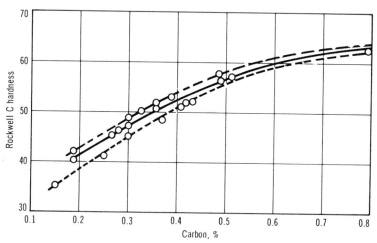

Fig. 13.4 Hardness of steel with structure of 95% martensite as a function of carbon content (from Hodge and Orehoski [5]).

Table 13.4 Approximate Savings[a] in Steel Costs if Annealing Operation is Eliminated [4]

| Currently Used Medium Carbon Grade | Q Temp Grade[b] | Saving | |
| | | Elimination of | |
		Regular Anneal (%)	Spheroidize Anneal (%)
1035, 1037, 1038	10B18Q	9	15
1039, 1040, 1042	10B18Q	9	15
1043, 1045, 1050	10B18Q	9	15
1024	10B18Q	23	24
1027, 1036	10B18Q	18	26
1041	10B23Q	15	21
1041	50B23Q	6	12

[a]Based on published mill prices for hot-rolled bars, 3/4-in. to 2 15/16 in. -diameter rounds.
[b]Q Temp is a trade name of U.S. Steel Corp.

For a highly competitive industry savings of this magnitude would be significant. They are, however, based on purchase of annealed stock; if a bolt manufacturer uses his own annealing furnaces, the savings may be substantially less. In any event, most heat-treated fastener steels will have nominal carbon contents of 0.18 to 0.40%.

Regardless of carbon content, hardenability should be such that all heats of steel harden to a microstructure of 95% min martensite throughout the section of the threaded length and to a similar structure at least in the surface of the body-to-head fillet. The remaining microstructure should be bainite or pearlite. The hardness of 95% martensite for various carbon contents is shown in Fig. 13.4 [5]. All of this must be accomplished while keeping the part straight within approximately 0.010 in./in. of bolt length, depending on drawing requirements. As pointed out in Chapter 3, one firm sets the limits for free water quenching of fasteners to a length of five times the diameter, for free oil quenching, eight diameters, and for austempering 10 diameters. The preferred heat treatment method for extremely long bolts is roller die quenching; a poor second is to straighten hot from the furnace and retemper.

With vigorously agitated water, brine, or caustic quenching properly selected carbon steels can be used in sections up to at least 0.625 in. and will give strengths up to 120,000 psi proof load. This requires that the steels have no more than 0.38% nominal carbon content, plus adequate manganese (1.40%

max) and/or high residual alloy content. The steel must be purchased with adequate hardenability guarantees or the source should be required to provide for significant testing* for hardenability. The Grange method of calculating the hardenability described in Appendix VII can be useful in the selection of carbon steel.

Because carbon steels have such low hardenability in terms of hardness and microstructural capability, the aqueous quenches must not only be highly vigorous but uniform as well. Belt hardening furnaces, which are standard in the industry, are sometimes loaded too heavily and this results in a conglomeration of fasteners falling into the quench at the same time. Those on the inside of this mass are sometimes slack quenched. The heat treater can make a quick check of the uniformity of his quench by etching as-quenched fasteners in picral and/or nital. Soft spots will show up as black streaks or spots. The ideal quench is by individual impingement; this can be mechanized quite readily with an induction heating unit.

To provide maximum hardenability of carbon steels at minimum cost coarse-grained material is sometimes used. For a 0.20% carbon steel the Grossmann factor for grain size 5, as related to grain size 8, is 1.25. At 0.30% C the factor ratio is 1.27. Except for steels of 0.25% nominal carbon content or less, we do not recommend this practice because of questionable toughness. This was vividly brought out in an epidemic of failures in dynamically loaded SAE grade 8 bolts. To solve the problem it was necessary to add a test requirement for a room-temperature V-notch Izod impact reading of 45 ft-lb at 32 Rc linearly down to 30 ft-lb at 38Rc. It was found that steels such as 1335 and 1340 could not consistently meet this requirement.

For high-strength oil-quenched alloy steel bolts it is especially necessary to select steels of known microstructural capability. Here grades such as 1335, 4037, 4042, 4137, 4140, 5140, 8637, 8640, and 8740 are widely used. Lean-alloy boron grades such as 81B35, 81B37, 94B30, 94B35, and 94B37 will provide even better microstructures. They also should be easier to anneal and

*Significant testing implies a procedure by which coils or bundles of bars are marked according to their position in the heat of steel. As a minimum, the testing locations should be at the bottom of the first ingot and at the top of the last ingot representing the product shipped. It is preferable to have testing locations at the bottom and top of the first ingot, the middle of the middle ingot, and the bottom and the top of the last ingot. Testing should consist of quenching samples of the steel (in production diameters) on each heat location in the fabricator's equipment (along with other bolts) and evaluating the microstructure at a magnification of at least 250X. Composition and surface quality checks should also be made on these samples. Heats showing a marked increase in phosphorus from the first to last ingot are highly questionable for use in high-strength fasteners.

less abrasive to boltmaker tooling than some other grades now used. A grade like 1340 is not recommended for fasteners because of erratic behavior both in heat treatment and engineering properties.

Carbon steels such as 1038, 1040, and 1045, oil-quenched, can be used for fasteners, particularly if they are high in manganese or residual alloy content, for SAE grade 5 strength levels. Better choices, however, are similar steels with boron added even though they increase the steel cost.

The "Q Temp" grades of U.S. Steel Corp. are mostly C-Mn-B or other low-alloy boron steels of nominal 0.20% C. Their excellent strength-toughness combination is typical of low-carbon martensite. A bend-test comparison of 10B21 and 4037 heat-treated to SAE grade 8 mechanical properties is made in Fig. 13.5. Hardenability ranges for these steels are discussed in Chapter 15. Typical mechanical properties of finished 10B21 bolts are listed in Table 13.5.

Table 13.5 Strength, Hardness and Notch Toughness of 10B21 Steel for Bolts

Source	Bolt Diameter (in.)	Temper Temperature (F)	Rc Core Hardness	Tensile Strength (psi)	Test Temperature	Charpy V-Notch (ft-lb)
(a)	7/8	400	41.2	206,900	Room	36
(a)	7/8	800	31.3	147,800	Room	104
(a)	1	400	45.8	209,500	Room	29
(a)	1	800	33.2	156,600	Room	92
(b)	7/8	400	41.0	. . .	–40 F	21
(b)	7/8	680	38.0	. . .	–40 F	25
(b)	7/8	850	29.0	. . .	–40 F	104

[a] Standard Pressed Steel Co.
[b] U.S. Steel Corp.

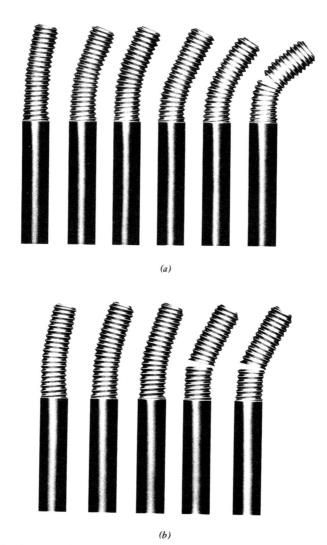

(a)

(b)

Fig. 13.5 Bend-test results show the superior toughness of low-carbon martensitic steel over that of 4037 at the same hardness; (a) 10B21 Q Temp; (b) standard 4037 (courtesy U.S. Steel Corp.).

264

HEAT TREATMENT

Fasteners are heat-treated (quenched and tempered) to restore carbon to the decarburized surfaces and to raise the strength to the desired level. Hardening therefore must be done in a furnace with an atmosphere maintained as close as possible to the mean carbon content of the steel. A lean endothermic gas or a richer carrier gas diluted with air to the correct carbon potential is generally used. Furnace time and temperature must be regulated as close as possible to restore the surface to the base carbon. A continuous furnace with a belt hearth is commonly used. Carburization beyond 0.45% C (depending on the steel grade) tends to embrittle bolts and must be avoided. Decarburization will lead to inadequate thread strength and a tendency to seize in assembly; the fatigue strength is also deleteriously affected. Quality fastener manufacturers have the latest in atmosphere-control equipment and also a staff of people who constantly check the product of their furnaces for carbon correction and as-quenched microstructure. An approximate check for decarburization or carburization can be made by comparing adjacent Rockwell C and 15N tests. A typical requirement for bolts specified to a hardness of 32 to 38 Rc is that the Rockwell 15N surface hardness tests converted to Rockwell C shall not be below 30 or above 40 Rc.

After austenitization the fasteners are quenched typically, by dropping them down a chute into an agitated medium. They are then elevated into a continuous furnace for tempering. After tempering they are cooled in a water-emulsifiable rustproofing oil and are ready for shipment or further metallurgical or mechanical processing.

POSTHEAT TREATMENT PROCESSING

Processing of fasteners after heat treatment is usually one or a combination of the following procedures:

1 Cold or warm rolling to improve fatigue qualities.
2 Surface treatment, such as phosphate coating, for ease of assembly and/or a moderate increase in corrosion resistance.
3 Plating for corrosion resistance and/or appearance.
4 Grinding the body for applications requiring extraordinary precision.

A typical rolling operation, after heat treatment, is the forming of the threads as specified; for example, in grade 7 of SAE J429g. This increases long-life fatigue strength, as shown in Fig. 13.6. This operation can be performed on fasteners having almost any degree of hardness, although the cost even at low hardness is substantial and becomes very high above a hardness of 50 Rc. The

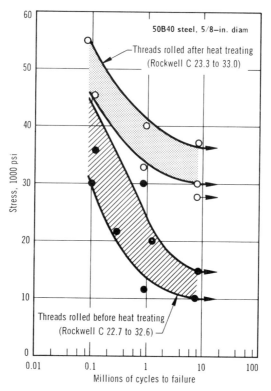

Fig. 13.6 Effect of rolling threads before and after heat treatment (from ASM *Metals Handbook,* Vol. 1. p. 176).

reasons are that it requires an additional operation on a machine other than the boltmaker, and of course the life of the rolling tool decreases sharply with increased hardness; for example, if the life at 15 Rc is 100,000 pieces, at 25 to 30 Rc it is 20,000 to 30,000 pieces, at 36 Rc 8000 pieces, and at 40 Rc only 3500 pieces. Rolling of the body-to-head fillet is also done to improve the long-life fatigue strength in this location.

The fatigue strength of axially loaded externally threaded fasteners with the threads rolled before heat treatment peaks out at approximately 35 Rc. Therefore, when a structure is fatigue sensitive, this bolt will not perform any better at a hardness of 40 or 45 Rc. This, however, is not true if the threads are rolled *after* heat treatment, where fatigue life shows improvement up to 50 Rc. Steels used at this strength level should also be vacuum-melted or vacuum-arc remelted. In regard to fatigue it should be remembered that seldom will an externally threaded fastener fail by fatigue if the proper clamping load is maintained.

SURFACE TREATMENT

The minimal surface treatment for heat-treated fasteners, other than a light coating of mildly rustproofing oil, is a coat of zinc phosphate. This increases the corrosion resistance somewhat, but the coefficient of friction is higher and more erratic than with zinc or cadmium plate.

Hot-dip galvanizing (see ASTM Standard A153) has found wide acceptance for its atmospheric corrosion resistance but it should be used only on coarse threads (½ in. or more).

Fasteners are plated for engineering purposes and appearance.

1 To prevent loss of section and thereby decreased load-carrying ability.
2 To prevent or at least minimize hydrogen embrittlement.
3 To prevent stress-corrosion cracking.

To prevent gross corrosion such as rusting the most popular plating for low-strength fasteners is electroplated cadmium (ASTM-A165) or zinc (ASTM-A164). The plating thickness can vary from 0.00015 to 0.0006 in., but 0.0002 in. is the most common. It must be remembered that cadmium plating reduces the torque necessary to obtain a specified clamping effect and zinc increases it. In fact, zinc tends to gall; therefore zinc-plated fasteners should be lubricated in assembly. Cadmium is widely used but has a duller finish. Assembly of cadmium-plated fasteners is also best done with a lubricant. For dynamically loaded high-strength fasteners where corrosion is quite severe, corrosion-resistant alloys such as 13-8 Mo, H950, and 718 are used and should preferably be vacuum-melted.

For most applications of constructional grades of alloy steel at tensile-strength levels up to 185,000 psi a 0.0003-in. cadmium plate, with a suitable bake after plating, will prevent hydrogen absorption or stress-corrosion cracking. (Threads rolled after heat treatment are preferred.) For ultimate strength levels of more than 200,000 psi vacuum cadmium plating 0.0005 in. thick is preferred. If a steel bolt is used in an aluminum forging, cadmium plating, plus assembly with zinc chromate paste, is recommended to prevent galvanic corrosion.

A fastener heat-treated to a tensile strength of more than 130,000 psi and fully torqued in use is nearly ideal as a test for hydrogen embrittlement. The effect of sharpness of notch radius is shown in Fig. 13.7, taken from data by Frohmberg, Barnett, and Troiano [6]. Atomic hydrogen from a humid atmosphere, from electroplating, or from pickling for hot-dip galvanizing is readily absorbed by the fasteners and concentrates at the points of highest stress; the result is embrittlement and sometimes sudden failure within a few hours of assembly. Hydrogen embrittlement failures usually occur within 500 hr after assembly, whereas with stress-corrosion cracking a time of at least 1000 hr is required, as shown schematically in Fig. 13.8.

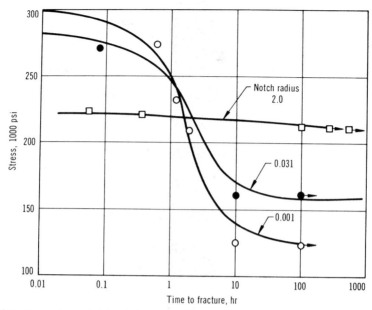

Fig. 13.7 Comparison of delayed fracture of heat treated steel specimens with notches of differing sharpness [6].

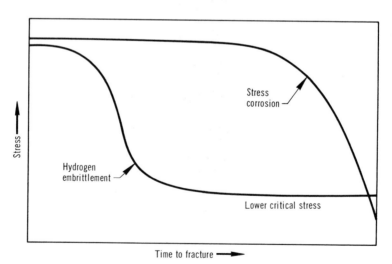

Fig. 13.8 Schematic representation of delayed fracture curves [8].

Stress-corrosion cracking is a different type of failure mechanism in which the corrosive environment creates corrosion pits and failure finally takes place abruptly. Stress-corrosion cracking (as well as hydrogen embrittlement) occurs frequently in steels heat-treated to tensile strengths above 165,000 psi and the sensitivity tends to rise sharply in proportion to the increase in strength. As might be expected, the time to fracture depends on the corrosiveness of the environment, as shown in Fig. 13.9.

A number of exhaustive studies have been made on delayed brittle fracture of bolts from which some conclusions may be drawn:

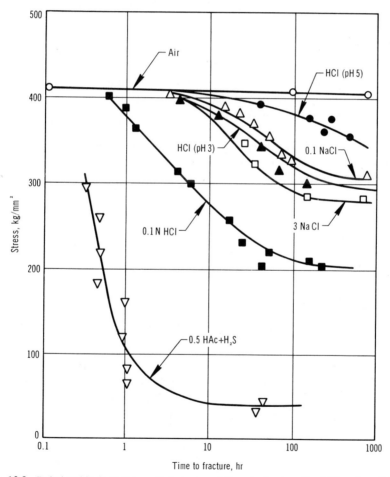

Fig. 13.9 Relationship between applied stress and time to fracture for 0.2% carbon Si-Mn-Cr steel in various environments (from Fukui and Asadi [7]).

1 The carbon content should be as low as possible to provide adequate strength with suitable tempering [8].

2 Cold working increases susceptibility to brittle fracture [9]. Cold working followed by a blue tempering treatment, however, provided better resistance to brittle fracture than that of parts given a regular quench-and-temper treatment.

3 Decarburization improves resistance to fracture; carburization lowers it [10].

4 Addition of titanium has a pronounced beneficial effect, as shown in Fig. 13.10 [8]. These tests were run on 51 different analyses of 0.20% carbon steels heat-treated to approximately 130,000 psi min proof load. Testing was done at 90% of notch tensile strength in a 95% relative humidity chamber at 108 F. The optimum composition was found to be as follows: 0.20-0.25% C; 0.60-0.85% Mn; 0.15-0.35% Si; 0.03% max S; 0.03% max P; 0.9-1.2% Cr; 0.04-0.06% Ti; and 0.001-0.003% B. Several hundred bolts made to this specification have been in test in Japan in a severe environment (at a proof stress level of more than 155,000 psi) for more than three years without failure.

The 0.04 to 0.06% Ti content is approximately that for most boron-treated steels made in the United States. When compositions like these are inadequate for resistance to stress-corrosion cracking, cadmium plating is sometimes effective.

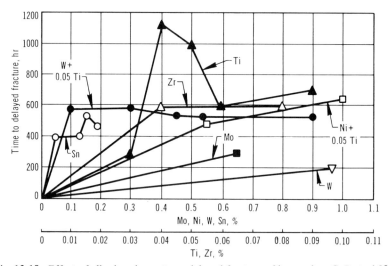

Fig. 13.10 Effect of alloying elements on delayed fracture of low-carbon Cr-B steel [8].

ASSEMBLY

The assembly of fasteners is as important in their successful use as geometry and metallurgical quality. Most failures are caused by improper assembly, which in turn, is caused by inadequate torquing. The result is a less-than-desired clamping stress. Because parts can move in relation to one another, fatigue failure of the bolt often occurs; for example, a farm-equipment manufacturer could not keep an assembly together with 1.25-in. SAE grade 5 bolts made of 1046 steel. It was suspected that the metallurgical quality of these bolts was questionable, even though a competitor was using similar fasteners successively. A change made to SAE grade 8, and 4140 steel, did not solve the problem. Called to assist with the problem, we found that the assembly torque was only 600 ft-lb. When tightening torque was increased to 1000 ft-lb, even the grade 5 bolts were satisfactory.

Functionally, torquing is a means to an end; namely, to create a desired level of tension stress in a fastener to obtain a required clamping force. Ideally, this effect of tightening is determined by accurately measuring the overall face length of a fastener and then torquing until a measured amount of stretch has occurred. Up to the proof strength the clamping force will be close to 30,000 psi for each 0.1% of stretch. In most instances to adopt this procedure would be highly uneconomical and in some, impossible because of inaccessibility. Accordingly, most tightening is done on the basis of torque applied to the head of a bolt or cap screw or to the nut if one is used. Typical recommended torques for unplated, high-strength fasteners are listed in Table 13.6.

To express tightness of fasteners only in terms of torque is a rather crude and difficult method for the following reasons:

1 To provide a given clamping force a required dynamic torque is needed, and a different value will result if the part is tested statically with a torque wrench.
2 The torque for a given tightening effect is sensitive to the coefficient of friction between the threads of the driven and stationary members and the driven member and its abutment. This coefficient of friction is, in turn, sensitive to such things as the finish of the thread surfaces, lubrication, plating, and bolt clearance.
3 Most air wrenches require constant auditing to make sure the torque setting has not changed.
4 Because many fasteners are hand-wrenched, the tightening quality by torquing can be quite erratic.

Another difficult problem in fastener tightening is the loss in clamping force of a properly torqued fastener as adjacent fasteners are torqued.

Table 13.6   Typical Tightening Specifications[a]

Thread Size (in.)	SAE Grade 5 Torque (lb-ft)	SAE Grade 8 Torque (lb-ft)
0.250	10	14
0.312	19	25
0.375	30	45
0.437	50	70
0.500	75	110
0.562	110	150
0.625	150	220
0.750	260	380
0.875	430	600
1.000	640	900
1.125	800	1200
1.250	1100	1800
1.375	1400	2400
1.500	1900	3100

[a] Unplated dry bolts ($K = 0.20$).

The solutions to most of these problems are obvious. However, because the number of fasteners used in a product like an automobile (approximately 3000) is so large, the investment in personnel and equipment may be difficult to justify. The first step is to make sure that the required fastener for a given application is, in fact, used. To prevent mix-ups some firms have standardized, for example, on SAE grade 8 for all except a few special applications. Relationships between static and dynamic torque can be established. Factors affecting the coefficient of friction can be controlled by calling for different torque values for dry, lubricated, or cadmium-plated fasteners. Air tools for tightening have become much more sophisticated in the effort to maintain constant torque setting and so has the equipment for checking the tools. Nevertheless, maintenance of clamping force of a group of cap screws, particularly with hand tightening, requires suitable drawing specifications, employee training, and supervision.

The tightening of externally threaded fasteners is done not only by the torque-control method but also by the turn-of-the-nut procedure. In the torque-control method the bolt tension producing the clamping force is below the yield point, whereas in the turn-of-the-nut method the tension occurs between the yield point and the tensile strength, (see Fig. 13.11). This curve shows that the latter method achieves a higher clamping force, but because bolt tension is above the yield point and in the plastic deformation region even with a relatively

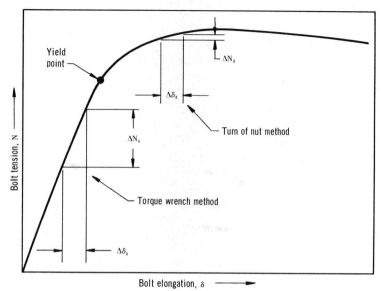

Fig. 13.11. Comparison of two bolt-tightening methods [8].

large variation in bolt elongation $(\Delta\delta_2)$ caused by variation in nut-rotation angle there will be only a small change in bolt tension (ΔN_2). Therefore a constant tension can be quite easily maintained (see Fig. 13.12). Commercial turn-of-the-nut tightening consists in torquing to a snug fit plus a nut rotation, as listed for ASTM A325 and A490 bolts:

Bolt Length	Nut Rotation
Not exceeding 8 × diameter or 8 in.	1/2 turn from snug position (180°)
Exceeding 8 × diameter or 8 in.	2/3 turn from snug position (240°)

The principle of this tightening practice is to pretorque part way to the yield point and then use the turn-of-the-nut method to exceed the yield point. Ideally, it should finish approximately halfway between the yield point and the ultimate tensile strength. Considerable use is made of a one-third turn of the nut from a pretorque, which will produce a tension stress in the bolt of one-fourth to one-third of the yield point.

The turn-of-the-nut method also has its disadvantages:

1 Highly ductile bolts are required.
2 Because the bolt has yielded, it usually cannot be reused.

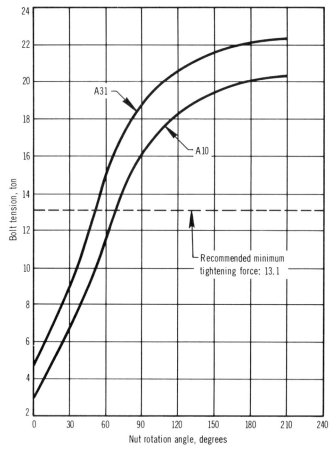

Fig. 13.12 Relationship between bolt tension and nut rotation angle [8].

3   When control of the rotation angle becomes lax, bolt breakage in as-
    sembly can necessitate a costly salvage operation.

Proper tightening sometimes is important enough to justify proprietary
Tel-Torq* bolts which have an optical indicator in the head that changes color
when the clamping force reaches a prescribed level.

Abutment relaxation due to insufficient compressive yield strength under
the fastener head can result in loss of clamping force. This can be a problem
when bolts are used to fasten structural steel to ASTM A325, A354, or A490
specifications. In this situation hardened washers can be used, especially under
the fastener component (nut or bolt) that is driven in. Loss of clamping load

*Manufactured by Modulus Corp., Mt. Pleasant, Pa.

and possible failure can also result from disintegration of rust and heat-treating scale.

An important adjunct to proper tightening of bolts is keeping them tight. This is done with various types of lock washers or wiring through drilled heads. Also, there are numerous mechanical gripping mechanisms in the threads of externally threaded fasteners. One air-wrench mechanism has hardened steel jaws that cold-forge indentations into the nut after it has been tightened to the prescribed torque. Chemically reactive, adhesive-locking materials such as Loctite, Dry-Loc, and Scotch Grip are coming into wide use.* Loctite and Dry-Loc are anaerobic materials, the latter being encased in tiny capsules that rupture as the fastener is tightened. Scotch Grip is a microencapsulated two-part epoxy adhesive. In tightening, the capsules rupture to mix the two parts and provide a strong locking action. Perhaps the simplest and most often overlooked method of maintaining tightness is to increase bolt length, thereby building more elasticity into the bolted joint.

SUPERBOLTS

Sometimes it is impossible to increase bolt size or revise a design to compensate for unexpected clamping forces. To salvage a design "superfasteners" can be used, one of which is the socket-head cap screw. This type of cap screw should have a forged head with flow lines closely following the contour of the part through the head-to-body fillet. The head sockets for driving are almost always hexagonal and of a size amenable to proper tightening. This type of cap screw is made of a medium-carbon alloy steel such as 4340, 8640, or 8740, quenched, and tempered to 36 to 42 Rc. At this hardness the minimum proof strength in diameters up to 1.5 in. is 130,000 psi and the minimum tensile strength is 160,000 psi. When the proper steel is selected, heat treating produces an essentially 100% tempered martensite or bainitic microstructure throughout the part; these bolts also show excellent toughness. The Izod V-notch impact strength at room temperature will be at least 35 ft-lb and at –20 F, 27 ft-lb. Steels with nominal carbon content higher than 0.40% should be avoided and hardness should preferably not exceed 42 Rc. Because of their high strength, these bolts are prone to hydrogen cracking. Accordingly, they should not be indiscriminately used in water or even highly humid environments nor subjected to contact with any material or electrical circuit that could release hydrogen at the bolt surface.

Other superbolts are made of H11, 300M, or similar steels, vacuum-melted with threads rolled after heat treatment in accordance with aircraft practice.

---

*Loctite and Dry-Loc are made by Loctite Service Products Co., Newington, Pa., and Scotch Grip, by 3M Co., St. Paul, Minn.

Bolts of this type now in use have tensile strengths up to 250,000 psi. Assemblies with all such fasteners require special care in preparing the underhead seat, maintaining parallel clamping surface, and proper torque.

REFERENCES

1 Mechanical and Quality Requirements for Externally Threaded Fasteners–SAE Standard J429g, *SAE Handbook,* New York, 1974, p. 119.

2 J. L. Hurley, A. L. Kitchin, and R. R. Crawford, *Metals Eng. Quart.,* American Society for Metals, **11,** No. 3, 36 (1971).

3 A. F. DeRetana and D. V. Doane, Predicting Hardenability of Carburizing Steels, *Metal Progr.,* **100,** No. 3, 65 (1971).

4 J. B. R. Anderson, Selection of Steels for Cold Forging, ASTME Pub. No. MF69-539, Society of Manufacturing Engineers, Detroit, 1969.

5 J. M. Hodge and M. A. Orehoski, Relationship Between Hardenability and Percentage of Martensite in Some Low Alloy Steels, Technical publication No. 1800, Figs. 15–17, 1945.

6 R. P. Frohmberg, W. J. Barnett, and A. R. Troiano, *Trans ASM,* **47,** 892 (1955).

7 S. Fukui and C. Asadi, *Tetsu to Hagane,* **54,** 1290 (1968).

8 I. Kimura, S. Watanabe, M. Honda, R. Hiroi, and M. Usa, Nippon Steel Technical Report Overseas, No. 3, June 1973.

9 H. H. Johnson, J. G. Morlet, and A. R. Troiano, *Trans AIME,* **212,** 528 (1958).

10 Y. Iwata, Y. Asayama, and A. Sakamoto, *J. Japan Inst. Metals,* **30,** 169 (1966).

CHAPTER FOURTEEN

SELECTION OF STEELS
FOR ABRASIVE-WEAR APPLICATIONS

THREE MAJOR CATEGORIES OF wear are discussed in Chapter 8. This chapter elaborates on abrasive wear because this phenomena can often be solved only by changes in material or heat treatment. In abrasive-wear situations sand, ore, or rock for example, contact the part surface or is located between two parts that tend to crush it. In the area of abrasive wear there are at least three subforms of this phenomenon [1,2]:

1 Low-stress abrasion.
2 High-stress abrasion.
3 Gouging abrasion.

Low-stress abrasion is defined [3] as a condition in which the stresses imposed on the abrasive particles do not exceed their crushing strength. This condition occurs typically in moving soft moist earth; for example, in plowing a field. It might also include pumping sand slurries from a dredge and even handling dusty air. Steels and heat treatments for resisting this type of wear are not always suitable for high-stress or gouging abrasion, which are defined in the following paragraphs.

High-stress abrasion occurs when two wearing surfaces, such as a grinding ball and the liner of a ball mill, come together in a gritty environment with enough force to crush the abrasive particles entrapped between them [4].

Gouging wear represents a condition in which rocks or other coarse abrasive materials cut into a wearing surface with considerable force, producing deep gouges and removing large particles from the surface [5].

The proper selection of steel for abrasion resistance first of all requires that the materials engineer identify the type of abrasion anticipated. This may be complicated by the fact that a bulldozer, for example, may on one day be pushing earth (low-stress abrasion) and on the next ripping up a broken concrete highway (gouging abrasion). It must also be recognized that although historically constructional grades of steel have been used, engineering characteristics available only with other materials are often required. The method of attachment of the wear part or incorporation into a machine or structure must be practical, and finally, but of great importance, its cost must be competitive.

In the selection of steels and heat treatments for abrasive wear, the four most important technical factors are the following:

1 Your own experience or the experience of someone who makes a similar part that has performed successfully.
2 Examination of the total environment and compatibilities of the materials in contact for adhesive or corrosive wear effects.
3 Almost without exception, the harder a steel part can be made without being excessively brittle, the longer its wear life will be.
4 As the abrasive conditions become increasingly severe (from low stress to high stress to gouging), the improvement that can be provided by different grades of constructional steels and types of heat treatment becomes less and less significant. In extreme cases the only answer is to provide more steel to wear away.

SELECTION OF STEELS FOR
LOW-STRESS ABRASION

Because this type of abrasive wear will not tear out hard carbides, the grade of steel can be quite significant. The complex carbides of chromium, molybdenum, and vanadium can add to the wear resistance (see Fig. 14.1). The superiority of 4140 steel is undoubtedly due to the presence of more hard carbides than exist in the low-carbon alloy steel plate. The slope of both curves in Fig. 14.1, however, shows the strong effect that hardness has in improving the abrasion resistance of heat-treated steel. The low unit stress and lack of impact in low stress abrasion permits the use of rather brittle materials, which, under high stress, might crumble or chip away. The results in Fig. 14.1 were derived from a rubber-wheel abrasion-test device developed by the Climax Molybdenum Company of Michigan. These results correlate well with the actual field tests

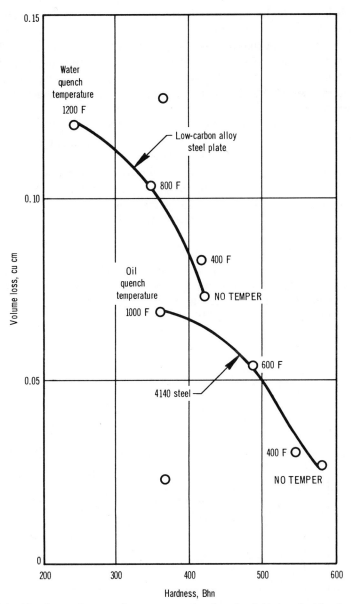

Fig. 14.1 Abrasion resistance of two constructional steels plotted against hardness (from Borik [1]).

shown in Fig. 14.2. (Relative wear in this study is based on the weight loss of 1090 make A steel rated as 100.)

One of the most extensive uses of heat-treated steel in low-stress abrasion is in farm-tillage tools. With the possible exception of plow moldboards, substantial toughness is required in these parts to prevent breakage when rocks and roots are struck by the tool in operation. Industry standards today are approximately as follows:

Plow shares	1080, 1085, 1095 quenched and tempered, 415–477 Bhn
Cultivator sweeps	1085, 1090 oil-quenched and tempered or austempered, 388–415 Bhn
Spring-tooth-harrow teeth	1085, 1090 oil-quenched and tempered, 388–429 Bhn

Plow moldboards present a problem that goes beyond mere abrasion resistance in that they must scour or polish to a smooth finish that will allow the soil to be turned over efficiently without sticking. Experience over many years has shown that steel with a soft center is the preferred material and that it must be quenched to a surface hardness of 60 Rc min. Soft-center steel is made by

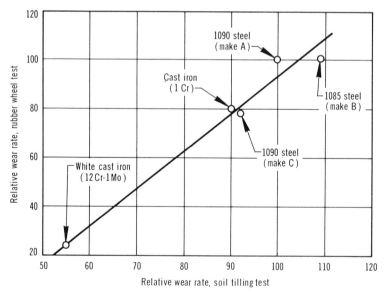

Fig. 14.2 Correlation of relative wear values of materials abraded in rubber-wheel abrasion test and in soil-tilling tests (from Stolk [12]).

casting 1020 steel into a hollow 1095 ingot and then rolling the composite ingot into plate. After the plate is hot-formed in a die of the moldboard shape it is hardened and polished.

Because of the many variables present in an abrasive-wear application, there is no reliable way of predicting the characteristics necessary to achieve a certain level of assured life. There are innumerable degrees of abrasiveness of the materials to be handled; for example, from cotton to crushed silicon carbide. There are also the questions of contact pressure and the velocity of the abrasive particles against the steel part. Finally, the expected life may permit the removal of only a few thousandths of an inch of metal, whereas under other circumstances the part can be useful until it is worn nearly completely away. As stated earlier, undoubtedly the most reliable rule is to relate the new part to something that has been in successful operation for a long time in a similar application; for example, if the new part is to be used to manipulate soft earth, the steels and heat treatments applicable to agricultural tillage tools would be a logical place to start.

At the prototype stage of a design it is good practice to think in terms of what could be done not only to improve wear resistance at a moderate increase in cost but also, perhaps, what could be done to improve toughness (e.g., if the 1090 used in tillage tools proved inadequate). The trend in Fig. 14.1 would indicate that a change to, say, 52100 would significantly improve wear life. Furthermore, austempering instead of a conventional quench and temper should improve toughness. From Fig. 14.3 it can be reasonably expected that the

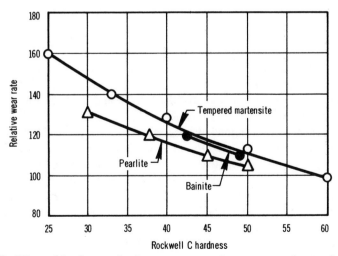

Fig. 14.3 Effect of hardness and microstructure on the wear rate of cast grinding balls (from *Metals Handbook*, Vol. 1, p. 245).

bainitic microstructure produced by austempering would provide essentially the same low-stress abrasion resistance as that of a martensitic structure.

The maximum resistance to low-stress abrasion that can be provided by the constructional steels is illustrated by the following examples of steel types and treatments:

1 4140 quenched and tempered to 32 to 38 Rc, nitrided to depth required; hardness, 86.5 Rockwell 15N min.
2 1060 to 1080 steel, induction hardened; hardness, 63 Rc min.
3 8822 or 9310 carburized to 1.80% min C; hardness, 62 Rc min.

For severe applications it is sometimes necessary to go beyond the constructional steels for a suitable material. Figure 14.4 shows the resistance to low-stress abrasion of a variety of materials measured by the Climax Molybdenum Company's rubber-wheel test apparatus. It should be pointed out, however, that the white irons are very brittle materials that must be assembled with care to prevent breakage. The carbides are expensive compared with the constructional steels and are usually assembled by brazing in place or making the entire part of solid carbide. In many instances, however, especially when downtime and/or installation costs are high, a carbide will comfortably pay its way. An alternative to making a part of solid carbide is to coat it with the carbide by any one of several commercial processes. For low-stress abrasion we have found tungsten carbide to be 40 to 80 times more wear resistant than 52100 steel at 60 Rc.

HIGH-STRESS ABRASION

The most widely known application in which this type of abrasion takes place is in a ball mill for grinding hard materials (e.g., taconite iron ore). A similar action takes place between a crawler-tractor track and the track rollers that run on it and support the weight of the vehicle.

A popular test to evaluate high-stress abrasion is the ball-mill procedure developed by Norman and Loeb [6,7]. Essentially, this test evaluates different materials in the form of 5-in.-diameter balls in a 9-ft-diameter production ball mill operated by the Climax Molybdenum Company at their plant in Climax, Colorado. The procedure consists in "wearing in" test balls to remove surface irregularities and decarburization. The properly worn-in balls are then sorted out, weighed, and charged into the production mill for enough time to wear off a layer of metal 0.125 to 0.250 in. thick. After the test the balls are carefully reweighed so that the weight loss per unit of surface area can be computed. This figure is divided by the results for the standard material, which is

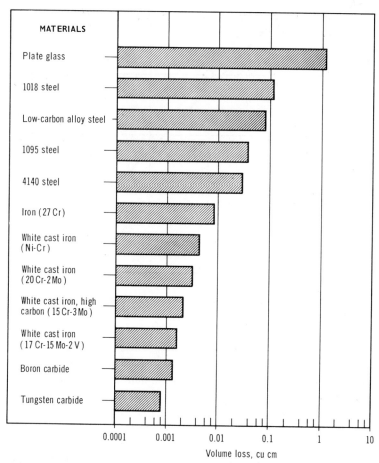

Fig. 14.4 Abrasion resistance of various materials as determined by the rubber-wheel abrasion test in terms of volume loss normalized to 55 Durometer hardness of the rubber wheel (from Borik [1]).

a martensitic 1% C, 6% Cr, 1% Mo steel, heat-treated by air quenching from 1900 F and tempering at 400 F. This quotient, multiplied by 100, is called the abrasion factor.

The ball-mill test yields results of high reproducibility; for example, for a set of five balls the abrasion factors have a typical range of 2% of the mean and a standard deviation of 1% of the mean. Typical results are shown in Fig. 14.5 for a 1% C, 9.6% Cr, 1.2% Mo steel. It is interesting to note that balls made of this highly alloyed test steel wore more rapidly than the somewhat leaner standard steel by a factor of approximately 5 to 10%.

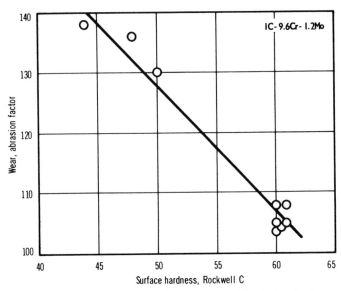

Fig. 14.5 Correlation between abrasion factor and work-hardening hardness of 1% C, 9.6% Cr, 1.2% Mo grinding balls (from Borik [4]).

It is obvious that for high-stress abrasion resistance the worked surface must be very hard. The presence of hard chromium and molybdenum carbides is quite beneficial. Because of the compressive nature of the loading of grinding balls, the beneficial effect of high carbon content on compressive yield point is significant. Because localized heating occurs in high-stress abrasion, steel compositions containing substantial amounts of chromium and molybdenum to provide high-temperature hardness and resistance to tempering are also desirable.

The use of grinding-ball steel for high-stress abrasion applications in complex shaped parts would, however, probably result in chipping, spalling, and breakage due to lack of toughness; for example, induction-hardened 1095 crawler-tractor track pins will break in only a few hours of operation, whereas 1049 or 1050 pins similarly treated seldom break. Again, because bainite has good high-stress abrasion characteristics (essentially equal to martensite), a 0.90% carbon steel of suitable hardenability might make excellent track pins when austempered.

High-stress abrasion almost always involves some degree of shock loading; for example, from the balls falling against each other and against the liner plates in a ball mill. Figure 14.6 illustrates this very well in that the brittle white cast irons, which have high wear resistance in low-stress abrasion, do not perform so

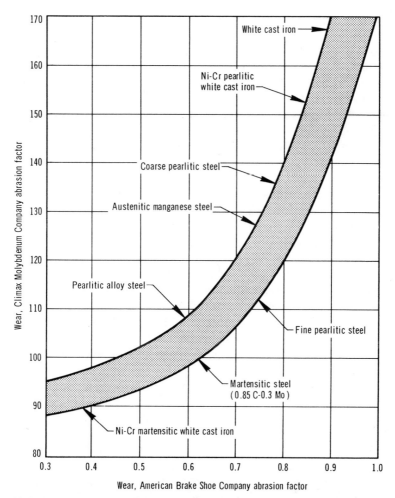

Fig. 14.6 Comparison of ball-mill-test results (Climax Molybdenum Co.) with laboratory results (American Brake Shoe Co.) on a high-strength abrasion-testing machine (from Borik [4]).

well in high-stress applications (with the notable exception of the Ni-Cr martensitic iron).

The great difference between ball-mill materials and those that must be used in more complex parts is cost; this includes not only the price of the steel but also the cost of suitable preparation for incorporation into a structure or mechanism. The standard ball steel used by Norman and Loeb [6,7] would cost approximately 25% more than 4340. Incorporation costs would also be very high because of the standard steel's poor machinability and its nearly unweldable composition.

The tests to establish Fig. 8.12 were under conditions of high-stress abrasion. It can be seen from these curves that as the carbon content is increased above 0.40% and the hardness above 55 Rc, the increments of improvement are quite small. The steels used in crawler-tractor track and track rollers, cited as an example of high-stress abrasion, are usually 0.35% to 0.40% C, plain carbon, or C-Mn-B steels induction-hardened or furnace-heated and impingement-quenched to approximately 55 Rc. For guidance to the materials engineer in selecting constructional grades of steel for parts subject to high-stress abrasion this steel might be a good place to start. Some improvements in wear resistance can be achieved with higher carbon contents and higher hardnesses but at a substantial sacrifice in toughness, machinability, and weldability.

Track bushings are another example of high-stress abrasion in a crawler-tractor that could serve as a guide in selecting steels for other types of part. It is these bushings that the drive sprocket engages to push the tractor along. Stones, sand, and cinders are crushed between the bushings and the sprocket as in a ball mill. Bushings are often made from silicon-killed, fine-grained 1018 or 10B18 steel, carburized 0.06 to 0.100 in. deep, depending on the wall thickness, and water-quenched to a hardness of 60 Rc min. The sprockets are made of 0.40 to 0.45% carbon steel induction-hardened or impingement-quenched to approximately 50 to 55 Rc. In quite highly abrasive environments such as crushed rock or hard angular sand these components will provide an operating life of about 2000 hr min. The bushings are carburized because through-hardened parts made of steels with 0.45% or higher carbon are excessively brittle. Case-carbon contents of 0.85 down to about 0.65% provide optimum crushing-fatigue strength. High case carbon, 1.10% or more, provides the best wear resistance but at a sacrifice in fatigue resistance. It has also been found that steels with 1.5 to 2.5% Cr and 0.30 to 0.50% Mo, carburized to case-carbon content of more than 1.80%, gave significantly longer wear life than 1018 or 10B18. Nitriding and carbonitriding were not satisfactory for these parts because of the tendency of the thin cases to spall off.

When weld fabricating is the preferred — and perhaps the only — practical method of incorporating abrasion-resistant steels into an assembly subject to high-stress abrasion, the use of low-carbon, heat-treated steels such as T-1*, T-1A*, or T-1B* at a hardness of 321 or 360 Bhn minimum is suggested. The wear resistance of this type of steel in chutes handling coarse dolomite rock for blast furnaces is 5 times the wear resistance of a .40–.50% C AR* steel (see tabulation).

*Registered trademarks, U.S. Steel Corp.

Type of steel	T1-321*	AR*
Hours of service	572	120
Total tonnage of dolomite	256,394	48,761
Plate gage, in.	0.504	0.494
Wear resistance, tons/0.001 in. thickness	509	99

Figure 14.6 shows that for high-stress abrasion pearlitic alloy steels often perform as well, for example, as a martensitic steel of 0.85% C and 0.30% Mo. A typical composition for a commercial pearlitic abrasion-resistant steel is as follows: 0.25–0.31% C, 1.65% max Mn, 0.15–0.35% Si, 1.20% Cr max, 0.35% Mo max, 0.0005% B. This type of steel is also quenched and tempered to ranges of 260 to 333, 300 to 340, 340 to 380, and 400 to 440 Bhn.

For crawler-tractor track bushings thin case hardening, as afforded by carbonitriding and nitriding, is generally not an acceptable heat-treating process for high-stress abrasion because of the tendency of the case to spall off. Deep case carburizing, on the other hand, is used not only for parts such as the track bushings but also on steels 4815, 4817, 4820, 9310, and 9317 steels for highly abusive applications such as the cutters in oil-well drill bits.

GOUGING ABRASION

This type of abrasion is costly to the mining and rock quarrying industries where the sharp edges of the strong materials handled cut into wearing surfaces with sufficient force to produce deep gouges and remove relatively large particles [5]. The gouging action may also be associated with high-velocity impact, as in hammer mills or breaker bars in pulverizers. It may also occur at relatively low velocities, as in a jaw or gyratory crusher. In some applications, the part subject to gouging wear is loaded only in compression. In others it is subject to bending, torsion, and impact forces or a combination thereof. Finally, some components can be most efficiently designed for bolting in place, whereas others may be welded in place or cast as an integral part of a machine. All of these parameters affect to an important degree the choice of materials used to resist gouging abrasion. The complexity of this problem, coupled with the cost of running meaningful tests to rate materials, resulted for many years in selection by experience. This limits optimum selection to those materials tested in specific environments.

Hall (8), Ksenofontov (9), Avery et al. (10), and Kawamura et al. (11) made intensive studies of the gouging-abrasion phenomenon with various laboratory devices. In general, tests with small jaw crushers showed that they could be used to study metallurgical variables and to rank materials for gouging-abrasion applications. Engineers of the Climax Molybdenum Research Laboratory modified a small production jaw crusher for these purposes and used it to develop new wear-resistant alloys [2].

The rock used in the Climax Molybdenum Company tests was a highly siliceous morainal material precrushed to a size range of 1.5 to 2 in. The test procedure consisted of crushing one ton of rock with one test plate and one plate of a reference composition and hardness. Results were reported as wear ratio, which is determined by dividing the weight loss of the test plate by the weight loss of the reference plate. Wear ratios of duplicate runs were averaged. With this test it was possible for the first time to rank alloys of engineering importance according to resistance to gouging wear [2]. Figure 14.7 shows the averages of 39 tests of various ferrous materials in which total carbon content is plotted against wear ratio. Note that up to approximately 0.80% C the gouging wear resistance increases (wear ratio decreases) sharply with increasing carbon

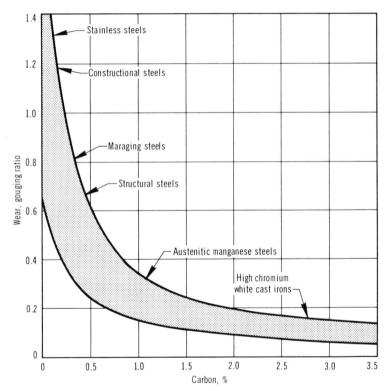

Fig. 14.7 Relationship between the gouging-wear ratio and the carbon content of various classes of ferrous materials (from Borik [4]).

content. Above this carbon content the improvement is at a much lower rate. Figure 14.8 is a curve of similar shape that indicates dramatic increases in gouging abrasion resistance (decreased in wear ratio) with hardness up to approximately 500 Bhn; a much reduced rate of improvement is shown for higher hardnesses.

By comparing Figs. 14.1, 14.5, and 14.8 it can be seen that as the abrasive conditions become more severe the effect of hardness on wear resistance is increasingly important — at least to 500 Bhn or about 52 Rc. Borik [1] found that, in common with high-stress abrasion, the gouging resistance of ferrous materials correlates with the hardness of the work-hardened surface. Little, if any, correlation was found between the wear ratio and the nonwork-hardened surface. What this means is that when gouging abrasion is a major engineering requirement steels with carbon content as close to 0.80% as possible should be used. Also, these steels should be heat-treated to approximately 52 to 58 Rc.

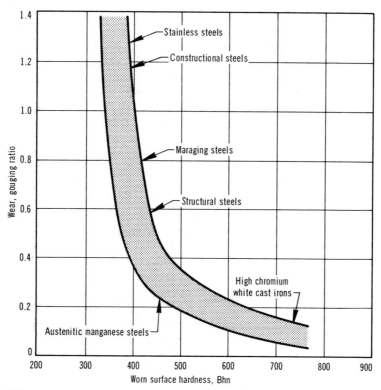

Fig. 14.8 Gouging-wear ratio plotted against work hardened surface hardness of a wide variety of steels and irons tested in the laboratory jaw crusher (from Borik [4]).

If welding must be done to incorporate a dynamically loaded part, the carbon should be no higher than a nominal 0.30% if the section is as much as 1 in. thick. For thicknesses of more than 1.0 in. the nominal carbon content should be 0.27% max, even when welded with low-hydrogen processes. Higher carbon contents require carefully controlled preheating and possibly postheating (see Chapter 7 for further discussion of weldability). High-quality production welding of materials like austenitic manganese steel is very difficult even with stainless steel electrodes.

When gouging-abrasion conditions exist, shock loading is often present as well. A typical example is the point of the ripper attached to the crawler-tractor shown in Fig. 14.9. Rippers are used to tear up rock in highway construction and in strip mines and often replace drilling and blasting. These ripper points are subject to impact loading and gouging abrasion. A widely used mater-

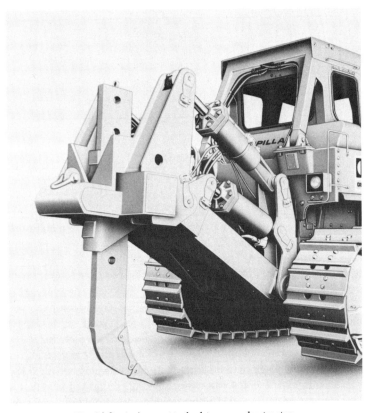

Fig. 14.9   A ripper attached to a crawler tractor.

ial is 4330, water-quenched and tempered at 450 F, with a final hardness of 47 to 53 Rc. Scarifier points on motor graders are made of similar steels heat-treated in the same manner. Steels with even 0.40% nominal carbon content are generally too brittle for this type of application.

Sometimes it is possible to design an assembly with easily replaceable plates to provide maximum gouging-abrasion resistance. A good example is the use of liner plates in off-the-highway truck bodies. The body itself is made of low-carbon steel of 80,000 to 100,000 psi minimum yield strength, which is readily weldable and tough. The liner plates are made of medium carbon steel heat-treated to more than 500 Bhn. Sometimes these liner plates are subjected to the impact of large rocks or pieces of ore falling from a power shovel or loader bucket, and impact strength then becomes an important characteristic; therefore the plates must be made of lower carbon steels (0.25 to 0.30% C) to provide adequate toughness.

The selection of steels and heat treatment for parts subjected to gouging abrasion and impact obviously must depend on a number of compromises. We suggest the following guidelines:

1 The steel must be as low in carbon as possible, yet have adequate hardenability to harden through all important sections, preferably to a microstructure of 99+% martensite.

2 The microstructure should contain no blocky ferrite resulting from inadequate austenitization. A boron steel is preferred to minimize the amount of proeutectoid ferrite present in the structure.

3 Even trace carburization during austenitization must be prevented.

4 Quench must be vigorous, with impingement preferred; cold water, brine, or 5% caustic should be used.

5 Tempering must be done below 475 F or at more than 800 F and for a enough time that retempering at the same temperature will not reduce the hardness at any point in the section more than three points Rockwell C.

6 Nickel-bearing steel is preferred even if it is the 0.35 to 0.75% content of the 8600 series. If operating temperatures are to be below -20F, the use of 1.20 to 1.50% Ni is suggested; if below -40 F, at least 1.65 to 2.00% Ni is better.

7 Phosphorus and sulfur should be below 0.025% and preferably below 0.010%.

When gouging abrasion becomes extremely severe, the only practical solution may be to provide more steel to wear away; for example, if a standard earth-moving scraper cutting edge is 1.50 in. thick, for extremely severe service the cutting edge should be made 2.25 or even 2.50 in. thick.

REFERENCES

1 F. Borik, Rubber Wheel Abrasion Test, SAE National Farm, Construction, and Industrial Machinery and Powerplant Meeting, Milwaukee, September 14, 1970 (SAE Reprint Number 700687).

2 Gouging Abrasion Test for Materials Used in Ore and Rock Crushing: Part I — Description of the Test; F. Borik and D. L. Sponseller; Part II — Effect of Metallurgical Variables on Gouging Wear, F. Borik and W. G. Scholz, *J. Mater,* 6, 576 (1971).

3 H. S. Avery, *Wear,* 4, 427 (1961).

4 F. Borik, *Metals Eng. Quart.,* 12, No. 2, 33 (1972).

5 T. E. Norman, *Handbook of Mechanical Wear,* University of Michigan Press, Ann Arbor, 1961, pp. 277–314.

6 T. E. Norman and C. M. Loeb, Jr., *Trans. AIME,* 176, 490 (1948).

7 T. E. Norman, *Modern Castings,* 33, 89 (1958).

8 J. H. Hall, *Proc. Am. Soc. Testing Mater.,* 28, 326 (1928).

9 V. P. Ksenofontov, *Russ. Cast Prod.,* No. 7, 310 (1966).

10 H. S. Avery et al., Surface Protection Against Wear and Corrosion, American Society for Metals, Metals Park, Ohio, 1954, pp. 22–24.

11 K. Kawamura, A. Harada, T. Kunitaki, and Y. Honda, *Sumitomo Kinzoku,* 20, No. 2, 79 (1968).

12 D. A. Stolk, Field and Laboratory Tests on Plowshares, SAE National Farm, Construction, and Industrial Machinery and Powerplant Meeting, Milwaukee, September 14, 1970.

CHAPTER FIFTEEN

SELECTION OF BORON STEEL
FOR HEAT-TREATED PARTS

BORON IS ADDED TO STEEL TO
increase hardenability; boron steels also have superior processing and engineering
characteristics, notably the following:

1 The hot and cold formability of boron steels is superior to the formability of equivalent grades made by conventional alloying with manganese, nickel, chromium, and molybdenum.
2 The machinability (as-rolled or annealed) for a given level of hardenability is superior to that of boron-free steels.
3 The weldability in the heat-treated condition is superior to that of boron-
   free steels of equal hardenability.
4 Because the M_s temperature is not significantly depressed by the boron
   addition, the susceptibility to quench cracking is reduced.
5 Boron steels can be readily quenched to a microstructure that is free of
   proeutectoid ferrite, thereby enhancing important engineering properties
   such as notch tensile strength, fracture toughness, and both short- and
   long-life fatigue strength.

Among the outstanding characteristics of steels containing boron is the exceptionally strong effect this element has in preventing the nucleation of ferrite
at the austenite grain boundaries in carbon and medium-hardening alloy steels.
In deep-hardening alloy steels that tend to nucleate bainite boron can be bene-

293

ficial by transferring this nucleation into the grains, which is less deleterious than at the grain boundaries.

In order to give boron steel high hardenability and all of the other processing and engineering advantages we have discussed, it must be properly made. This correctness of manufacture can be determined by calculating the Grossmann boron factor (see Chapter 6 and Appendix VII). Boron has a strong affinity for oxygen and nitrogen; consequently the content of these gases in molten steel must be held as low as possible and must then be reduced to very small amounts (less than 0.008%) by treatment with aluminum, titanium, zirconium, and vanadium, singly or in combination. The typical ladle additions of boron are made with the following:

1 Titanium and ferroboron.
2 Grainal 1, which contains approximately 0.20% B, 15% Ti, 10% Al, and 25% V.
3 Grainal 79, which contains approximately 0.50% B, 8% Mn, 5% Si, 20% Ti, 13% Al, and 4% Zr.
4 Grainal 100, which contains 1.00% B, 8% Mn, 5% Si, 20% Ti, and 4% Zr.

Kapadia, Brown, and Murphy [1] found that the optimum boron content was 0.001% when denitrification to 0.006% N was performed with titanium, as shown in Fig. 15.1. Grange et al. [2,3] concluded that the boron effect was due to segregation of boron atoms at austenite grain boundaries. It was also concluded by Grange that boron contents above the optimum amount tend to promote the precipitation of a boron-rich grain-boundary constituent that depletes the austenite grain boundaries of boron atoms and results in an overall loss of effectiveness. The maximum boron effect reported by Kapadia, Brown, and Murphy on a high side heat of 8822 was a Grossmann factor of 2.80, whereas the formula we suggested for this steel in Chapter 6 of $B_f = 1 + 1.6(1.01 - \%C)$ would indicate 2.34. The typical effect of boron on end-quench hardenability is shown in Fig. 15.2. Although the effect of boron is strong in increasing hardenability measured to 50% martensite, it is even more effective when measured in terms of D_I to 95% martensite; for example, the actual boron factor in terms of 50% martensite for the ingots shown in Fig. 15.3 is 1.33. Using the technique described by Hodge and Orehoski [4] for 95% martensite, we find that the boron factor is 1.69. What this means in a practical sense is that if a part requires high percentages of martensite, particularly near the surface, a boron-containing steel is the obvious choice. This is seen by comparing the minimum Jominy distances to 99.9, 95, and 90% martensite for 4140H, 8620H, and 94B30H, shown in Table 6.4.

Grange's hot-brine hardenability test, which predicts the diameter in which a steel will harden to 90% min martensite at the center, can be used to test

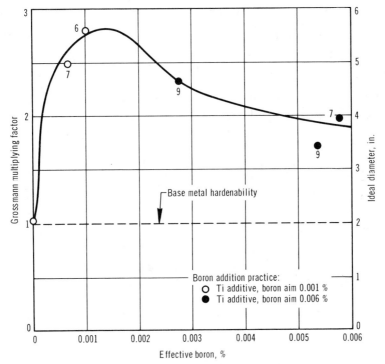

Fig. 15.1 Effect of boron content on hardenability of steel. Numerals adjcent to data points represent nitrogen content × 0.001% (from Kapadia, Brown, and Murphy [1]).

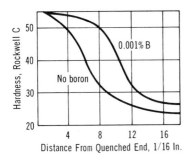

Fig. 15.2 Typical effect of boron on hardenability in a 0.36% C, 1.45% Mn steel.

round bars of steels with and without various alloys to determine the increase in diameter (ΔD) attributed to the addition of boron. Table 15.1 shows the values of increased critical diameter achieved by boron additions to steels of four grain sizes and the amounts of alternative alloy required to give values of ΔD equal to those of the boron addition. As shown in this table, boron is extremely potent in terms of hardenability to 90% martensite.

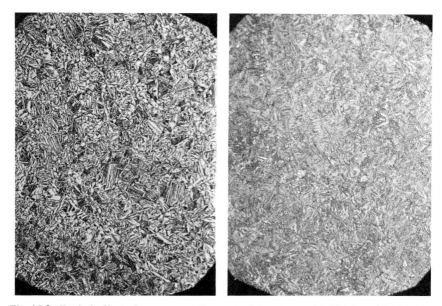

Fig. 15.3 Typical effect of boron on refinement of structure and inhibition of ferrite formation in 8620 steel. Same heat of steel, center of 1 1/8-in. round, oil-quenched from 1700°F (courtesy U.S. Steel Corp.).

Table 15.1 Increase in Critical Diameter to 90% Martensite Due to Boron Addition for .40% C Steel. Alloy Addition Required for Equivalent Effect Is Also Shown[a]

Grain Size	ΔD_B (in.)	Addition for Equal ΔD_B%			
		Mn	Ni	Cr	Mo
4	0.15	0.26	1.46	0.35	0.20
6	0.18	0.32	1.70	0.40	0.24
8	0.25	0.48	2.00	0.54	0.34
10	0.32	0.66	2.00	0.66	0.44

[a] From Grange et al. [2–3].

The attainment of high percentages of martensite in hardening is necessary for many engineering characteristics such as maximum ratio of yield strength to tensile strength, reduction in area (for short-life fatigue strength), and toughness (for shock loading). The exceptional microstructural capability of boron steels is believed to be responsible for their unusual fracture toughness (Fig. 15.4). Typical notch tensile strengths are shown in Table 15.2.

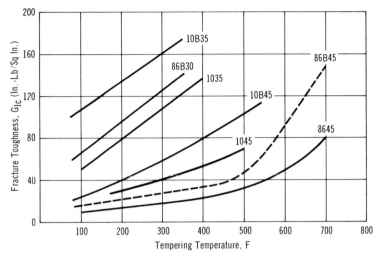

Fig. 15.4 Typical effect of boron in improving fracture toughness of several direct-hardening steels (*Metal Progress,* March 1973, Fig. 3, p. 93).

Table 15.2 Notch Tensile Strength of Boron and Boron-Free 1036 Steel [6]

	Notch Tensile Strength (psi)	
	45 Rc	52 Rc
1036[a] (no boron)	275,000	255,000
1036[a] (with boron)	300,000	345,000

[a] 1.50% Mn.

The production of high-quality boron steels is relatively new. As late as 1969 0.30% C boron steels were being made, for example, with boron factors ranging from 1 (no effect) to more than 2. Boron-factor studies of many hundreds of heats showed similar variations in quality (see Fig. 15.5). A major breakthrough occurred when titanium was found to be an outstanding de-nitrifier to protect the boron addition. Today it is common for boron steels to be consistently made with boron factors varying by ±0.2. This spread is the equivalent of approximately 0.20% Cr or 0.25% Mn and presents no difficulties to the heat treater. In fact, boron steels are often more consistent in heat-treat response because boron exerts a leveling effect on hardenability. When the carbon content of a heat of steel is at the high limit, the boron effect is less than when the carbon is at the low limit. As cited in Table 15.1, the boron

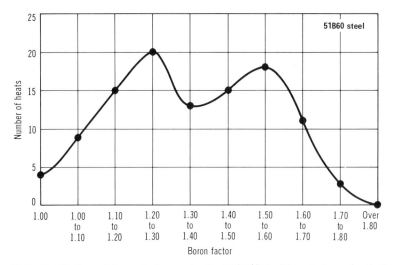

Fig. 15.5 Distribution of boron factors on heats of 51B60 steel (two tests per heat). Boron factor is the quotient of actual D_I with boron, divided by the calculated D_I without boron (per Grossman factors: see Appendix VII).

effect is in direct proportion to the grain size which further tends to minimize variations in hardenability caused by grain-size variations.

Sometimes plain carbon boron steels exhibit an erratic nature in forming or cold extruding. This is perhaps a reflection on the fact that today all types of carbon steel seem more erratic than ever. The reason may be that some producers who use taconite pellets have such an abundance of hot metal that little scrap is used. The result is low residuals of nickel, chromium, and molybdenum. At the same time other producers are employing electric-furnace melting and a 100% scrap charge, which produces a high residual alloy content. Boron cannot level out these wide differences in composition, and specifications that require cold extrusion should be written with suitable control of residual alloy content, hardenability, or source restrictions.

The characteristics exhibited by boron steels in heat treatment depend on one or a combination of the following factors:

1 Expertise in setting up specifications.
2 Recognizing that the boron factor decreases with carbon content up to 0.85% and beyond that can even be negative in carbon grades, as found by Grange. The boron effect has been negative in carburized alloy steels with 1% case carbon when direct-quenched but slightly positive when reheated for hardening. The reason is probably that on slow cooling more free carbide is formed which depletes the martix not only of carbon to hypoeutectoid levels but also the alloy content.

With proper upper and lower limits on hardenability (and attention to boron factor) direct hardening of boron steels is more trouble free than the boron-free grades.

The fact that boron steels are cold-extruded and used for important parts such as piston pins and heavy-duty crankshafts is evidence that they can be made clean. This is especially true when carbon deoxidation and vacuum degassing are employed. Vacuum degassing reduces the oxygen and nitrogen contents to such low levels that the amounts of deoxidizers and denitrifiers which must be added to protect the boron can also be substantially reduced.

Machinability has not been a problem with boron steels. The machinability before quenching and tempering is substantially better than that of steels alloyed to the same hardenability. In the heat-treated condition it must be recognized that carbon-boron steels can harden deeply to a fine and tough microstructure. Cutting a tough steel is usually more difficult than machining a brittle material like a cold-drawn resulfurized grade.

AVAILABILITY OF BORON STEELS

The major issue in regard to boron steels is availability. As shown in Appendix VI, there are only eight standard carbon-boron H steels; we hesitate to recommend the use of four because of their high manganese content which causes erratic transverse properties and erratic machinability with or without boron. All of these steels, with the possible exception of 50B60H, require excellent boron practice to achieve the hardenability shown with the specified minimum compositions. In addition to these grades, one company markets the steels listed in Table 15.3 with guaranteed hardenabilities. At least two other American companies sell weldable carbon-boron plates up to 1.50 in. thick, quenched and tempered up to 100,000-psi minimum yield point.

Table 15.3   Available Nonstandard Boron Steels

| Steel | Carbon (%) | Manganese (%) | End-Quench Hardenability (Rc) | | | | | | |
			1	2	3	4	5	6	7
0B18	0.15–0.20	0.80–1.10	39–46	37–45	30–44	20–41	...	...	...
0B21	0.18–0.23	0.80–1.10	40–46	40–45	35–44	20–42	...	...	...
0B22	0.17–0.23	1.00–1.30	42–48	42–47	41–46	36–46	20–44		
0B23	0.17–0.23	1.10–1.40	43–46	43–46	41–46	37–46	22–45	25–43	
0B24	0.19–0.25	1.35–1.65	44–50	44–49	44–48	43–46	38–45	33–47	...

The cost-reduction possibilities of carbon-boron steels are substantial because, with the exception of structural shapes, they are priced on a carbon steel base. Boron-containing alloy steel can also reduce cost. As an example, the cost per ton to double the 50% martensite hardenability of an alloy steel is as follows: boron, $5; manganese, $8; chromium, $12; molybdenum, $26; and nickel, $69.

As supplies become critical, conventional alloying elements will become increasingly expensive; therefore the advantage of boron will become even more pronounced. In the United States there is a nearly inexhaustible supply. Beyond the economic considerations, the failure to utilize boron in steel would be a waste of natural resources.

SELECTION OF BORON STEEL FOR
DIRECT-HARDENED PARTS

The selection of boron steels can usually be made on the basis of hardenability, just as for any other steel. It usually works out that selection is based on a combination of the following reasons:

1   To provide deeper hardening than can be provided by plain carbon steels.
2   To reduce costs of processing, such as forging, extruding, and machining.
3   To take advantage of the unusual toughness and/or notch tensile strength of these steels.
4   To reduce quench cracking.
5   To reduce costs by replacing full-alloy steel with carbon-boron or a leaner alloy boron grade.

Low-residual plain-carbon steels (0.60 to 0.90% Mn) can be hardened only to about 70% martensite at a 0.125-in. depth on parts 1.25 to 2.00-in. round or equivalent, even by induction, when the cold core assists in quenching. A typical example is the crawler-tractor track link shown in Fig. 15.6. When greater depths are required, the first step is to increase the manganese. When heats are received with manganese on the high side (1.40%), machinability problems and quench cracking may develop. This situation can often be solved with carbon-boron steels and today a widely popular material for crawler-tractor links is 15B37H. This steel can be cold-sheared into forging blanks; it is easier to forge than more highly alloyed boron-free steels and it heat-treats with ease.

Boron steels are excellent selections for shafts, especially with shell or induction hardening. For maximum residual compressive stress, however, it should be remembered that through hardening must be avoided. Carbon-boron steels austenitize rapidly, a characteristic that make them particularly well suited to induction hardening.

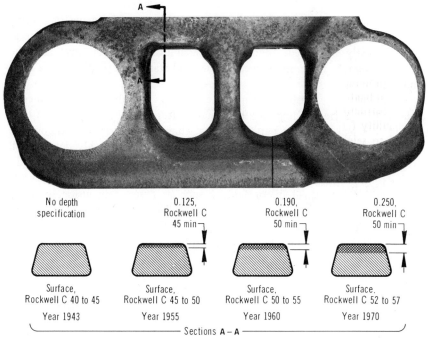

No depth specification	0.125, Rockwell C 45 min	0.190, Rockwell C 50 min	0.250, Rockwell C 50 min
Surface, Rockwell C 40 to 45	Surface, Rockwell C 45 to 50	Surface, Rockwell C 50 to 55	Surface, Rockwell C 52 to 57
Year 1943	Year 1955	Year 1960	Year 1970

Sections **A – A**

Fig. 15.6 Carbon-manganese-boron steels are used extensively to provide the hardenability and other qualities necessary for crawler-tractor track links.

A major difference between carbon-boron steel and some alloy steels lies in the tempering procedure. When replacing 4140 in thin sections, for example, the tempering temperature for carbon-boron steels must be reduced approximately 50 F to achieve the same final hardness. The inherently good toughness of carbon-boron steel permits the use of lower tempering temperatures without the risk of brittle fracture. When toughness is a major engineering requirement, however, all steels under 0.40% C should preferably *not* be tempered between 450 and 750 F. Tempering times, to reduce costs, can be shorter with carbon-boron steels than with some alloy grades.

Typical parts made from direct-hardening grades of carbon-boron steel are as follows:

Axles and shafts	Heat-treated springs
Wheel spindles	Torsion bars
Steering knuckles	Crankshafts
Induction hardened gears	Connecting rods
Hand tools such as wrenches and chisels	High-strength welded structures

Precise selection of boron steel for its capability of providing high-quality microstructure, outstanding fracture toughness, and unusually high notch tensile strength is a matter of highly sophisticated materials engineering. First, the engineering requirements for these characteristics must be known or at least accurately estimated. Second, the capabilities of boron steel must be known, preferably in terms of mean values and standard deviation, for a large number of heats.

Microstructural capability can be estimated roughly from end-quench hardenability (corresponding to 99.9% martensite). This information can then be used to select steels for parts requiring what might be called ultralong-life fatigue strength, in which the presence of ferrite can be highly detrimental. A typical part is a high-performance internal-combustion-engine connecting rod.

Fracture toughness is important, of course, on dynamically loaded parts that might contain small discontinuities such as a seam, inclusion, or (in a welded structure) an undercut or fusion defect. The superiority of boron steels in this situation is well known; however, possible flaw sizes and specific capabilities of candidate steels must be known in order to apply the principles of fracture mechanics for quality-control purposes.

The same reasoning suggested for fracture toughness applies to the characteristic of notch tensile strength. Engineering requirements and steel capabilities must be known for correct steel selection. As is often the case, materials-engineering decisions should be evaluated at the test facility or at least by laboratory testing of parts or prototypes. If a part fails and the failure indicates lack of microstructural quality, fracture toughness, or notch tensile strength, then changing to a boron steel may be a simple solution. It must be remembered, however, that the demands of the design may be beyond the capabilities even of boron steel.

A typical example of a part changed from 1541 to 10B40 to eliminate quench cracking is shown in Fig. 15.7. With 1541 it was necessary on some heats to use Ucon* concentrations as high as 30% to control cracking; however, all had to be magnetic-particle inspected for cracks. A change was made to 10B40 and no further cracking occurred.

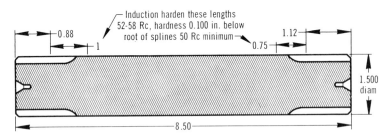

Fig. 15.7 An involute splined shaft of 1541 steel that cracked excessively in hardening.

*A quenching compound made by Union Carbide Corporation.

Fig. 15.8 Welded components for heavy equipment, such as this bulldozer blade, can be made of C-Mn-B steels. Sections can be welded without preheating or postheating (*Metal Progress,* March 1973, Fig. 4 p. 99).

An outstanding application of direct-hardening carbon-boron steels is in parts that require welding. Underbead cracking is much easier to control compared with alloy steels of equivalent strength. A bull-dozer blade such as that shown in Fig. 15.8 is a complex welded structure, especially in the moldboard reinforcements. Delayed cracking in the welds was a constant problem until the alloy steel was replaced with carbon-boron steel. Today such critical parts as crane booms and excavator dipper sticks are being made of carbon-boron steels quenched and tempered to 100,000-psi minimum yield point. Weld efficiency and fatigue qualities are the same as those of the alloy steels formerly used. Toughness is excellent; for example, at 90,000-psi yield point the transverse Charpy V-notch value will exceed 20 ft-lb at –20 F.

For best results heat-treated carbon-boron steel plates should be through-hardened to at least 80% min martensite. Because of the limits of hardenability of carbon-boron steels, the maximum plate thickness is 1.25 in. at the 100,000-psi yield level and 1.50 in. at 80,000- and 90,000-psi strength. Above these thicknesses boron-treated alloy grades are generally necessary.

SELECTION OF BORON STEEL FOR
CASE-HARDENED PARTS

Perhaps the most practical use of boron in case hardening steels is to provide higher hardenability in the core and to minimize the need for conventional alloying elements such as manganese, nickel, chromium, and molybdenum. Beyond material cost the reason is that boron has almost no effect on the compressive yield point. This is in sharp contrast to other alloying elements, as shown in the formula (13-1) developed for $\sigma_{0.2}$ by Hurley, Kitchin, and Crawford (see Chapter 13). This factor permits cold extrusion of boron steels with core hardenability at the level of 8617 to 8620 with the same ease as 1016 or 1018. Annealing before cold forming is much simpler for the boron steels, as also noted in Chapter 13; for example, when cold extruding a 4-in. OD by 0.625-in. wall by 12-in.-long bushing of 10B18 steel from a slug 4½ in. in diameter by 5 in. long, only simple normalizing before extrusion and heating to 1250 F between extrusion steps are required.

In case-hardened parts boron is useful for engineering qualities in that it provides higher hardenability not only in the core but also in the hypoeutectoid zone of a carburized or carbonitrided case. This higher hardenability is useful to ensure adequate case support and/or a microstructure of more than 90% low-carbon martensite for maximum toughness.

By examining the minimum hardenability limits and the minimum carbon contents of the Q Temp* steels we note that the approximate minimums for relationship of J distance to various martensite levels are as follows:

	J Distance to Martensite Content of		
Steel	99.9%	95%	90%
10B18	1	2	3
10B21	1	2.5	3
10B22	1	3.5	4+
10B23	3	4	4.5+
15B24	4	4+	5.5+
8620H (for comparison)	1	2	3

The effect of boron on hardenability is quite strongly influenced by carbon content in hypoeutectoid steels, and the gradient of carbon from eutectoid carbon (0.85%) down to core carbon must be closely controlled for maximum uniformity of response in heat treatment. Wide variations in carbon content from batch to batch at various depths below the eutectoid carbon may cause wide variations in microstructure, mechanical properties, and dimensional accuracy of the hardened part.

Table 15.1 indicates that boron effect is enhanced by finer grain size. We have no knowledge of the commercial use of coarse-grained boron steel. The

*Registered trademark, U.S. Steel Corporation.

average grain size of boron steels lies between ASTM 8 and 12. However, any processing that will coarsen the austenite grain size, such as excessive forging or heat-treating temperatures (e.g., carburizing at 2000 F), may have a seriously deleterious effect on boron effectiveness.

The materials engineer must recognize that the presence of boron will have only a slight effect on hardenability in the portions of a carburized or carbonitrided case containing more than 0.85% C. Therefore adequate conventional alloying elements must be present or the quench must be drastic (e.g., water, brine, or caustic soda) to ensure that proper microstructure and hardness will result. Particular caution should be exercised on parts such as piston pins. It is nearly impossible to water-quench the internal surface of such tubular sections without a strong flush through the bore or an inserted spray head.

The selection of boron steel for case-hardened parts should proceed about the same as for a born-free steel, beginning with provision for adequate case hardenability at the surface. The core hardness (and strength) for case support and structural strength are determined by selection on the basis of core hardenability.

When precision and/or highly loaded parts are involved, test lots should be employed to check for microstructural and dimensional quality.

GEARS

The only alloy boron steel that is being used today in significant tonnages for case-hardened gears is 94B17. For some manufacturers it has been successful in highly loaded hypoid and spiral bevel pinions.

Although we are not aware of their use, it would seem that boron-treated alloy steels could be useful for nitriding; for example, in heavy sections a grade such as 41B40 could be used to provide the high-quality tempered martensite microstructures necessary for optimum nitriding response. The small amounts (0.02 to 0.06%) of titanium and/or vanadium commonly added with boron would be expected to enhance nitriding qualities.

REFERENCES

1 B. M. Kapadia, R. M. Brown, and W. J. Murphy, The Influence of Nitrogen, Titanium, and Zirconium on the Boron Hardenability Effect in Constructional Alloy Steels. Transactions of the Metallurgical Society of the American Institute of Mining, Metallurgical, and Petroleum Engineers, 242, 1689 (August 1968).
2 R. A. Grange and T. M. Garvey, Trans ASM, 37, 136 (1946).
3 R. A. Grange and J. B. Mitchell, Trans ASM, 53, 157 (1961).

4 J. M. Hodge and M. A. Orehoski, Trans, (American Institute of Mining, Metallurgical, and Petroleum Engineers, **167**, 627 (1946).

5 R. A. Grange and H. R. Hribal, Hardenability Effect of Boron in Carbon Steels, Research Laboratory, U.S. Steel Corp., Monroeville, Pennsylvania, June 20, 1972.

6 R. F. Kern, Heat Treating Carbon-Manganese-Boron Steel Parts for Heavy Equipment, *Metal Progr.,* March 1973.

APPENDIXES

I Hardness Conversion Chart

C Scale	B Scale	15-N Scale	30-N Scale	Diameter (mm)	Hardness No.	Knoop Hardness No. (500-g load)	Shore Scleroscope Hardness No.	Approximate Tensile Strength (ksi)
68	...	93.2	84.4	...	...	858	97	...
67	...	92.9	83.6	...	...	835	95	...
66	...	92.5	82.8	...	...	810	92	...
65	...	92.2	81.9	2.26	739	790	91	...
64	...	91.8	81.1	2.28	722	768	88	...
63	...	91.4	80.1	2.31	705	747	87	...
62	...	91.1	79.3	2.34	688	726	85	...
61	...	90.7	78.4	2.37	670	707	83	...
60	...	90.2	77.5	2.40	654	687	81	...
59	...	89.8	76.6	2.44	634	670	80	351
58	...	89.3	75.7	2.47	615	653	78	337
57	...	88.9	74.8	2.51	595	637	76	324
56	...	88.3	73.9	2.55	577	620	75	313
55	...	87.9	73.0	2.59	560	605	74	301
54	...	87.4	72.0	2.63	543	587	72	292
53	...	86.9	71.2	2.67	525	571	71	283
52	...	86.4	70.2	2.71	512	557	69	273
51	...	85.9	69.4	2.75	496	543	68	264
50	...	85.5	68.5	2.79	481	530	67	255
49	...	85.0	67.6	2.83	469	516	66	246

I Hardness Conversion Chart (Continued)

C Scale	B Scale	15-N Scale	30-N Scale	Brinell 3000-kg Load, 10-mm Ball Diameter (mm)	Brinell Hardness No.	Knoop Hardness No. (500-g load)	Shore Sclerescope Hardness No.	Approximate Tensile Strength (ksi)
48	...	84.5	66.7	2.87	455	503	64	237
47	...	83.9	65.8	2.90	443	490	63	229
46	...	83.5	64.8	2.94	432	477	62	222
45	...	83.0	64.0	2.98	421	465	60	215
44	...	82.5	63.1	3.02	409	454	58	208
43	...	82.0	62.2	3.05	400	440	57	201
42	...	81.5	61.3	3.09	390	430	56	194
41	...	80.9	60.4	3.13	381	417	55	188
40	...	80.4	59.5	3.17	371	406	54	181
39	...	79.9	58.6	3.21	362	395	52	176
38	...	79.4	57.7	3.25	353	385	51	171
37	...	78.8	56.8	3.29	344	375	50	168
36	109.0	78.3	55.9	3.33	336	365	49	162
35	108.5	77.7	55.0	3.37	327	357	48	157
34	108.0	77.2	54.2	3.41	319	347	47	153
33	107.5	76.6	53.3	3.45	311	337	46	149
32	107.0	76.1	52.1	3.50	301	327	44	145
31	106.0	75.6	51.3	3.54	294	318	43	142
30	105.5	75.0	50.4	3.59	286	310	42	138
29	104.5	74.5	49.5	3.64	279	302	41	135
28	104.0	73.9	48.6	3.69	271	295	41	132

27	103.0	73.3	47.7	3.73	264	287	40	128
26	102.5	72.8	46.8	3.77	258	280	38	125
25	101.5	72.2	45.9	3.81	253	273	38	122
24	101.0	71.6	45.0	3.85	247	265	37	120
23	100.0	71.0	44.0	3.89	243	259	36	117
22	99.0	70.5	43.2	3.93	237	253	35	114
21	98.5	69.9	42.3	3.98	231	247	35	112
20	97.8	69.4	41.5	4.02	226	240	34	110
18.8	97.3	⋮	⋮	4.05	223	⋮	⋮	⋮
17.5	96.4	⋮	⋮	4.10	217	⋮	33	105
16.0	95.5	⋮	⋮	4.15	212	⋮	⋮	102
15.2	94.6	⋮	⋮	4.20	207	⋮	32	100
13.8	93.8	⋮	⋮	4.25	201	⋮	31	98
12.7	92.8	⋮	⋮	4.30	197	⋮	30	95
11.5	91.9	⋮	⋮	4.35	192	⋮	29	93
10.0	90.7	⋮	⋮	4.40	187	⋮	⋮	90
9.0	90.0	⋮	⋮	4.45	183	⋮	28	89
8.0	89.0	⋮	⋮	4.50	179	⋮	27	87
6.4	87.8	⋮	⋮	4.55	174	⋮	⋮	85
5.4	86.8	⋮	⋮	4.60	170	⋮	26	83
4.4	86.0	⋮	⋮	4.65	167	⋮	⋮	81
3.3	85.0	⋮	⋮	4.70	163	⋮	25	79
0.9	82.9	⋮	⋮	4.80	156	⋮	⋮	76
⋮	80.8	⋮	⋮	4.90	149	⋮	23	73
⋮	78.7	⋮	⋮	5.00	143	⋮	22	71
⋮	76.4	⋮	⋮	5.10	137	⋮	21	67
⋮	74.0	⋮	⋮	5.20	131	⋮	⋮	65

II Conversion Factors to SI Units

Quantity	To Convert	Multiply by
Length	in. to m	2.54 E – 02
	ft. to m	3.048 E – 01
	in./ft to m/m	8.333 E – 02
Mass	lb to kg	4.5359 E – 01
	cwt to kg	4.5359 E + 01
	ton (short) to kg	9.0718 E + 02
Area	$in.^2$ to m^2	6.4516 E – 04
	ft^2 to m^2	9.2903 E – 02
Stress	psi to Pa	6.8947 E + 03
	ksi to Pa	6.8947 E + 06
Stress Intensity	ksi $\sqrt{in}$ to Pa $\sqrt{m}$	1.0988 E + 06
Bending Moment or Torque	in.-lbf to Nm	1.1298 E – 01
	ft.-lbf to Nm	1.3558
Impact Energy	ft-lbf to J	1.3558
Flow	gal/min to m^3/min	3.7854 E – 03
	gal/min to m^3/sec	6.309 E – 05
Temperature	F = 1.8C + 32	
	C = (F – 32)/1.8	

III   Estimated Minimum Mechanical Properties of As-Rolled Steel

Data are from *SAE Handbook*. Properties are those that can generally be expected in bar sizes of 0.75 to 1.25 in. diameter, based on the standard round tensile-test specimen with 2-in. gage length. Properties for cold-finished bars are those normally expected when drawn with normal draft. Heavier drafts can be used to produce higher strengths.

Table III.1   Estimated Mechanical Properties and Machinability of SAE Carbon Steel Bars

SAE and AISI No.	Type of Processing	Estimated Minimum Values					Average Machinability Rating (cold-drawn 1112 = 100%)
		Tensile Strength (psi)	Yield Strength (psi)	Elongation in 2 (in., %)	Reduction in area, (%)	Brinell Hardness	
1006	Hot-rolled	43,000	24,000	30	55	86	
	Cold-drawn	48,000	41,000	20	45	95	50
1008	Hot-rolled	44,000	24,500	30	55	86	
	Cold-drawn	49,000	41,500	20	45	95	55
1010	Hot-rolled	47,000	26,000	28	50	95	
	Cold-drawn	53,000	44,000	20	40	105	55
1012	Hot-rolled	48,000	26,500	28	50	95	
	Cold-drawn	54,000	45,000	19	40	105	55
1015	Hot-rolled	50,000	27,500	28	50	101	
	Cold-drawn	56,000	47,000	18	40	111	60
1016	Hot-rolled	55,000	30,000	25	50	111	

Table III.1 Estimated Mechanical Properties and Machinability of SAE Carbon Steel Bars (Continued)

SAE and AISI No.	Type of Processing	Estimated Minimum Values					Average Machinability Rating (cold-drawn 1112 = 100%)
		Tensile Strength (psi)	Yield Strength (psi)	Elongation in 2 (in., %)	Reduction in area (%)	Brinell Hardness	
1017	Cold-drawn	61,000	51,000	18	40	121	70
	Hot-rolled	53,000	29,000	26	50	105	
1018	Cold-drawn	59,000	49,000	18	40	116	65
	Hot-rolled	58,000	32,000	25	50	116	
1019	Cold-drawn	64,000	54,000	15	40	126	70
	Hot-rolled	59,000	32,500	25	50	116	
1020	Cold-drawn	66,000	55,000	15	40	131	70
	Hot-rolled	55,000	30,000	25	50	111	
1021	Cold-drawn	61,000	51,000	15	40	121	65
	Hot-rolled	61,000	33,000	24	48	116	
1022	Cold-drawn	68,000	57,000	15	40	131	70
	Hot-rolled	62,000	34,000	23	47	121	
1023	Cold-drawn	69,000	58,000	15	40	137	70
	Hot-rolled	56,000	31,000	25	50	111	
1024[a]	Cold-drawn	62,000	52,500	15	40	121	65
	Hot-rolled	74,000	41,000	20	42	149	
1025	Cold-drawn	82,000	69,000	12	35	163	60
	Hot-rolled	58,000	32,000	25	50	116	
	Cold-drawn	64,000	54,000	15	40	126	65

1026	Hot-rolled	64,000	35,000	24	49	126	75
	Cold-drawn	71,000	60,000	15	40	143	65
1027[a]	Hot-rolled	75,000	41,000	18	40	149	70
	Cold-drawn	83,000	70,000	12	35	163	65
1030	Hot-rolled	68,000	37,500	20	42	137	55
	Cold-drawn	76,000	64,000	12	35	149	65
1035	Hot-rolled	72,000	39,500	18	40	143	65
	Cold-drawn	80,000	67,000	12	35	163	
1036[a]	Hot-rolled	83,000	45,500	16	40	163	60
	Cold-drawn	92,000	77,500	12	35	187	60
1037	Hot-rolled	74,000	40,500	18	40	143	45
	Cold-drawn	82,000	69,000	12	35	167	
1038	Hot-rolled	75,000	41,000	18	40	149	60
	Cold-drawn	83,000	70,000	12	35	163	60
1039	Hot-rolled	79,000	43,500	16	40	156	70
	Cold-drawn	88,000	74,000	12	35	179	
1040	Hot-rolled	76,000	42,000	18	40	149	
	Cold-drawn	85,000	71,000	12	35	170	60
1041[a]	Hot-rolled	92,000	51,000	15	40	187	45
	Cold-drawn	102,500	87,000	10	30	207	
	ACD[b]	94,000	80,000	10	45	184	
1042[a]	Hot-rolled	80,000	44,000	16	40	163	60
	Cold-drawn	89,000	75,000	12	35	179	60
	NCD[c]	85,000	73,000	12	45	179	
1043[a]	Hot-rolled	82,000	45,000	16	40	163	70
	Cold-drawn	91,000	77,000	12	35	179	
	NCD[c]	87,000	75,000	12	45	179	60
1044	Hot-rolled	80,000	44,000	16	40	163	
1045	Hot-rolled	82,000	45,000	16	40	163	70
	Cold-drawn	91,000	77,000	12	35	179	55
	ACD[b]	85,000	73,000	12	45	170	65

Table III.1   Estimated Mechanical Properties and Machinability of SAE Carbon Steel Bars (Continued)

SAE and AISI No.	Type of Processing	Estimated Minimum Values				Brinell Hardness	Average Machinability Rating (cold-drawn 1112 = 100%)
		Tensile Strength (psi)	Yield Strength (psi)	Elongation in 2 (in., %)	Reduction in area, (%)		
1046	Hot-rolled	85,000	47,000	15	40	170	
	Cold-drawn	94,000	79,000	12	35	187	55
	ACD[b]	90,000	75,000	12	45	179	65
1047[a]	Hot-rolled	94,000	52,000	15	30	192	
	Cold-drawn	103,000	88,000	10	28	207	40
	ACD[b]	95,000	85,000	10	35	187	45
1048[a]	Hot-rolled	96,000	53,000	14	33	197	
	Cold-drawn	106,500	89,500	10	28	217	45
	ACD[b]	93,500	78,500	10	35	192	50
1049	Hot-rolled	87,000	48,000	15	35	179	
	Cold-drawn	97,000	81,500	10	30	197	45
	ACD[b]	92,000	77,000	10	40	187	55
1050	Hot-rolled	90,000	49,500	15	35	179	
	Cold-drawn	100,000	84,000	10	30	197	45
	ACD[b]	95,000	80,000	10	40	189	55
1052[a]	Hot-rolled	108,000	59,500	12	30	217	
	ACD[b]	98,000	83,000	10	40	193	50
1055	Hot-rolled	94,000	51,500	12	30	192	
	ACD[b]	96,000	81,000	10	40	197	55

Grade	Condition					
1060	Hot-rolled	98,000	54,000	12	30	201	60
	SACD[d]	90,000	70,000	10	45	183	
1064	Hot-rolled	97,000	53,500	12	30	201	60
	SACD[d]	89,000	69,000	10	45	183	
1065	Hot-rolled	100,000	55,000	12	30	207	60
	SACD[d]	92,000	71,000	10	45	187	
1070	Hot-rolled	102,000	56,000	12	30	212	55
	SACD[d]	93,000	72,000	10	45	192	
1074	Hot-rolled	105,000	58,000	12	30	217	55
	SACD[d]	94,500	73,000	10	40	192	
1078	Hot-rolled	100,000	55,000	12	30	207	55
	SACD[d]	94,000	72,500	10	40	192	
1080	Hot-rolled	112,000	61,500	10	25	229	45
	SACD[d]	98,000	75,000	10	40	192	
1084	Hot-rolled	119,000	65,500	10	25	241	45
	SACD[d]	100,000	77,000	10	40	192	
1085	Hot-rolled	121,000	66,500	10	25	248	45
	SACD[d]	100,500	78,000	10	40	192	
1086	Hot-rolled	112,000	61,500	10	25	229	45
	SACD[d]	97,000	74,000	10	40	192	
1090	Hot-rolled	122,000	67,000	10	25	248	45
	SACD[d]	101,000	78,000	10	40	197	
1095	Hot-rolled	120,000	66,000	10	25	248	45
	SACD[d]	99,000	76,000	10	40	197	

[a] These grades, with max Mn in excess of 1%, have been renumbered 1500 series. See Table 3, SAE J403.
[b] ACD represents annealed cold-drawn.
[c] NCD represents normalized cold-drawn.
[d] SACD represents spheroidized annealed cold-drawn.

Table III.2 Estimated Mechanical Properties and Machinability Ratings of Resulfurized Carbon Steel Bars[a]

SAE and AISI No.	Type of Processing	Estimated Minimum Values					Average Machinability Rating (cold-drawn 1112 = 100%)
		Tensile Strength (psi)	Yield Strength (psi)	Elongation in 2 (in., %)	Reduction in area, (%)	Brinell Hardness	
1111	Hot-rolled	55,000	33,000	25	45	121	
	Cold-drawn	75,000	58,000	10	35	163	95
1112	Hot-rolled	56,000	33,500	25	45	121	
	Cold-drawn	78,000	60,000	10	35	167	100
1113	Hot-rolled	56,000	33,500	25	45	121	
	Cold-drawn	78,000	60,000	10	35	167	135
12L14	Hot-rolled	57,000	34,000	22	45	121	
	Cold-drawn	78,000	60,000	10	35	163	160
1108	Hot-rolled	50,000	27,500	30	50	101	
	Cold-drawn	56,000	47,000	20	40	121	80
1109	Hot-rolled	50,000	27,500	30	50	101	
	Cold-drawn	56,000	47,000	20	40	121	80
1117	Hot-rolled	62,000	34,000	23	47	121	
	Cold-drawn	69,000	58,000	15	40	137	90
1118	Hot-rolled	65,000	36,000	23	47	131	
	Cold-drawn	72,000	61,000	15	40	143	85
1119	Hot-rolled	62,000	34,000	23	47	121	
	Cold-drawn	69,000	58,000	15	40	137	100
1132	Hot-rolled	83,000	45,500	16	40	167	

1137	Cold-drawn	92,000	77,000	12	35	183	75
	Hot-rolled	88,000	48,000	15	35	179	70
1140	Cold-drawn	98,000	82,000	10	30	197	70
	Hot-rolled	79,000	43,500	16	40	156	
1141	Cold-drawn	88,000	74,000	12	35	170	70
	Hot-rolled	94,000	51,500	15	35	187	
1144	Cold-drawn	105,100	88,000	10	30	212	70
	Hot-rolled	97,000	53,000	15	35	197	
1145	Cold-drawn	108,000	90,000	10	30	217	80
	Hot-rolled	85,000	47,000	15	40	170	
1146	Cold-drawn	94,000	80,000	12	35	187	65
	Hot-rolled	85,000	47,000	15	40	170	
1151	Cold-drawn	94,000	80,000	12	35	187	70
	Hot-rolled	92,000	50,500	15	35	187	
	Cold-drawn	102,000	86,000	10	30	207	65

[a] All SAE 1100 series steels are rated on the basis of 0.10% max or coarse-grain melting practice.

IV Size Tolerances for Carbon and Alloy Steel Bars (per AISI)

Table IV.1 Cold-Drawn

Specified Size		Size Tolerances							
		Carbon Range (max)							
		Carbon Steels				Alloy Steels			
Over	Incl	Carbon 0.28% or Less	0.29% to 0.55% Incl	Carbon to 0.55%; Incl, Stress Relieved or Annealed After Cold Finish	Carbon Over 0.55%; All Carbons Quenched and Tempered, or Normalized Before Cold Finish	Carbon 0.28% or Less	Carbon 0.29% to 0.55% Incl	Carbon to 0.55% Incl, Stress Relieved or Annealed After Cold Finish	Over 0.55%; All Carbons Quenched and Tempered or Normalized Before Cold Finish[a]
		Variations from size (undersize only)							
		Rounds (cold-drawn)							
	To 1.50	0.002	0.003	0.004	0.005	0.003	0.004	0.005	0.006

1.50 to 2.50	0.003	0.004	0.005	0.006	0.004	0.005	0.006	0.007
2.50 to 4.00	0.004	0.005	0.006	0.007	0.005	0.006	0.007	0.008
Squares								
To 0.75	0.002	0.004	0.005	0.003	0.005	0.006	0.007	0.008
0.75 to 1.50	0.003	0.005	0.006	0.004	0.006	0.007	0.008	0.009
1.50 to 2.50	0.004	0.006	0.007	0.005	0.007	0.008	0.009	0.010
2.50 to 4.00	0.006	0.008	0.009	0.007	0.009	0.010	0.010	0.012
4.00 to 5.00	···	···	···	0.011	···	···	···	···
Hexagons								
To 0.75	0.002	0.003	0.004	0.003	0.004	0.005	0.006	0.007
0.75 to 1.50	0.003	0.004	0.005	0.004	0.005	0.006	0.007	0.008
1.50 to 2.50	0.004	0.005	0.006	0.005	0.006	0.007	0.008	0.009
2.50 to 3.12	0.005	0.006	0.007	0.006	0.007	0.008	0.009	0.010
3.12 to 4.00	···	···	···	0.006	···	···	···	···
Flats (thickness or width variation)								
Width[b]								
To 0.75	0.003	0.004	0.006	0.004	0.006	0.007	0.007	0.009
0.75 to 1.50	0.004	0.005	0.008	0.005	0.008	0.009	0.009	0.011
1.50 to 3.00	0.005	0.006	0.010	0.006	0.010	0.011	0.011	0.013
3.00 to 4.00	0.006	0.008	0.011	0.007	0.011	0.012	0.012	0.017
4.00 to 6.00	0.008	0.010	0.012	0.009	0.012	0.013	0.013	0.021
Over 6.00	0.013	0.015	···	0.014	···	···	···	···

[a]Also applies to bars with or without stress relieving or annealing after cold finishing.
[b]Width governs tolerances for both width and thickness. Dimensions in inches.

Appendix IV (Continued) Carbon and Alloy Steel Bars

Table IV.2 Turned and Polished

Specified Diameter		Carbon Steels				Alloy Steels			
		Carbon Range (max)							
Over	Inc	Carbon 0.28% or Less	Carbon 0.29% to 0.55% Incl	Carbon to 0.55%; Incl, Stress Relieved or Annealed After Cold Finish	Carbon Over 0.55%; All Carbons Quenched and Tempered, or Normalized Before Cold Finish	Carbon 0.28% or Less	Carbon 0.29% to 0.55% Incl	Carbon to 0.55% Incl, Stress Relieved or Annealed After Cold Finish	Over 0.55%; All Carbons Quenched and Tempered or Normalized Before Cold Finish[a]
		Variations from Size (undersize only)							
To 1.50		0.002	0.003	0.004	0.005	0.003	0.004	0.005	0.006
1.50 to 2.50		0.003	0.004	0.005	0.006	0.004	0.005	0.006	0.007
2.50 to 4.00		0.004	0.005	0.006	0.007	0.005	0.006	0.007	0.008
4.00 to 6.00		0.005	0.006	0.007	0.008	0.006	0.007	0.008	0.009
6.00 to 8.00		0.006	0.007	0.008	0.009	0.007	0.008	0.009	0.010
8.00 to 9.00		0.007	0.008	0.009	0.010	0.008	0.009	0.010	0.011
Over 9.00		0.008	0.009	0.010	0.011	0.009	0.010	0.011	0.012

[a] Applies to bars with or without stress relieving or annealing after cold finishing.

Table IV.3 Cold Drawn, Ground and Polished

Specified Size		Variation from Specified Size (undersize only)
Over	Incl	
To 1.50		0.001
1.50 to 2.50		0.0015
2.50 to 3.00		0.002
3.00 to 4.00		0.003

Table IV.4 Turned, Ground, and Polished

Size Tolerances			

		Alloy Steels	
Specified Size	Carbon Steels	Not Heat-Treated	Heat-Treated[a] Before Turn
Over Incl	Variations from Size (Undersize Only)		
To 1.50	0.001	0.001	0.001
1.50 to 2.499	0.0015	0.0015	0.0015
2.499 to 3.00	0.002	0.002	0.002
3.00 to 4.00	0.003	0.003	0.003
4.00 to 6.00	0.004[b]	0.004	0.005
Over 6.00	0.005[b]	0.005	0.006

[a] Heat-treated applies to quenched and tempered, normalized, and tempered, or any similar double treatment before turning.

[b] For Nonresulfurized steels (steels specified to maximum sulfur limits under 0.08%) or for steels thermally treated, the tolerance is increased by 0.001 in. Dimensions in inches.

Table IV.5 Rounds, Squares, and Round-Cornered Squares. Hot-Rolled Carbon and Alloy Steel Bar Tolerances.

Specified Size		Size Tolerance		Out-of-Round[a] or Out-of-Square[b]
Over	Incl	Plus	Minus	
To 0.31		0.005	0.005	0.008
0.31 to 0.44		0.006	0.006	0.009
0.44 to 0.62		0.007	0.007	0.010
0.62 to 0.88		0.008	0.008	0.012
0.88 to 1.00		0.009	0.009	0.013
1.00 to 1.12		0.010	0.010	0.015
1.12 to 1.25		0.011	0.011	0.016
1.25 to 1.38		0.012	0.012	0.018
1.38 to 1.50		0.014	0.014	0.021
1.50 to 2.00		0.016	0.016	0.023
2.00 to 2.50		0.031	0	0.023
2.50 to 3.50		0.047	0	0.035
3.50 to 4.50		0.062	0	0.046
4.50 to 5.50		0.078	0	0.058
5.50 to 6.50		0.125	0	0.070
6.50 to 8.25		0.156	0	0.085
8.25 to 9.50		0.188	0	0.100
9.50 to 10.00[c]		0.250	0	0.120

[a] Out-of-round–the difference between the maximum and minimum diameters of the bar measured at the same cross section.
[b] Out-of-square — the difference in the two dimensions at the same cross section of a square bar between opposite faces.
[c] Carbon steel only.

Table IV.6 Hexagons

Specified Size (between flats)		Size Tolerance		Out of Hexagon[a]
Over	Inc.	Plus	Minus	
To 0.50		0.007	0.007	0.011
0.50 to 1.00		0.010	0.010	0.015
1.00 to 1.50		0.021	0.013	0.025
1.50 to 2.00		0.031	0.016	0.031
2.00 to 2.50		0.047	0.016	0.047
2.50 to 3.50		0.062	0.016	0.062

[a] Out-of-hexagon — the difference between any two dimensions at the same cross section between opposite faces. Dimensions in inches.

324

Table IV.7 Hot-Rolled Alloy Steel. Thickness and Width Tolerances (Square-Edge and Round-Edge Flats)

Width		Thickness Tolerances for Thicknesses Given (over and under in inches)							Width Tolerances (in inches)	
Over	Incl	0.203 to 0.230 Excl	0.230 to 0.250 Excl	0.250 to 0.500 Incl	Over 0.500 to 1.000 Incl	Over 1.000 to 2.000 Incl	Over 2.000 to 3.000 Incl	Over 3.000	Over	Under
To	1.00	0.007	0.007	0.008	0.010	...	...	...	0.016	0.016
1.00	2.00	0.007	0.007	0.012	0.015	0.031	...	...	0.031	0.031
2.00	4.00	0.008	0.008	0.015	0.020	0.031	0.047	0.047	0.062	0.031
4.00	6.00	0.009	0.009	0.015	0.020	0.031	0.062	0.062[b]	0.094	0.062
6.00	8.00	[a]	0.015	0.016	0.025	0.031	0.062	[b]	0.125[b]	0.094[b]

[a] Not produced as bar flats.

[b] Thickness and width tolerances for flats more than 6 to 8 in. incl, in width and more than 3 in. in thickness shall be negotiated with the supplier.

Note. In the above table, when the thickness equals the width, the product is a square. Size and out-of-square tolerances for squares are shown in size tolerance table for hot-rolled carbon and alloy steel rounds, squares, and round-cornered squares.

Table IV.8 Hot-Rolled Carbon Steel

Width Over	Width Incl	Thickness Tolerances for Thicknesses Given (over and under in inches)							Width Tolerances (in inches)	
		0.203 to 0.230 Excl	0.230 to 0.250 Excl	0.250 to 0.500 Incl	Over 0.500 to 1.000 Incl	Over 1.000 to 2.000 Incl	Over 2.000 to 3.000 Incl	Over 3.000	Over	Under
	To. 100	0.007	0.007	0.008	0.010	. . .	. . .	. . .	0.016	0.016
1.00	2.00	0.007	0.007	0.012	0.015	0.031	. . .	0.047	0.031	0.031
2.00	4.00	0.008	0.008	0.015	0.020	0.031	0.047	0.047	0.062	0.031
4.00	6.00	0.009	0.009	0.015	0.020	0.031	0.047	0.047	0.094	0.062
6.00	8.00	[a]	0.015	0.016	0.025	0.031	0.047	. . .	0.125[b]	0.094[b]

[a] Flats more than 6 to 8 in. incl in width are not available as HR carbon steel bars in thicknesses under 0.230 in.
[b] For flats more than 6 to 8 inches incl in width, and to 3 in. incl, in thickness.

Table IV.9   Bar Size Angles[a] Hot-Rolled Carbon Steel Bar Size Section Tolerances

Specified Length of Leg		Thickness Tolerances for Thicknesses Given (over and under in inches)			Tolerances for Length of Leg (over and under in inches)
		To 0.188 Incl	Over 0.188 to 0.375 Incl	Over 0.375	
Over	Incl				
To 1.00		0.008	0.010	. . .	0.031
1.00 to 2.00		0.010	0.010	012	0.047
2.00 to 3.00 excl		0.012	0.015	015	0.062

[a]The longer leg of an unequal angle determines the size for tolerance. The out-of-square tolerance in either direction is 1.5 degrees.

Table IV.10   Bar Size Channels[a]

	Size Tolerances (over and under in inches)				
Size of Channel	Depth of Section	Width of Flanges	Thickness of Web For Thicknesses Given		Out-of-Square of Either Flange per Inch of Flange Width
			To 0.188 Incl	Over 0.188	
To 1.50 incl	031	0.031	0.010	0.015	0.031
Over 1.50 to to 3.00 excl	062	0.062	0.015	020	0.031

[a]Measurements for depth of section and width of flange are overall. For channels 0.625 in. or under in depth the out-of-square tolerance is 0.047 in. per in. of depth.

Table IV.11 Bar Size Tees

Size of Tee	Width or Depth		Thickness of Flange		Thickness of Stem		Stem Out-of-Square[a]
	Over	Under	Over	Under	Over	Under	
To 1.25 incl	0.047	0.047	0.010	0.010	0.005	0.020	0.031
Over 1.25 to 2.00 incl	0.062	0.062	0.012	0.012	0.010	0.020	0.062
Over 2.00 to 3.00 excl	0.094	0.094	0.015	0.015	0.015	0.020	0.094

[a]Stem out-of-square is the variation from its true position of the centerline of stem measured at the point. The longer member of an unequal tee determines the size for tolerances. Measurements for both width and depth are overall.

Table IV.12 Hot-Rolled Carbon Steel Bars (Hot Cutting). Rounds, Squares, Hexagons, Bar Size Sections, Flats. (Hot-Rolled Carbon and Alloy Steel Bar Length Tolerances

Size of Rounds, Squares, and Hexagons		Size of Flats		Length Tolerances, Over Only				
Over	Incl	Thickness	Width	5 to 10 Ft Excl	10 to 20 Ft Excl	20 to 30 Ft Excl	30 to 40 Ft Excl	40 to 60 Ft Incl
				Mill Cutting				
To 1.00		To 1.00 incl	To 3.00 incl	0.50	0.75	1.25	1.75	2.25
1.00 to 2.00		Over 1.00	To 3.00 incl	0.62	1.00	1.50	2.00	2.50
... ...		To 1.00 incl	Over 3.00 to 6.00 incl	0.62	1.00	1.50	2.00	2.50
2.00 to 5.00		Over 1.00	Over 3.00 to 6.00 incl	1.00	1.50	1.75	2.25	2.75
5.00 to 10.00		... ...	... ...	2.00	2.50	2.75	3.00	3.25
... ...		23 to 1.00 incl	Over 6.00 to 8.00 incl	0.75	1.25	1.75	3.50	4.00
... ...		Over 1.00 to 3.00 incl	Over 6.00 to 8.00 incl	1.25	1.75	2.00	3.50	4.00
Bar size sections				0.62	1.00	1.50	2.00	2.50
				Hot Sawing				
2.00 to 5.00[a]		1.00 and over	3.00 and over	a	1.50	1.75	2.25	2.75
5.00 to 10.00		... ...	... ...	a	2.50	2.75	3.00	3.25

[a]Smaller sizes and shorter lengths are not hot-sawed.

Table IV.13 Hot-Rolled Alloy Steel Bars (Hot Cutting). Rounds, Squares, Hexagons, Octagons, Round-Edge and Square-Edge Flats

Size of Rounds, Squares, Hexagons and Octagons	Size of Flats		Length Tolerances, Over Only				
	Thickness	Width	5 to 10 Ft Excl	10 to 20 Ft Excl	20 to 30 Ft Excl	30 to 40 Ft Excl	40 to 60 Ft Incl
Rounds, Squares, Hexagons, Octagons							
Hot Shearing							
To 1.00 incl			.50	.75	1.25	1.75	2.25
Over 1.00 to 2.00 incl			.62	1.00	1.50	2.00	2.50
Over 2.00 to 5.00 incl			1.00	1.50	1.75	2.25	2.75
Over 5.00 to 9.50 incl			2.00	2.50	2.75	3.00	3.25
Hot Sawing							
2.00[a] to 5.00 incl			1.50	1.50	1.75	2.25	2.75
Over 5.00 to 9.50 incl			1.50	2.50	2.75	3.00	3.25

330

Round Edge and Square Edge Flats

Hot Shearing

Thickness	Width					
To 1.00 incl	To 3.00 incl	0.50	0.75	1.25	1.75	2.25
Over 1.00	To 3.00 incl	0.62	1.00	1.50	2.00	2.50
To 1.00 incl	Over 3.00 to 6.00 incl	0.62	1.00	1.50	2.00	2.50
Over 1.00	Over 3.00 to 6.00 incl	1.00	1.50	1.75	2.25	2.75
To 1.00 incl	Over 6.00 to 8.00 incl	0.75	1.25	1.75	3.50	4.00
Over 1.00 to 3.00 incl	Over 6.00 to 8.00 incl	1.25	1.75	2.00	3.50	4.00

Hot Sawing

Thickness	Width					
1.00[a] and over	3.00[a] and over		1.50	1.75	2.25	2.75

[a]Smaller sizes are not hot-sawed

Note 1. Length tolerances for cutting apply also to hot-rolled bars that are specified to be machine cut on one end, including bars specified to special straightness.

Note 2. In the lower portion of the above table, when the thickness equals the width the product is a square; length tolerances for squares are given in upper portion of table.

Table IV.14 Hot-Rolled Carbon and Alloy Steel Bars (Machine Cut After Machine Straightening), Rounds, Squares, Hexagons, and Flats

Size of Rounds, Squares, Hexagons, Width of Flats, and Max Dimension of Other Sections	Length Tolerances			
	To 12 Ft incl		Over 12 Ft	
	Over	Under	Over	Under
To 3.00 incl	0.19	0.06	0.25	0.06
Over 3.00 to 6.00 incl	0.25	0.06	0.38	0.06
Over 6.00 to 8.00 incl	0.38	0.06	0.50	0.06
Over 8.00 to 10.00 incl	0.50	0.06	0.62	0.06

Note. Tolerances are sometimes required all over or all under the specified length, in which case the sum of the two tolerances apply.

Table IV.15

Standard Straightness Tolerances

1 Hot-rolled carbon steel bars and bar size sections 0.250 in. in any 5.00 ft or
$0.250 \times \dfrac{\text{number of feet in length}}{5.00}$ because of warpage. Straightness tolerances do not apply to HR bars if a subsequent heating operation or controlled cooling has been performed.

2 Hot-rolled alloy steel bars 0.250 in. in any 5.00 ft or 0.250
$\times \dfrac{\text{number of feet in length}}{5.00}$

Special Straightness Tolerances

1 Hot-rolled carbon steel bars and bar size sections 0.125 in. in any 5.00 ft or
$0.125 \times \dfrac{\text{number of feet in length}}{5.00}$ because of warpage. Straightness tolerances do not apply to HR bars if a subsequent heating operation or controlled cooling has been performed.

2 Hot-rolled alloy steel bars 0.125 in. in any 5.00 ft or 0.125
$\times \dfrac{\text{number of feet in length}}{5.00}$.

V Recommended Stock Removal from Steel Products

Table V.1 Hot-Rolled Carbon Steel Bars. Recommended Minimum Machining Allowances for Turning on Centers[a]

Specified Hot-Rolled Diameter (in.)	Minimum Machining Allowance for Radius (in.)
To 7/8 incl	0.025
Over 7/8 to 1 incl	0.028
Over 1 to 1 1/8 incl	0.031
Over 1 1/8 to 1 1/4 incl	0.034
Over 1 1/4 to 1 3/8 incl	0.037
Over 1 3/8 to 1 1/2 incl	0.040
Over 1 1/2 to 2 incl	0.053
Over 2 to 2 1/2 incl	0.065
Over 2 1/2 to 3 1/2 incl	0.090
Over 3 1/2 to 4 1/2 incl	0.115
Over 4 1/2 to 5 1/2 incl	0.140
Over 5 1/2 to 6 1/2 incl	0.165
Over 6 1/2 to 8 1/4 incl	0.209
Over 8 1/4 to 9 1/2 incl	0.240
Over 9 1/2 to 10 incl	0.253

[a]Per AISI Steel Products Manual: carbon steel, semifinished for forging, hot-rolled and cold-finished bars, hot-rolled deformed concrete reinforcing bars, May 1972.

Table V.2 Alloy Steel, Hot-Rolled, or Thermally Treated Bars Recommended Allowances for Machining[a]

Specified Sizes (in.)	Minimum Allowance from the Surface, (in.)[b]
To 5/8 incl	0.016
Over 5/8 to 7/8 incl	0.021
Over 7/8 to 1 incl	0.023
Over 1 to 1 1/8 incl	0.025
Over 1 1/8 to 1 1/4 incl	0.028
Over 1 1/4 to 1 3/8 incl	0.030
Over 1 3/8 to 1 1/2 incl	0.033
Over 1 1/2 to 2 incl	0.042
Over 2 to 2 1/2 incl	0.052
Over 2 1/2 to 3 1/2 incl	0.072
Over 3 1/2 to 4 1/2 incl	0.090
Over 4 1/2 to 5 1/2 incl	0.110
Over 5 1/2 to 6 1/2 incl	0.125
Over 6 1/2 to 8 1/4 incl	0.155
Over 8 1/4 to 9 1/2 incl	0.203

[a]Per AISI Steel Products Manual: alloy steel, semifinished, hot-rolled and cold-finished bars, August 1970.
[b]For example, the minimum reduction in diameter of rounds is twice the minimum allowance from the surface.

VI Compositions and Hardenability Requirements of Heat Treating Grades of Steel Made in the United States[a, b]

Chemical composition limits or ranges of hardenability bands and hardness limits for specification purposes are shown on the following pages. Other compositional aspects are as follows:

Table VI.1 Standard Alloy H Steels

1 The phosphorus and sulfur limitations for each steelmaking process are as follows:

	Maximum (%) Phosphorus	Sulfur
Basic electric	0.025	0.025
Basic oxygen or basic open hearth	0.035	0.040
Acid electric	0.050	0.050
Acid open hearth	0.050	0.050

2 Minimum silicon limit for acid open-hearth or acid electric-furnace alloy steel is 0.15%.
3 Small quantities of certain elements are present in alloy steels that are not specified or required. These elements are considered incidental and may be present to the following maximum amounts: copper, 0.35%; nickel, 0.25%; chromium, 0.20%; molybdenum, 0.06%.
4 Standard H-steels can be produced with a lead content of 0.15 to 0.35%. Such steels are identified by insertion of the letter "L" between the second and third numbers of the AISI designation; e.g. 41L40H. Lead is reported only as a range of 0.15 to 0.35% because it is added to the ladle stream as the steel is being poured.

[a]The hardenability data are taken from AISI Steel Products Manual: alloy steel, semifinished, hot-rolled, and cold-finished bars, August 1970.
[b]For compositions of constructional steels manufactured or available in Belgium, Bulgaria, West Germany, France, Great Britain, Italy, Japan, Yugoslavia, Norway, Austria, Poland, Rumania, Russia, Sweden, Switzerland, Spain, Czechoslovakia, and Hungary see Stahlschlussel, 1974 edition, available from American Society for Metals, Metals Park, Ohio 44073.

Table VI.2 Compositions of AISI-SAE Standard Carbon Steels

Free-Machining Grades

Corr:

Resulfurized

AISI No.[a]	Composition (%)[b] C	Composition (%)[b] Mn	Composition (%)[b] P	Composition (%)[b] S	SAE No.
1109	0.08 to 0.13	0.60 to 0.90	0.040 max	0.08 to 0.13	1109
1110	0.08 to 0.13	0.30 to 0.60	0.040 max	0.08 to 0.13	...
B1111	0.13 max	0.60 to 0.90	0.07 to 0.12	0.10 to 0.15	1211
B1112	0.13 max	0.70 to 1.00	0.07 to 0.12	0.16 to 0.23	1212
B1113	0.13 max	0.70 to 1.00	0.07 to 0.12	0.24 to 0.33	1213
1116	0.14 to 0.20	1.10 to 1.40	0.040 max	0.16 to 0.23	...
1117	0.14 to 0.20	1.00 to 1.30	0.040 max	0.08 to 0.13	1117
1118	0.14 to 0.20	1.30 to 1.60	0.040 max	0.08 to 0.13	1118
1119	0.14 to 0.20	1.00 to 1.30	0.040 max	0.24 to 0.33	1119
1132	0.27 to 0.34	1.35 to 1.65	0.040 max	0.08 to 0.13	1132
1137	0.32 to 0.39	1.35 to 1.65	0.040 max	0.08 to 0.13	1137
1139	0.35 to 0.43	1.35 to 1.65	0.040 max	0.13 to 0.20	...
1140	0.37 to 0.44	0.70 to 1.00	0.040 max	0.08 to 0.13	1140
1141	0.37 to 0.45	1.35 to 1.65	0.040 max	0.08 to 0.13	1141

Nonresulfurized Grades

AISI No.[a]	Composition (%)[b] C	Composition (%)[b] Mn	Composition (%)[b] P (max)	Composition (%)[b] S (max)	SAE No.
—	0.06 max	0.35 max	0.040	0.050	1005
1006	0.08 max	0.25 to 0.40	0.040	0.050	1006
1008	0.10 max	0.30 to 0.50	0.040	0.050	...
1010	0.08 to 0.13	0.30 to 0.60	0.040	0.050	1010
1012	0.10 to 0.15	0.30 to 0.60	0.040	0.050	1012
1513	0.10 to 0.16	1.10 to 1.40	0.040	0.050	1513
1015	0.13 to 0.18	0.30 to 0.60	0.040	0.050	1015
1016	0.13 to 0.18	0.60 to 0.90	0.040	0.050	1016
1017	0.15 to 0.20	0.30 to 0.60	0.040	0.050	1017
1018	0.15 to 0.20	0.60 to 0.90	0.040	0.050	1018
1518	0.15 to 0.21	1.10 to 1.40	0.040	0.050	1518
1019	0.15 to 0.20	0.70 to 1.00	0.040	0.050	1019
1020	0.18 to 0.23	0.30 to 0.60	0.040	0.050	1020
1021	0.18 to 0.23	0.60 to 0.90	0.040	0.050	1021
1022	0.18 to 0.23	0.70 to 1.00	0.040	0.050	1022

336

1144	0.40 to 0.48	1.35 to 1.65	0.040 max	0.24 to 0.33	1144
1145	0.42 to 0.49	0.70 to 1.00	0.040 max	0.04 to 0.07	1145
1146	0.42 to 0.49	0.70 to 1.00	0.040 max	0.08 to 0.13	1146
1151	0.48 to 0.55	0.70 to 1.00	0.040 max	0.08 to 0.13	1151

Resulfurized and Rephosphorized

1211	0.13 max	0.60 to 0.90	0.07 to 0.12	0.10 to 0.15	...
1212	0.13 max	0.70 to 1.00	0.07 to 0.12	0.16 to 0.23	1112
1213	0.13 max	0.70 to 1.00	0.07 to 0.12	0.24 to 0.33	1113
1215	0.09 max	0.75 to 1.05	0.04 to 0.09	0.26 to 0.35	...
12L14[a]	0.15 max	0.85 to 1.15	0.04 to 0.09	0.26 to 0.35	...

1522	0.18 to 0.24	1.10 to 1.40	0.040	0.050	1522
1023	0.20 to 0.25	0.30 to 0.60	0.040	0.050	1023
1524	0.19 to 0.25	1.35 to 1.65	0.040	0.050	1524
1025	0.22 to 0.28	0.30 to 0.60	0.040	0.050	1025
1525	0.23 to 0.29	0.80 to 1.10	0.040	0.050	1525
1026	0.22 to 0.28	0.60 to 0.90	0.040	0.050	1026
1526	0.22 to 0.29	1.10 to 1.40	0.040	0.050	1526
1527	0.22 to 0.29	1.20 to 1.50	0.040	0.050	1527
1029	0.25 to 0.31	0.60 to 0.90	0.040	0.050	...
1030	0.28 to 0.34	0.60 to 0.90	0.040	0.050	1030
1035	0.32 to 0.38	0.60 to 0.90	0.040	0.050	1035
1536	0.30 to 0.37	1.20 to 1.50	0.040	0.050	1536
1037	0.32 to 0.38	0.70 to 1.00	0.040	0.050	1037
1038	0.35 to 0.42	0.60 to 0.90	0.040	0.050	1038
1039	0.37 to 0.44	0.70 to 1.00	0.040	0.050	1039
1040	0.37 to 0.44	0.60 to 0.90	0.040	0.050	1040
1541	0.36 to 0.44	1.35 to 1.65	0.040	0.050	1541
1042	0.40 to 0.47	0.60 to 0.90	0.040	0.050	1042
1043	0.40 to 0.47	0.70 to 1.00	0.040	0.050	1043
1044	0.43 to 0.50	0.30 to 0.60	0.040	0.050	1044
1045	0.43 to 0.50	0.60 to 0.90	0.040	0.050	1045
1046	0.43 to 0.50	0.70 to 1.00	0.040	0.050	1046
1547	0.45 to 0.51	1.35 to 1.65	0.040	0.050	1547
1548	0.44 to 0.52	1.10 to 1.40	0.040	0.050	1548
1049	0.46 to 0.53	0.60 to 0.90	0.040	0.050	1049
1050	0.48 to 0.55	0.60 to 0.90	0.040	0.050	1050
1551	0.45 to 0.56	0.85 to 1.15	0.040	0.050	1551
1552	0.47 to 0.55	1.20 to 1.50	0.040	0.050	1552
1053	0.48 to 0.55	0.70 to 1.00	0.040	0.050	...
1055	0.50 to 0.60	0.60 to 0.90	0.040	0.050	1055
1060	0.55 to 0.65	0.60 to 0.90	0.040	0.050	1060

Table VI.2 Compositions of AISI-SAE Standard Carbon Steels

| | Free-Machining Grades | | | | | | Nonresulfurized Grades | | | | |
| | Composition (%)[b] | | | | | | Composition (%)[b] | | | | |
AISI No.[a]	C	Mn	P	S	SAE No.	AISI No.[a]	C	Mn	P (max)	S (max)	SAE No.
						1561	0.55 to 0.65	0.75 to 1.05	0.040	0.050	1561
						1566	0.60 to 0.71	0.85 to 1.15	0.040	0.050	1566
						1070	0.65 to 0.75	0.60 to 0.90	0.040	0.050	1070
						1572	0.65 to 0.76	1.00 to 1.30	0.040	0.050	1572
						1078	0.72 to 0.85	0.30 to 0.60	0.040	0.050	1078
						1080	0.75 to 0.88	0.60 to 0.90	0.040	0.050	1080
						1084	0.80 to 0.93	0.60 to 0.90	0.040	0.050	1084
						1090	0.85 to 0.98	0.60 to 0.90	0.040	0.050	1090
						1095	0.90 to 1.03	0.30 to 0.50	0.040	0.050	1095

[a]Prefix B indicates acid bessemer steel; others are made by the electric furnace, open hearth, or basic oxygen furnace processes.

[b]When silicon is required, the following ranges and limits are commonly used: for nonresulfurized steels: up to 1015, 0.10% max; 1015 to 1025, 0.10% max, or 0.10 to 0.20%, or 0.15 to 0.30%; over 1025, 0.10 to 0.20%, or 0.15 to 0.30%; 1513 to 1524, 0.10% max, or 0.10 to 0.20%, or 0.15 to 0.30%; 1525 and over, 0.10 to 0.20%, or 0.15 to 0.30%. For resulfurized steels: to 1110, 0.10% max; 1116 and over, 0.10% max, or 0.10 to 0.20%, or 0.15 to 0.30%. It is not common practice to produce resulfurized and rephosphorized steels to specified limits for silicon because of its adverse effect on machinability. Copper can be added to a standard steel.

[c]0.15 to 0.35% Pb. When lead is required as an added element to a standard steel, a range of 0.15 to 0.35%, inclusive, is generally used. Such a steel is identified by inserting the letter "L" between the second and third numeral of the AISI number.

Sources: American Iron & Steel Institute, Washington, D.C.; SAE Standard J403c.

Table VI.3 Compositions of Standard Carbon H-Steels

SAE or AISI	Carbon (%)	Manganese (%)	Max Phosphorus (%)	Max Sulfur (%)	Silicon (%)
1038H	0.34/0.43	0.50/1.00	0.040	0.050	0.15/0.30
1045H	0.42/0.51	0.50/1.00	0.040	0.050	0.15/0.30
1522H	0.17/0.25	1.00/1.50	0.040	0.050	0.15/0.30
1524H	0.18/0.26	1.25/1.75	0.040	0.050	0.15/0.30
1526H	0.21/0.30	1.00/1.50	0.040	0.050	0.15/0.30
1541H	0.35/0.45	1.25/1.75	0.040	0.050	0.15/0.30

Table VI.4 Compositions of Standard Carbon Boron H-Steels[a]

15B21H	0.17/0.24	0.70/1.20	0.035	0.045	0.15/0.30
15B28H	0.25/0.34	1.00/1.50	0.035	0.045	0.15/0.30
15B30H	0.26/0.34	0.70/1.00	0.035	0.045	0.15/0.30
15B35H	0.31/0.39	0.70/1.20	0.035	0.045	0.15/0.30
15B37H	0.30/0.39	1.00/1.50	0.035	0.045	0.15/0.30
15B41H	0.36/0.44	1.25/1.75	0.035	0.045	0.15/0.30
15B46H	0.42/0.51	0.70/1.20	0.035	0.045	0.15/0.30
15B62H	0.54/0.67	1.00/1.50	0.035	0.045	0.15/0.30

[a]These steels can be expected to have 0.0005% minimum boron content.

Table VI.5 Hardness Limits for Specification Purposes, Carbon H Steels

Hardness - Rockwell C

Jominy Distance (in./16)	1038 H		1045 H		1522 H		1524 H		1526 H		1541 H	
	Min	Max	Min	Max	Min	Max	Min	Max	Min	Max	Min	Max
1.0	51	58	55	62	41	50	42	51	44	53	53	60
1.5	42	56	52	61	41	48	42	49	42	50	52	59
2.0	34	55	42	59	32	47	38	48	38	49	50	59
2.5	29	53	34	56	27	46	34	47	33	47	47	58
3.0	26	49	31	52	22	45	29	45	26	46	44	57
3.5	24	43	29	46	21	42	25	43	25	42	41	56
4.0	23	37	28	38	20	39	22	39	21	39	38	55
4.5	22	33	27	34	...	37	20	38	20	37	35	53
5.0	22	30	26	33	...	34	...	35	...	33	32	52
5.5	21	29	26	32	...	32	...	34	...	31	29	50
6.0	21	28	25	32	...	30	...	32	...	30	27	48
6.5	20	27	25	31	...	28	...	30	...	28	26	46
7.0	...	27	25	31	...	27	...	29	...	27	25	44
7.5	...	26	24	30	...	...	...	28	...	26	24	41
8.0	...	26	24	30	...	...	...	27	...	26	23	39
9.0	...	25	23	29	...	...	...	26	...	24	23	35
10.0	...	25	22	29	...	...	...	25	...	24	22	33
12.0	...	25	21	28	...	...	...	23	...	23	21	32
14.0	...	24	20	27	...	...	...	22	...	...	20	31
16.0	...	21	...	26	...	...	...	...	...	...	...	30

Table VI.6 Hardness Limits for Specification Purposes, Carbon Boron H Steels

Jominy Distance (in./16)	15B21 H		15B28 H		15B30 H		15B35 H		15B37 H		15B41 H		15B46 H		15B62 H	
	Max	Min	Max	Min	Max	Min	Max	Min	Max	Min	Max	Min	Max	Min	Max	Min
1	49	42	56	49	56	48	58	51	58	50	61	53	63	56	65	60
2	48	41	56	48	56	47	56	50	56	50	61	53	62	54	65	60
3	47	38	55	47	55	46	55	49	55	49	60	52	61	53	65	60
4	45	32	54	45	53	42	54	48	54	48	60	51	59	51	65	60
5	40	24	53	43	51	28	53	39	53	43	59	51	57	36	65	60
6	35	...	51	39	45	22	51	28	52	37	59	50	53	30	65	59
7	30	...	49	29	38	...	47	24	51	33	58	48	49	28	65	58
8	27	...	45	22	33	...	41	22	50	26	57	45	43	26	64	57
10	23	...	36	...	27	...	30	20	45	22	56	32	38	23	63	48
12	20	...	29	...	24	...	27	...	40	21	53	28	34	21	63	38
14	...	...	27	...	21	...	26	...	33	20	50	25	32	...	57	34
16	...	...	25	...	20	...	25	...	29	...	47	23	30	...	47	31
20	...	...	22	...	...	...	24	...	27	...	40	20	28	...	41	28
24	...	...	20	...	...	...	22	...	25	...	35	...	26	...	39	26
28	...	...	...	...	...	...	20	...	23	...	30	...	24	...	37	25
32	...	...	...	...	...	...	...	...	...	...	28	...	22	...	35	24

HARDENABILITY BAND 1330 H

C	Mn	Si
.27/.33	1.45/2.05	.20/.35

DIAMETERS OF ROUNDS WITH SAME AS QUENCHED HARDNESS

Diameters									LOCATION IN ROUND	QUENCH
3.8									SURFACE	MILD WATER QUENCH
1.1	2.0	2.9	3.8	4.8	5.8	6.7			3/4 RADIUS FROM CENTER	
0.7	1.2	1.6	2.0	2.4	2.8	3.2	3.6	3.9	CENTER	
0.8	1.8	2.5	3.0	3.4	3.8				SURFACE	MILD OIL QUENCH
0.5	1.0	1.6	2.0	2.4	2.8	3.2	3.6	4.0	3/4 RADIUS FROM CENTER	
0.2	0.6	1.0	1.4	1.7	2.0	2.4	2.8	3.1	CENTER	

ROCKWELL HARDNESS C SCALE

DISTANCE FROM QUENCHED END — SIXTEENTHS OF AN INCH

HARDNESS LIMITS FOR SPECIFICATION PURPOSES

"J" DISTANCE SIXTEENTHS OF AN INCH	1330 H MAX.	1330 H MIN.
1	56	49
2	56	47
3	55	44
4	53	40
5	52	35
6	50	31
7	48	28
8	45	26
9	43	25
10	42	23
11	40	22
12	39	21
13	38	20
14	37	-
15	36	-
16	35	-
18	34	-
20	33	-
22	32	-
24	31	-
26	31	-
28	31	-
30	30	-
32	30	-

HEAT TREATING TEMPERATURES
RECOMMENDED BY SAE

*NORMALIZE 1650 °F
AUSTENITIZE 1600 °F

*For forged or rolled specimens only.

342

HARDENABILITY BAND ___ 1335 H

C	Mn	Si
.32 / .38	1.45 / 2.05	.20 / .35

HARDNESS LIMITS FOR SPECIFICATION PURPOSES

"J" DISTANCE SIXTEENTHS OF AN INCH	1335 H MAX.	MIN.
1	58	51
2	57	49
3	56	47
4	55	44
5	54	38
6	52	34
7	50	31
8	48	29
9	46	27
10	44	26
11	42	25
12	41	24
13	40	23
14	39	22
15	38	22
16	37	21
18	35	20
20	34	-
22	33	-
24	32	-
26	31	-
28	31	-
30	30	-
32	30	-

HEAT TREATING TEMPERATURES RECOMMENDED BY SAE
*NORMALIZE 1600 °F
AUSTENITIZE 1550 °F

*For forged or rolled specimens only.

DIAMETERS OF ROUNDS WITH SAME AS QUENCHED HARDNESS

LOCATION IN ROUND	QUENCH								
SURFACE	MILD WATER QUENCH	3.8							
3/4 RADIUS FROM CENTER		1.1	2.0	2.9	3.8	4.8	5.8	6.7	
CENTER		0.7	1.2	1.6	2.0	2.4	2.8	3.2	3.6	3.9
SURFACE	MILD OIL QUENCH	0.8	1.8	2.5	3.0	3.4	3.8		
3/4 RADIUS FROM CENTER		0.5	1.0	1.6	2.0	2.4	2.8	3.2	3.6	4.0
CENTER		0.2	0.6	1.0	1.4	1.7	2.0	2.4	2.8	3.1

ROCKWELL HARDNESS C SCALE (20 – 65)

DISTANCE FROM QUENCHED END — SIXTEENTHS OF AN INCH (2 4 6 8 10 12 14 16 18 20 22 24 26 28 30 32)

Constructional Steels for Heat Treating — United States (Alloy Bars, Blooms, Billets, and Slabs)

343

HARDENABILITY BAND 1340 H

C	Mn	Si
.37 / .44	1.45 / 2.05	.20 / .35

DIAMETERS OF ROUNDS WITH SAME AS QUENCHED HARDNESS

Diameter values	LOCATION IN ROUND	QUENCH
3.8	SURFACE	MILD WATER QUENCH
1.1 2.0 2.9 3.8 4.8 5.8 6.7	3/4 RADIUS FROM CENTER	
0.7 1.2 1.6 2.0 2.4 2.8 3.2 3.6 3.9	CENTER	
3.8	SURFACE	MILD OIL QUENCH
0.8 1.8 2.5 3.0 3.4 3.8	3/4 RADIUS FROM CENTER	
0.5 1.0 1.6 2.0 2.4 2.8 3.2 3.6 4.0	CENTER	
0.2 0.6 1.0 1.4 1.7 2.0 2.4 2.8 3.1		

ROCKWELL HARDNESS C SCALE (65 – 20)

DISTANCE FROM QUENCHED END – SIXTEENTHS OF AN INCH (2 – 32)

HARDNESS LIMITS FOR SPECIFICATION PURPOSES

"J" DISTANCE SIXTEENTHS OF AN INCH	1340 H MAX.	MIN.
1	60	53
2	60	52
3	59	51
4	58	49
5	57	46
6	56	40
7	55	35
8	54	33
9	52	31
10	51	29
11	50	28
12	48	27
13	46	26
14	44	25
15	42	25
16	41	24
18	39	23
20	38	23
22	37	22
24	36	22
26	35	21
28	35	21
30	34	20
32	34	20

HEAT TREATING TEMPERATURES
RECOMMENDED BY SAE
*NORMALIZE 1600 °F
AUSTENITIZE 1550 °F

*For forged or rolled specimens only.

344

HARDENABILITY BAND 1345 H

C	Mn	Si	Cr	Ni	Mo
.42/.49	1.45/2.05	.20/.35			

DIAMETERS OF ROUNDS WITH SAME AS QUENCHED HARDNESS

Diameters									LOCATION IN ROUND	QUENCH
3.8									SURFACE	MILD WATER QUENCH
1.1	2.0	2.9	3.8	4.8	5.8	6.7			3/4 RADIUS FROM CENTER	
0.7	1.2	1.6	2.0	2.4	2.8	3.2	3.9		CENTER	
0.8	1.8	2.5	3.0	3.4	3.8				SURFACE	MILD OIL QUENCH
0.5	1.0	1.6	2.0	2.4	2.8	3.2	3.6	4.0	3/4 RADIUS FROM CENTER	
0.2	0.6	1.0	1.4	1.7	2.0	2.4	2.8	3.1	CENTER	

ROCKWELL HARDNESS C SCALE

DISTANCE FROM QUENCHED END — SIXTEENTHS OF AN INCH

HARDNESS LIMITS FOR SPECIFICATION PURPOSES

"J" DISTANCE SIXTEENTHS OF AN INCH	1345 H MAX.	MIN.
1	63	56
2	63	56
3	62	55
4	61	54
5	61	51
6	60	44
7	60	38
8	59	35
9	58	33
10	57	32
11	56	31
12	55	30
13	54	29
14	53	29
15	52	28
16	51	28
18	49	27
20	47	27
22	45	26
24	44	26
26	47	25
28	46	25
30	45	24
32	45	24

HEAT TREATING TEMPERATURES RECOMMENDED BY SAE

*NORMALIZE 1600 °F

AUSTENITIZE 1550 °F

*For forged or rolled specimens only.

Constructional Steels for Heat Treating — United States (Alloy Bars, Blooms, Billets, and Slabs)

HARDENABILITY BAND 4027 H & 4028 H*

C	Mn	Si	Mo
.24 / .30	.60 / 1.00	.20 / .35	.20 / .30

*SULPHUR CONTENT 0.035/0.050

DIAMETERS OF ROUNDS WITH SAME AS QUENCHED HARDNESS									LOCATION IN ROUND	QUENCH
3.8									SURFACE	MILD
1.1	2.0	2.9	3.8	4.8	5.8	6.7			3/4 RADIUS FROM CENTER	WATER
0.7	1.2	1.6	2.0	2.4	2.8	3.2	3.6	3.9	CENTER	QUENCH
0.8	1.8	2.5	3.0	3.4	3.8				SURFACE	MILD
0.5	1.0	1.6	2.0	2.4	2.8	3.2	3.6	4.0	3/4 RADIUS FROM CENTER	OIL
0.2	0.6	1.0	1.4	1.7	2.0	2.4	2.8	3.1	CENTER	QUENCH

ROCKWELL HARDNESS C SCALE — DISTANCE FROM QUENCHED END — SIXTEENTHS OF AN INCH

HARDNESS LIMITS FOR SPECIFICATION PURPOSES

"J" DISTANCE SIXTEENTHS OF AN INCH	4027H & 4028H MAX.	MIN.
1	52	45
2	50	40
3	46	31
4	40	25
5	34	22
6	30	20
7	28	–
8	26	–
9	25	–
10	25	–
11	24	–
12	23	–
13	23	–
14	22	–
15	22	–
16	21	–
18	21	–
20	20	–
22	–	–
24	–	–
26	–	–
28	–	–
30	–	–
32	–	–

HEAT TREATING TEMPERATURES
RECOMMENDED BY SAE
*NORMALIZE 1650 °F
AUSTENITIZE 1600 °F

*For forged or rolled specimens only.

346

HARDENABILITY BAND 4037 H

C	Mn	Si	Mo
.34 / .41	.60 / 1.00	.20 / .35	.20 / .30

DIAMETERS OF ROUNDS WITH SAME AS QUENCHED HARDNESS									LOCATION IN ROUND	QUENCH
3.8									SURFACE	MILD
1.1	2.0	2.9	3.8	4.8	5.8	6.7			3/4 RADIUS FROM CENTER	WATER QUENCH
0.7	1.2	1.6	2.0	2.4	2.8	3.2	3.6	3.9	CENTER	
0.8	1.8	2.5	3.0	3.4	3.8				SURFACE	MILD
0.5	1.0	1.6	2.0	2.4	2.8	3.2	3.6	4.0	3/4 RADIUS FROM CENTER	OIL QUENCH
0.2	0.6	1.0	1.4	1.7	2.0	2.4	2.8	3.1	CENTER	

ROCKWELL HARDNESS C SCALE

DISTANCE FROM QUENCHED END —SIXTEENTHS OF AN INCH

HARDNESS LIMITS FOR SPECIFICATION PURPOSES

"J" DISTANCE SIXTEENTHS OF AN INCH	4037 H MAX.	4037 H MIN.
1	59	52
2	57	49
3	54	42
4	51	35
5	45	30
6	38	26
7	34	23
8	32	22
9	30	21
10	29	20
11	28	-
12	27	-
13	26	-
14	26	-
15	26	-
16	25	-
18	25	-
20	25	-
22	25	-
24	24	-
26	24	-
28	24	-
30	23	-
32	23	-

HEAT TREATING TEMPERATURES RECOMMENDED BY SAE
*NORMALIZE 1600 °F
AUSTENITIZE 1550 °F

*For forged or rolled specimens only.

Constructional Steels for Heat Treating – United States (Alloy Bars, Blooms, Billets, and Slabs)

347

HARDENABILITY BAND 4047 H

C	Mn	Si	Mo
.44/.51	.60/1.00	.20/.35	.20/.30

DIAMETERS OF ROUNDS WITH SAME AS QUENCHED HARDNESS	LOCATION IN ROUND	QUENCH
3.8	SURFACE	MILD WATER QUENCH
1.1 2.0 2.9 3.8 4.8 5.8 6.7	3/4 RADIUS FROM CENTER	
0.7 1.2 1.6 2.0 2.4 2.8 3.2 3.6 3.9	CENTER	
0.8 1.8 2.5 3.0 3.4 3.8	SURFACE	MILD OIL QUENCH
0.5 1.0 1.6 2.0 2.4 2.8 3.2 3.6 4.0	3/4 RADIUS FROM CENTER	
0.2 0.6 1.0 1.4 1.7 2.0 2.4 2.8 3.1	CENTER	

ROCKWELL HARDNESS C SCALE — (20 to 65)

DISTANCE FROM QUENCHED END —SIXTEENTHS OF AN INCH (2 to 32)

HARDNESS LIMITS FOR SPECIFICATION PURPOSES

"J" DISTANCE SIXTEENTHS OF AN INCH	4047 H MAX.	4047 H MIN.
1	64	57
2	62	55
3	60	50
4	58	42
5	55	35
6	52	32
7	47	30
8	43	28
9	40	28
10	38	27
11	37	26
12	35	26
13	34	25
14	33	25
15	33	25
16	32	25
18	31	24
20	30	24
22	30	23
24	30	23
26	30	22
28	29	22
30	29	21
32	29	21

HEAT TREATING TEMPERATURES
RECOMMENDED BY SAE
*NORMALIZE   1600 °F
AUSTENITIZE 1550 °F

*For forged or rolled specimens only.

HARDENABILITY BAND — 4118 H

C	Mn	Si	Cr	Mo
.17 / .23	.60 / 1.00	.20 / .35	.30 / .70	.08 / .15

DIAMETERS OF ROUNDS WITH SAME AS QUENCHED HARDNESS

DIAMETERS OF ROUNDS WITH SAME AS QUENCHED HARDNESS									LOCATION IN ROUND	QUENCH
3.8										
1.1	2.0	2.9	3.8	4.8	5.8	6.7			SURFACE	MILD WATER QUENCH
0.7	1.2	1.6	2.0	2.4	2.8	3.2	3.6	3.9	3/4 RADIUS FROM CENTER	
									CENTER	
0.8	1.8	2.5	3.0	3.4	3.8				SURFACE	MILD OIL QUENCH
0.5	1.0	1.8	2.0	2.4	2.8	3.2	3.6	4.0	3/4 RADIUS FROM CENTER	
0.2	0.6	1.0	1.4	1.7	2.0	2.4	2.8	3.1	CENTER	

ROCKWELL HARDNESS C SCALE: 65 60 55 50 45 40 35 30 25 20

DISTANCE FROM QUENCHED END — SIXTEENTHS OF AN INCH: 2 4 6 8 10 12 14 16 18 20 22 24 26 28 30 32

HARDNESS LIMITS FOR SPECIFICATION PURPOSES

"J" DISTANCE SIXTEENTHS OF AN INCH	4118 H MAX.	4118 H MIN.
1	48	41
2	46	36
3	41	27
4	35	23
5	31	20
6	28	–
7	27	–
8	25	–
9	24	–
10	23	–
11	22	–
12	21	–
13	21	–
14	20	–
15	–	–
16	–	–
18	–	–
20	–	–
22	–	–
24	–	–
26	–	–
28	–	–
30	–	–
32	–	–

HEAT TREATING TEMPERATURES RECOMMENDED BY SAE
*NORMALIZE 1700 °F
AUSTENITIZE 1700 °F

*For forged or rolled specimens only.

Constructional Steels for Heat Treating — United States (Alloy Bars, Blooms, Billets, and Slabs)

HARDENABILITY BAND   4130 H

C	Mn	Si	Cr	Mo
.27 / .33	.30 / .70	.20 / .35	.75 / 1.20	.15 / .25

DIAMETERS OF ROUNDS WITH SAME AS QUENCHED HARDNESS

									LOCATION IN ROUND	QUENCH
3.8										MILD
1.1	2.0	2.9	3.8	4.8	5.8	6.7			SURFACE	WATER
0.7	1.2	1.6	2.0	2.4	2.8	3.2	3.6	3.9	3/4 RADIUS FROM CENTER	QUENCH
0.8	1.8	2.5	3.0	3.4	3.8				CENTER	
0.8	1.8	2.5	3.0	3.4	3.8				SURFACE	MILD
0.5	1.0	1.6	2.0	2.4	2.8	3.2	3.6	4.0	3/4 RADIUS FROM CENTER	OIL
0.2	0.6	1.0	1.4	1.7	2.0	2.4	2.8	3.1	CENTER	QUENCH

ROCKWELL HARDNESS C SCALE

DISTANCE FROM QUENCHED END – SIXTEENTHS OF AN INCH

HARDNESS LIMITS FOR SPECIFICATION PURPOSES

"J" DISTANCE SIXTEENTHS OF AN INCH	4130 H MAX.	4130 H MIN.
1	56	49
2	55	46
3	53	42
4	51	38
5	49	34
6	47	31
7	44	29
8	42	27
9	40	26
10	38	26
11	36	25
12	35	25
13	34	24
14	34	24
15	33	23
16	33	23
18	32	22
20	32	21
22	32	20
24	31	-
26	31	-
28	30	-
30	30	-
32	29	-

HEAT TREATING TEMPERATURES
RECOMMENDED BY SAE
*NORMALIZE 1650 °F
AUSTENITIZE 1600 °F

*For forged or rolled specimens only.

HARDENABILITY BAND ___ 4137 H

C	Mn	Si	Cr	Mo
.34 / .41	.60 / 1.00	.20 / .35	.75 / 1.20	.15 / .25

HARDNESS LIMITS FOR SPECIFICATION PURPOSES		
"J" DISTANCE SIXTEENTHS OF AN INCH	4137 H MAX.	MIN.
1	59	52
2	59	51
3	58	50
4	58	49
5	57	49
6	57	48
7	56	45
8	55	43
9	55	40
10	54	39
11	53	37
12	52	36
13	51	35
14	50	34
15	49	33
16	48	33
18	46	32
20	45	31
22	44	30
24	43	30
26	42	30
28	42	29
30	41	29
32	41	28

HEAT TREATING TEMPERATURES RECOMMENDED BY SAE
*NORMALIZE 1600 °F
AUSTENITIZE 1550 °F

*For forged or rolled specimens only.

Constructional Steels for Heat Treating – United States (Alloy Bars, Blooms, Billets, and Slabs)

351

HARDENABILITY BAND — 4140 H

C	Mn	Si	Cr	Mo
.37 / .44	.65 / 1.10	.20 / .35	.75 / 1.20	.15 / .25

DIAMETERS OF ROUNDS WITH SAME AS QUENCHED HARDNESS

LOCATION IN ROUND	QUENCH
SURFACE	MILD WATER QUENCH
3/4 RADIUS FROM CENTER	
CENTER	
SURFACE	MILD OIL QUENCH
3/4 RADIUS FROM CENTER	
CENTER	

3.8								
1.1	2.0	2.9	3.8	4.8	5.8	6.7		
0.7	1.2	1.8	2.0	2.4	2.8	3.2	3.6	3.9
0.8	1.8	2.5	3.0	3.4	3.8			
0.5	1.0	1.6	2.0	2.4	2.8	3.2	3.6	4.0
0.2	0.6	1.0	1.4	1.7	2.0	2.4	2.8	3.1

ROCKWELL HARDNESS C SCALE vs. DISTANCE FROM QUENCHED END — SIXTEENTHS OF AN INCH

HARDNESS LIMITS FOR SPECIFICATION PURPOSES

"J" DISTANCE SIXTEENTHS OF AN INCH	4140 H MAX.	4140 H MIN.
1	60	53
2	60	53
3	60	52
4	59	51
5	59	51
6	58	50
7	58	48
8	57	47
9	57	44
10	56	42
11	56	40
12	55	39
13	55	38
14	54	37
15	54	36
16	53	35
18	52	34
20	51	33
22	49	33
24	48	32
26	47	32
28	46	31
30	45	31
32	44	30

HEAT TREATING TEMPERATURES RECOMMENDED BY SAE

*NORMALIZE 1600 °F
AUSTENITIZE 1550 °F

*For forged or rolled specimens only.

352

HARDENABILITY BAND — 4142 H

C	Mn	Si	Cr	Mo
.39 / .46	.65 / 1.10	.20 / .35	.75 / 1.20	.15 / .25

DIAMETERS OF ROUNDS WITH SAME AS QUENCHED HARDNESS

									LOCATION IN ROUND	QUENCH
3.8									SURFACE	MILD WATER QUENCH
1.1	2.0	2.9	3.8	4.8	5.8	6.7			3/4 RADIUS FROM CENTER	
0.7	1.2	1.6	2.0	2.4	2.8	3.2	3.6	3.9	CENTER	
0.8	1.8	2.5	3.0	3.4	3.8				SURFACE	MILD OIL QUENCH
0.5	0.8	1.6	2.0	2.4	2.8	3.2	3.6	4.0	3/4 RADIUS FROM CENTER	
0.2	0.6	1.0	1.4	1.7	2.0	2.4	2.8	3.1	CENTER	

ROCKWELL HARDNESS C SCALE (20–65) vs DISTANCE FROM QUENCHED END — SIXTEENTHS OF AN INCH (2–32)

HARDNESS LIMITS FOR SPECIFICATION PURPOSES

"J" DISTANCE SIXTEENTHS OF AN INCH	4142 H MAX.	MIN.
1	62	55
2	62	55
3	62	54
4	61	53
5	61	53
6	61	52
7	60	51
8	60	50
9	60	49
10	59	47
11	59	46
12	58	44
13	58	42
14	57	41
15	57	40
16	56	39
18	55	37
20	54	36
22	53	35
24	53	34
26	52	34
28	51	34
30	51	33
32	50	33

HEAT TREATING TEMPERATURES RECOMMENDED BY SAE
*NORMALIZE 1600 °F
AUSTENITIZE 1550 °F

*For forged or rolled specimens only.

Constructional Steels for Heat Treating — United States (Alloy Bars, Blooms, Billets, and Slabs)

353

HARDENABILITY BAND ___ 4145 H

C	Mn	Si	Cr	Mo
.42 / .49	.65 / 1.10	.20 / .35	.75 / 1.20	.15 / .25

DIAMETERS OF ROUNDS WITH SAME AS QUENCHED HARDNESS

	LOCATION IN ROUND	QUENCH
	SURFACE	MILD
	3/4 RADIUS FROM CENTER	WATER
	CENTER	QUENCH
	SURFACE	MILD
	3/4 RADIUS FROM CENTER	OIL
	CENTER	QUENCH

3.8

1.1 2.0 2.9 3.8 4.8 5.8 6.7
0.7 1.2 1.6 2.0 2.4 2.8 3.2 3.6 3.9
0.8 1.8 2.5 3.0 3.4 3.8
0.5 1.0 1.6 2.0 2.4 2.8 3.2 3.6 4.0
0.2 0.6 1.0 1.4 1.7 2.0 2.4 2.8 3.1

ROCKWELL HARDNESS C SCALE (20–65)

DISTANCE FROM QUENCHED END — SIXTEENTHS OF AN INCH (2–32)

HARDNESS LIMITS FOR SPECIFICATION PURPOSES

"J" DISTANCE SIXTEENTHS OF AN INCH	4145 H MAX.	4145 H MIN.
1	63	56
2	63	55
3	62	55
4	62	54
5	62	53
6	61	53
7	61	52
8	61	52
9	60	51
10	60	50
11	60	49
12	59	48
13	59	46
14	59	45
15	58	43
16	58	42
18	57	40
20	57	38
22	56	37
24	55	36
26	55	35
28	55	35
30	55	34
32	54	34

HEAT TREATING TEMPERATURES RECOMMENDED BY SAE

*NORMALIZE 1600 °F
AUSTENITIZE 1550 °F

*For forged or rolled specimens only.

354

HARDENABILITY BAND 4147 H

C	Mn	Si	Cr	Mo
.44 / .51	.65 / 1.10	.20 / .35	.75 / 1.20	.15 / .25

DIAMETERS OF ROUNDS WITH SAME AS QUENCHED HARDNESS

Diameters									LOCATION IN ROUND	QUENCH
3.8									SURFACE	MILD WATER QUENCH
1.1	2.0	2.9	3.8	4.8	5.8	6.7			3/4 RADIUS FROM CENTER	
0.7	1.2	1.6	2.0	2.4	2.8	3.2	3.6	3.9	CENTER	
0.8	1.8	2.5	3.0	3.4	3.8				SURFACE	MILD OIL QUENCH
0.5	1.0	1.6	2.0	2.4	2.8	3.2	3.6	4.0	3/4 RADIUS FROM CENTER	
0.2	0.6	1.0	1.4	1.7	2.0	2.4	2.8	3.1	CENTER	

ROCKWELL HARDNESS C SCALE

DISTANCE FROM QUENCHED END — SIXTEENTHS OF AN INCH

HARDNESS LIMITS FOR SPECIFICATION PURPOSES

"J" DISTANCE SIXTEENTHS OF AN INCH	4147 H MAX.	4147 H MIN.
1	64	57
2	64	57
3	64	56
4	64	56
5	63	55
6	63	55
7	63	55
8	63	54
9	63	54
10	62	53
11	62	52
12	62	51
13	61	49
14	61	48
15	60	46
16	60	45
18	59	42
20	59	40
22	58	39
24	57	38
26	57	37
28	57	37
30	56	37
32	56	36

HEAT TREATING TEMPERATURES RECOMMENDED BY SAE

*NORMALIZE 1600 °F
AUSTENITIZE 1550 °F

*For forged or rolled specimens only.

Constructional Steels for Heat Treating — United States (Alloy Bars, Blooms, Billets, and Slabs)

355

HARDENABILITY BAND — 4150 H

C	Mn	Si	Cr	Mo
.47 / .54	.65 / 1.10	.20 / .35	.75 / 1.20	.15 / .25

DIAMETERS OF ROUNDS WITH SAME AS QUENCHED HARDNESS

							LOCATION IN ROUND	QUENCH
3.8								
1.1	2.0	2.9	3.8	4.8	5.8	6.7	SURFACE	MILD WATER QUENCH
0.7	1.2	1.6	2.0	2.4	2.8	3.2	3.6	3.9	3/4 RADIUS FROM CENTER	
0.8	1.8	2.5	3.0	3.4	3.8		CENTER	
0.5	1.0	1.6	2.0	2.4	2.8	3.2	3.6	4.0	SURFACE	MILD OIL QUENCH
0.2	0.6	1.0	1.4	1.7	2.0	2.4	2.8	3.1	3/4 RADIUS FROM CENTER	
							CENTER	

DISTANCE FROM QUENCHED END — SIXTEENTHS OF AN INCH

ROCKWELL HARDNESS C SCALE

HARDNESS LIMITS FOR SPECIFICATION PURPOSES

"J" DISTANCE SIXTEENTHS OF AN INCH	4150 H MAX.	4150 H MIN.
1	65	59
2	65	59
3	65	59
4	65	58
5	65	58
6	65	57
7	65	57
8	64	56
9	64	56
10	64	55
11	64	54
12	63	53
13	63	51
14	62	50
15	62	48
16	62	47
18	61	45
20	60	43
22	59	41
24	59	40
26	58	39
28	58	38
30	58	38
32	58	38

HEAT TREATING TEMPERATURES RECOMMENDED BY SAE

*NORMALIZE   1600 °F
AUSTENITIZE 1550 °F

*For forged or rolled specimens only.

356

HARDENABILITY BAND   4161 H

C	Mn	Si	Ni	Cr	Mo
.55/.65	.65/1.10	.20/.35		.65/.95	.25/.35

DIAMETERS OF ROUNDS WITH SAME AS QUENCHED HARDNESS

LOCATION IN ROUND	QUENCH
SURFACE	MILD WATER QUENCH
3/4 RADIUS FROM CENTER	
CENTER	
SURFACE	MILD OIL QUENCH
3/4 RADIUS FROM CENTER	
CENTER	

Diameter values:

3.8									
1.1	2.0	2.9	3.8	4.8	5.8	6.7			
0.7	1.2	1.6	2.0	2.4	2.8	3.2	3.6	3.9	
0.8	1.8	2.5	3.0	3.4	3.8				
0.8	1.0	1.6	2.0	2.4	2.8	3.2	3.6	4.0	
0.5	1.0	1.6	2.0	2.4	2.8	2.4	2.8	3.1	
0.2	0.6	1.0	1.4	1.7	2.0	2.0	2.4	2.8	3.1

ROCKWELL HARDNESS C SCALE (65 60 55 50 45 40 35 30 25 20)

DISTANCE FROM QUENCHED END —SIXTEENTHS OF AN INCH (2 4 6 8 10 12 14 16 18 20 22 24 26 28 30 32)

HARDNESS LIMITS FOR SPECIFICATION PURPOSES

"J" DISTANCE SIXTEENTHS OF AN INCH	4161 H MAX.	MIN.
1	65	60
2	65	60
3	65	60
4	65	60
5	65	60
6	65	60
7	65	60
8	65	60
9	65	59
10	65	59
11	65	59
12	64	59
13	64	58
14	64	58
15	64	57
16	64	56
18	64	55
20	63	53
22	63	50
24	63	48
26	63	45
28	63	43
30	63	42
32	63	41

HEAT TREATING TEMPERATURES RECOMMENDED BY SAE
*NORMALIZE   1600 °F
AUSTENITIZE 1550 °F
*For forged or rolled specimens only.

Constructional Steels for Heat Treating – United States (Alloy Bars, Blooms, Billets, and Slabs)

357

HARDENABILITY BAND 4320 H

C	Mn	Si	Ni	Cr	Mo
.17/.23	.40/.70	.20/.35	1.55/2.00	.35/.65	.20/.30

DIAMETERS OF ROUNDS WITH SAME AS QUENCHED HARDNESS

									LOCATION IN ROUND	QUENCH
3.8										
1.1	2.0	2.9	3.8	4.8	5.8	6.7			SURFACE	MILD WATER QUENCH
0.7	1.2	1.6	2.0	2.4	2.8	3.2	3.6	3.9	3/4 RADIUS FROM CENTER	
									CENTER	
0.8	1.8	2.5	3.0	3.4	3.8				SURFACE	MILD OIL QUENCH
0.5	1.0	1.6	2.0	2.4	2.8	3.2	3.6	4.0	3/4 RADIUS FROM CENTER	
0.2	0.6	1.0	1.4	1.7	2.0	2.4	2.8	3.1	CENTER	

ROCKWELL HARDNESS C SCALE

DISTANCE FROM QUENCHED END — SIXTEENTHS OF AN INCH

HARDNESS LIMITS FOR SPECIFICATION PURPOSES

"J" DISTANCE SIXTEENTHS OF AN INCH	4320 H MAX.	MIN.
1	48	41
2	47	38
3	45	35
4	43	32
5	41	29
6	38	27
7	36	25
8	34	23
9	33	22
10	31	21
11	30	20
12	29	20
13	28	-
14	27	-
15	27	-
16	26	-
18	25	-
20	25	-
22	24	-
24	24	-
26	24	-
28	24	-
30	24	-
32	24	-

HEAT TREATING TEMPERATURES RECOMMENDED BY SAE

*NORMALIZE 1700 °F
AUSTENITIZE 1700 °F

*For forged or rolled specimens only.

358

HARDENABILITY BAND E4340 H

C	Mn	Si	Ni	Cr	Mo
.37/.44	.60/.95	.20/.35	1.55/2.00	.65/.95	.20/.30

DIAMETERS OF ROUNDS WITH SAME AS QUENCHED HARDNESS

Diameters									LOCATION IN ROUND	QUENCH
3.8									SURFACE	MILD
1.1	2.0	2.9	3.8	4.8	5.8	6.7			3/4 RADIUS FROM CENTER	WATER
0.7	1.2	1.6	2.0	2.4	2.8	3.2	3.6	3.9	CENTER	QUENCH
0.8	1.8	2.5	3.0	3.4	3.8				SURFACE	MILD
0.5	1.0	1.6	2.0	2.4	2.8	3.2	3.6	4.0	3/4 RADIUS FROM CENTER	OIL
0.2	0.6	1.0	1.4	1.7	2.0	2.4	2.8	3.1	CENTER	QUENCH

ROCKWELL HARDNESS C SCALE vs. DISTANCE FROM QUENCHED END — SIXTEENTHS OF AN INCH

HARDNESS LIMITS FOR SPECIFICATION PURPOSES

"J" DISTANCE SIXTEENTHS OF AN INCH	E4340 H MAX.	MIN.
1.	60	53
2	60	53
3	60	53
4	60	53
5	60	53
6	60	53
7	60	53
8	60	53
9	60	53
10	60	53
11	60	53
12	60	53
13	60	52
14	59	52
15	59	52
16	59	51
18	58	51
20	58	50
22	58	49
24	57	48
26	57	47
28	57	46
30	57	45
32	57	44

HEAT TREATING TEMPERATURES
RECOMMENDED BY SAE
*NORMALIZE 1600 °F
AUSTENITIZE 1550 °F

*For forged or rolled specimens only.

Constructional Steels for Heat Treating – United States (Alloy Bars, Blooms, Billets, and Slabs)

HARDENABILITY BAND 4340 H

C	Mn	Si	Ni	Cr	Mo
.37/.44	.55/.90	.20/.35	1.55/2.00	.65/.95	.20/.30

DIAMETERS OF ROUNDS WITH SAME AS QUENCHED HARDNESS

									LOCATION IN ROUND	QUENCH
3.8									SURFACE	MILD WATER QUENCH
1.1	2.0	2.9	3.8	4.8	5.8	6.7			3/4 RADIUS FROM CENTER	
0.7	1.2	1.6	2.0	2.4	2.8	3.2	3.6	3.9	CENTER	
0.8	1.8	2.5	3.0	3.4	3.8				SURFACE	MILD OIL QUENCH
0.5	1.0	1.6	2.0	2.4	2.8	3.2	3.6	4.0	3/4 RADIUS FROM CENTER	
0.2	0.6	1.0	1.4	1.7	2.0	2.4	2.8	3.1	CENTER	

ROCKWELL HARDNESS C SCALE

DISTANCE FROM QUENCHED END — SIXTEENTHS OF AN INCH

HARDNESS LIMITS FOR SPECIFICATION PURPOSES

"J" DISTANCE SIXTEENTHS OF AN INCH	4340 H MAX.	MIN.
1	60	53
2	60	53
3	60	53
4	60	53
5	60	53
6	60	53
7	60	53
8	60	52
9	60	52
10	60	52
11	59	51
12	59	51
13	59	50
14	58	49
15	58	49
16	58	48
18	58	47
20	57	46
22	57	45
24	57	44
26	57	43
28	56	42
30	56	41
32	56	40

HEAT TREATING TEMPERATURES
RECOMMENDED BY SAE
*NORMALIZE 1600 °F
AUSTENITIZE 1550 °F

*For forged or rolled specimens only.

360

HARDENABILITY BAND 4419H

C	Mn	Si	Ni	Cr	Mo
.17 / .23	.35 / .75	.20 / .35			.45 / .60

LOCATION IN ROUND	QUENCH
SURFACE	MILD
3/4 RADIUS FROM CENTER	WATER
CENTER	QUENCH
SURFACE	MILD
3/4 RADIUS FROM CENTER	OIL
CENTER	QUENCH

DIAMETERS OF ROUNDS WITH SAME AS QUENCHED HARDNESS

3.8								
1.1	2.0	2.9	3.8	4.8	5.8	6.7		
0.7	1.2	1.6	2.0	2.4	2.8	3.2	3.6	3.9
0.8	1.8	2.5	3.0	3.4	3.8			
0.5	1.0	1.6	2.0	2.4	2.8	3.2	3.6	4.0
0.2	0.6	1.0	1.4	1.7	2.0	2.4	2.8	3.1

ROCKWELL HARDNESS C SCALE

DISTANCE FROM QUENCHED END – SIXTEENTHS OF AN INCH

HARDNESS LIMITS FOR SPECIFICATION PURPOSES

"J" DISTANCE SIXTEENTHS OF AN INCH	4419 H MAX.	4419 H MIN.
1	48	40
2	45	33
3	41	27
4	34	23
5	30	21
6	28	20
7	27	-
8	25	-
9	25	-
10	24	-
11	24	-
12	23	-
13	23	-
14	22	-
15	22	-
16	21	-
18	21	-
20	20	-
22	-	-
24	-	-
26	-	-
28	-	-
30	-	-
32	-	-

HEAT TREATING TEMPERATURES RECOMMENDED BY SAE

*NORMALIZE 1700 °F
AUSTENITIZE 1700 °F

*For forged or rolled specimens only.

Constructional Steels for Heat Treating – United States (Alloy Bars, Blooms, Billets, and Slabs)

361

HARDENABILITY BAND ___4620 H

C	Mn	Si	Ni	Mo
.17 / .23	.35 / .75	.20 / .35	1.55 / 2.00	.20 / .30

DIAMETERS OF ROUNDS WITH SAME AS QUENCHED HARDNESS									LOCATION IN ROUND	QUENCH
3.8									SURFACE	MILD WATER QUENCH
1.1	2.0	2.9	3.8	4.8	5.8	6.7			3/4 RADIUS FROM CENTER	
0.7	1.2	1.6	2.0	2.4	2.8	3.2	3.6	3.9	CENTER	
0.8	1.8	2.5	3.0	3.4	3.8				SURFACE	MILD OIL QUENCH
0.5	1.0	1.6	2.0	2.4	2.8	3.2	3.6	4.0	3/4 RADIUS FROM CENTER	
0.2	0.6	1.0	1.4	1.7	2.0	2.4	2.8	3.1	CENTER	

ROCKWELL HARDNESS C SCALE: 20 25 30 35 40 45 50 55 60 65

DISTANCE FROM QUENCHED END — SIXTEENTHS OF AN INCH: 2 4 6 8 10 12 14 16 18 20 22 24 26 28 30 32

HARDNESS LIMITS FOR SPECIFICATION PURPOSES

"J" DISTANCE SIXTEENTHS OF AN INCH	4620 H MAX.	4620 H MIN.
1	48	41
2	45	35
3	42	27
4	39	24
5	34	21
6	31	-
7	29	-
8	27	-
9	26	-
10	25	-
11	24	-
12	23	-
13	22	-
14	22	-
15	22	-
16	21	-
18	21	-
20	20	-
22	-	-
24	-	-
26	-	-
28	-	-
30	-	-
32	-	-

HEAT TREATING TEMPERATURES RECOMMENDED BY SAE
*NORMALIZE 1700 °F
AUSTENITIZE 1700 °F

*For forged or rolled specimens only.

362

HARDENABILITY BAND 4621 H

C	Mn	Si	Ni	Mo
.17/.23	.60/1.00	.20/.35	1.55/2.00	.20/.30

DIAMETERS OF ROUNDS WITH SAME AS QUENCHED HARDNESS

LOCATION IN ROUND									QUENCH
SURFACE									MILD WATER QUENCH
3/4 RADIUS FROM CENTER									
CENTER									
SURFACE									MILD OIL QUENCH
3/4 RADIUS FROM CENTER									
CENTER									

3.8								
1.1	2.0	2.9	3.8	4.8	5.8	6.7		
0.7	1.2	1.6	2.0	2.4	2.8	3.2	3.6	3.9
0.8	1.0	1.8	2.5	3.0	3.4	3.8		
0.5	1.0	1.6	2.0	2.4	2.8	3.2	3.6	4.0
0.2	0.6	1.0	1.4	1.7	2.0	2.4	2.8	3.1

ROCKWELL HARDNESS C SCALE

DISTANCE FROM QUENCHED END — SIXTEENTHS OF AN INCH

HARDNESS LIMITS FOR SPECIFICATION PURPOSES

"J" DISTANCE SIXTEENTHS OF AN INCH	4621 H MAX.	MIN.
1	48	41
2	47	38
3	46	34
4	44	30
5	41	27
6	37	25
7	34	23
8	32	22
9	30	20
10	28	-
11	27	-
12	26	-
13	26	-
14	25	-
15	25	-
16	24	-
18	24	-
20	23	-
22	23	-
24	22	-
26	22	-
28	22	-
30	21	-
32	21	-

HEAT TREATING TEMPERATURES RECOMMENDED BY SAE.

*NORMALIZE 1700 °F
AUSTENITIZE 1700 °F

*For forged or rolled specimens only.

Constructional Steels for Heat Treating — United States (Alloy Bars, Blooms, Billets, and Slabs)

HARDENABILITY BAND 4626 ___H

C	Mn	Si	Ni	Cr	Mo
.23/.29	.40/.70	.20/.35	.65/1.05		.15/.25

DIAMETERS OF ROUNDS WITH SAME AS QUENCHED HARDNESS

									LOCATION IN ROUND	QUENCH
3.8										
1.1	2.0	2.9	3.8	4.8	5.8	6.7			SURFACE	MILD
0.7	1.2	1.6	2.0	2.4	2.8	3.2	3.6	3.9	3/4 RADIUS FROM CENTER	WATER QUENCH
0.8	1.8	2.5	3.0	3.4	3.8				CENTER	
0.8	1.8	2.5	3.0	3.4	3.8				SURFACE	MILD
0.5	1.0	1.6	2.0	2.4	2.8	3.2	3.6	4.0	3/4 RADIUS FROM CENTER	OIL QUENCH
0.2	0.6	1.0	1.4	1.7	2.0	2.4	2.8	3.1	CENTER	

ROCKWELL HARDNESS C SCALE

DISTANCE FROM QUENCHED END —SIXTEENTHS OF AN INCH

HARDNESS LIMITS FOR SPECIFICATION PURPOSES

"J" DISTANCE SIXTEENTHS OF AN INCH	4626 H MAX.	MIN.
1	51	45
2	48	36
3	41	29
4	33	24
5	29	21
6	27	-
7	25	-
8	24	-
9	23	-
10	22	-
11	22	-
12	21	-
13	21	-
14	20	-
15	-	-
16	-	-
18	-	-
20	-	-
22	-	-
24	-	-
26	-	-
28	-	-
30	-	-
32	-	-

HEAT TREATING TEMPERATURES RECOMMENDED BY SAE
*NORMALIZE 1700 °F
AUSTENITIZE 1700 °F

*For forged or rolled specimens only.

HARDENABILITY BAND 4718H

C	Mn	Si	Ni	Cr	Mo
.15/.21	.60/.95	.20/.35	.85/1.25	.30/.60	.30/.40

DIAMETERS OF ROUNDS WITH SAME AS QUENCHED HARDNESS

Diameters									LOCATION IN ROUND	QUENCH
3.8									SURFACE	MILD WATER QUENCH
1.1	2.0	2.9	3.8	4.8	5.8	6.7			3/4 RADIUS FROM CENTER	MILD WATER QUENCH
0.7	1.2	1.6	2.0	2.4	2.8	3.2	3.6	3.9	CENTER	MILD WATER QUENCH
0.8	1.8	2.5	3.0	3.4	3.8				SURFACE	MILD OIL QUENCH
0.5	1.0	1.6	2.0	2.4	2.8	3.2	3.6	4.0	3/4 RADIUS FROM CENTER	MILD OIL QUENCH
0.2	0.6	1.0	1.4	1.7	2.0	2.4	2.8	3.1	CENTER	MILD OIL QUENCH

Chart axes:
- Vertical: ROCKWELL HARDNESS C SCALE (20, 25, 30, 35, 40, 45, 50, 55, 60, 65)
- Horizontal: DISTANCE FROM QUENCHED END — SIXTEENTHS OF AN INCH (2, 4, 6, 8, 10, 12, 14, 16, 18, 20, 22, 24, 26, 28, 30, 32)

HARDNESS LIMITS FOR SPECIFICATION PURPOSES

"J" DISTANCE SIXTEENTHS OF AN INCH	4718 H MAX.	4718 H MIN.
1	47	40
2	47	40
3	45	38
4	43	33
5	40	29
6	37	27
7	35	25
8	33	24
9	32	23
10	31	22
11	30	22
12	29	21
13	29	21
14	28	21
15	27	20
16	27	20
18	27	-
20	26	-
22	28	-
24	25	-
26	25	-
28	24	-
30	24	-
32	24	-

HEAT TREATING TEMPERATURES RECOMMENDED BY SAE

*NORMALIZE	1700 °F
AUSTENITIZE	1700 °F

*For forged or rolled specimens only.

Constructional Steels for Heat Treating — United States (Alloy Bars, Blooms, Billets, and Slabs)

365

HARDENABILITY BAND ___ 4720 H

C	Mn	Si	Ni	Cr	Mo
.17/.23	.45/.75	.20/.35	.85/1.25	.30/.60	.15/.25

DIAMETERS OF ROUNDS WITH SAME AS QUENCHED HARDNESS									LOCATION IN ROUND	QUENCH
3.8										
1.1	2.0	2.9	3.8	4.8	5.8	6.7			SURFACE	MILD WATER QUENCH
0.7	1.2	1.8	2.0	2.4	2.8	3.2	3.6	3.9	3/4 RADIUS FROM CENTER	
0.8	1.8	2.5	3.0	3.4	3.8				CENTER	
0.8	1.0	1.6	2.0	2.4	2.8	3.2	3.6	4.0	SURFACE	MILD OIL QUENCH
0.5	1.0	1.4	1.7	2.0	2.4	2.8	3.6	3.1	3/4 RADIUS FROM CENTER	
0.2	0.6	1.0		2.4	2.8	3.1			CENTER	

Hardenability band graph — ROCKWELL HARDNESS C SCALE (vertical, 20–65) vs DISTANCE FROM QUENCHED END — SIXTEENTHS OF AN INCH (horizontal, 2–32).

HARDNESS LIMITS FOR SPECIFICATION PURPOSES

"J" DISTANCE SIXTEENTHS OF AN INCH	4720 H MAX.	MIN.
1	48	41
2	47	39
3	43	31
4	39	27
5	35	23
6	32	21
7	29	-
8	28	-
9	27	-
10	26	-
11	25	-
12	24	-
13	24	-
14	23	-
15	23	-
16	22	-
18	21	-
20	21	-
22	21	-
24	20	-
26	-	-
28	-	-
30	-	-
32	-	-

HEAT TREATING TEMPERATURES
RECOMMENDED BY SAE
*NORMALIZE 1700 °F
AUSTENITIZE 1700 °F

*For forged or rolled specimens only.

366

HARDENABILITY BAND ____ 4815 H

C	Mn	Si	Ni	Mo
.12 / .18	.30 / .70	.20 / .35	3.20 / 3.80	.20 / .30

DIAMETERS OF ROUNDS WITH SAME AS QUENCHED HARDNESS

		LOCATION IN ROUND	QUENCH
3.8			
1.1	2.0 2.9 3.8 4.8 5.8 6.7	SURFACE	MILD WATER QUENCH
0.7	1.2 1.6 2.0 2.4 2.8 3.2 3.6 3.9	3/4 RADIUS FROM CENTER	
		CENTER	
0.8	1.8 2.5 3.0 3.4 3.8	SURFACE	MILD OIL QUENCH
0.5	1.0 1.6 2.0 2.4 2.8 3.2 3.6 4.0	3/4 RADIUS FROM CENTER	
0.2	0.6 1.0 1.4 1.7 2.0 2.4 2.8 3.1	CENTER	

Vertical axis: ROCKWELL HARDNESS C SCALE (65, 60, 55, 50, 45, 40, 35, 30, 25, 20)

Horizontal axis: DISTANCE FROM QUENCHED END — SIXTEENTHS OF AN INCH (2, 4, 6, 8, 10, 12, 14, 16, 18, 20, 22, 24, 26, 28, 30, 32)

HARDNESS LIMITS FOR SPECIFICATION PURPOSES

"J" DISTANCE SIXTEENTHS OF AN INCH	4815 H MAX.	MIN.
1	45	38
2	44	37
3	44	34
4	42	30
5	41	27
6	39	24
7	37	22
8	35	21
9	33	20
10	31	–
11	30	–
12	29	–
13	28	–
14	28	–
15	27	–
16	27	–
18	26	–
20	25	–
22	24	–
24	24	–
26	24	–
28	23	–
30	23	–
32	23	–

HEAT TREATING TEMPERATURES RECOMMENDED BY SAE

*NORMALIZE 1700 °F
AUSTENITIZE 1550 °F

*For forged or rolled specimens only.

Constructional Steels for Heat Treating — United States (Alloy Bars, Blooms, Billets, and Slabs)

367

HARDENABILITY BAND ___ 4817 H

C	Mn	Si	Ni	Mo
.14 / .20	.30 / .70	.20 / .35	3.20 / 3.80	.20 / .30

DIAMETERS OF ROUNDS WITH SAME AS QUENCHED HARDNESS

									LOCATION IN ROUND	QUENCH
3.8									SURFACE	MILD
1.1	2.0	2.9	3.8	4.8	5.8	6.7			3/4 RADIUS FROM CENTER	WATER
0.7	1.2	1.6	2.0	2.4	2.8	3.2	3.6	3.9	CENTER	QUENCH
0.8	1.8	2.5	3.0	3.4	3.8				SURFACE	MILD
0.5	1.0	1.6	2.0	2.4	2.8	3.2	3.6	4.0	3/4 RADIUS FROM CENTER	OIL
0.2	0.6	1.0	1.4	1.7	2.0	2.4	2.8	3.1	CENTER	QUENCH

ROCKWELL HARDNESS C SCALE

DISTANCE FROM QUENCHED END — SIXTEENTHS OF AN INCH

HARDNESS LIMITS FOR SPECIFICATION PURPOSES

"J" DISTANCE SIXTEENTHS OF AN INCH	4817 H MAX.	MIN.
1	46	39
2	46	38
3	45	35
4	44	32
5	42	29
6	41	27
7	39	25
8	37	23
9	35	22
10	33	21
11	32	20
12	31	20
13	30	–
14	29	–
15	28	–
16	28	–
18	27	–
20	26	–
22	25	–
24	25	–
26	25	–
28	25	–
30	24	–
32	24	–

HEAT TREATING TEMPERATURES RECOMMENDED BY SAE

*NORMALIZE 1700 °F
AUSTENITIZE 1550 °F

*For forged or rolled specimens only.

368

HARDENABILITY BAND 4820 H

C	Mn	Si	Ni	Mo
.17/.23	.40/.80	.20/.35	3.20/3.80	.20/.30

ROCKWELL HARDNESS C SCALE

DISTANCE FROM QUENCHED END — SIXTEENTHS OF AN INCH

DIAMETERS OF ROUNDS WITH SAME AS QUENCHED HARDNESS

Diameters									LOCATION IN ROUND	QUENCH
3.8										
1.1	2.0	2.9	3.8	4.8	5.8	6.7			SURFACE	MILD WATER
0.7	1.2	1.6	2.0	2.4	2.8	3.2	3.9		3/4 RADIUS FROM CENTER	WATER QUENCH
									CENTER	
0.8	1.8	2.5	3.0	3.4	3.8				SURFACE	MILD OIL
0.5	1.0	1.6	2.0	2.4	2.8	3.2	3.6	4.0	3/4 RADIUS FROM CENTER	OIL QUENCH
0.2	0.6	1.0	1.4	1.7	2.0	2.4	2.8	3.1	CENTER	QUENCH

HARDNESS LIMITS FOR SPECIFICATION PURPOSES

"J" DISTANCE SIXTEENTHS OF AN INCH	4820 H MAX.	MIN.
1	48	41
2	48	40
3	47	39
4	46	38
5	45	34
6	43	31
7	42	29
8	40	27
9	39	26
10	37	25
11	36	24
12	35	23
13	34	22
14	33	22
15	32	21
16	31	21
18	29	20
20	28	20
22	28	–
24	27	–
26	27	–
28	26	–
30	26	–
32	25	–

HEAT TREATING TEMPERATURES RECOMMENDED BY SAE
*NORMALIZE   1700 °F
AUSTENITIZE   1550 °F

*For forged or rolled specimens only.

Constructional Steels for Heat Treating – United States (Alloy Bars, Blooms, Billets, and Slabs)

HARDENABILITY BAND   5120 H

C	Mn	Si	Cr
.17/.23	.60/1.00	.20/.35	.60/1.00

DIAMETERS OF ROUNDS WITH SAME AS QUENCHED HARDNESS

Diameters									LOCATION IN ROUND	QUENCH
3.8										
1.1	2.0	2.9	3.8	4.8	5.8	6.7			SURFACE	MILD WATER QUENCH
0.7	1.2	1.6	2.0	2.4	2.8	3.2	3.6	3.9	3/4 RADIUS FROM CENTER	
0.8	1.8	2.5	3.0	3.4	3.8				CENTER	
0.8	1.0	1.6	2.0	2.4	2.8	3.2	3.6	4.0	SURFACE	MILD OIL QUENCH
0.5	1.0	1.4	1.7	2.0	2.4	2.8	3.1		3/4 RADIUS FROM CENTER	
0.2	0.6	1.0							CENTER	

ROCKWELL HARDNESS C SCALE

DISTANCE FROM QUENCHED END — SIXTEENTHS OF AN INCH

HARDNESS LIMITS FOR SPECIFICATION PURPOSES

"J" DISTANCE SIXTEENTHS OF AN INCH	5120H MAX.	MIN.
1	48	40
2	46	34
3	41	28
4	36	23
5	33	20
6	30	-
7	28	-
8	27	-
9	25	-
10	24	-
11	23	-
12	22	-
13	21	-
14	21	-
15	20	-
16	-	-
18	-	-
20	-	-
22	-	-
24	-	-
26	-	-
28	-	-
30	-	-
32	-	-

HEAT TREATING TEMPERATURES RECOMMENDED BY SAE

*NORMALIZE 1700 °F
AUSTENITIZE 1700 °F

*For forged or rolled-specimens only.

HARDENABILITY BAND 5130 H

C	Mn	Si	Cr
.27/.33	.60/1.00	.20/.35	.75/1.20

HARDNESS LIMITS FOR SPECIFICATION PURPOSES

"J" DISTANCE SIXTEENTHS OF AN INCH	5130H MAX.	MIN.
1	56	49
2	55	46
3	53	42
4	51	39
5	49	35
6	47	32
7	45	30
8	42	28
9	40	26
10	38	25
11	37	23
12	36	22
13	35	21
14	34	20
15	34	–
16	33	–
18	32	–
20	31	–
22	30	–
24	29	–
26	27	–
28	26	–
30	25	–
32	24	–

HEAT TREATING TEMPERATURES RECOMMENDED BY SAE

*NORMALIZE 1650 °F
AUSTENITIZE 1600 °F

*For forged or rolled specimens only.

DIAMETERS OF ROUNDS WITH SAME AS QUENCHED HARDNESS

Values	LOCATION IN ROUND	QUENCH
3.8	SURFACE	MILD WATER QUENCH
1.1 2.0 2.9 3.8 4.8 5.8 6.7	3/4 RADIUS FROM CENTER	
0.7 1.2 1.6 2.0 2.4 2.8 3.2 3.6 3.9	CENTER	
0.8 1.8 2.5 3.0 3.4 3.8	SURFACE	MILD OIL QUENCH
0.5 1.0 1.6 2.0 2.4 2.8 3.2 3.6 4.0	3/4 RADIUS FROM CENTER	
0.2 0.6 1.0 1.4 1.7 2.0 2.4 2.8 3.1	CENTER	

ROCKWELL HARDNESS C SCALE: 65 60 55 50 45 40 35 30 25 20

DISTANCE FROM QUENCHED END — SIXTEENTHS OF AN INCH: 2 4 6 8 10 12 14 16 18 20 22 24 26 28 30 32

Constructional Steels for Heat Treating — United States (Alloy Bars, Blooms, Billets, and Slabs)

HARDENABILITY BAND 5132H

C	Mn	Si	Cr
.29/.35	.50/.90	.20/.35	.65/1.10

DIAMETERS OF ROUNDS WITH SAME AS QUENCHED HARDNESS

LOCATION IN ROUND		QUENCH
SURFACE	3.8	MILD WATER QUENCH
3/4 RADIUS FROM CENTER	1.1 2.0 2.9 3.8 4.8 5.8 6.7	
CENTER	0.7 1.2 1.6 2.0 2.4 2.8 3.2 3.6 3.9	
SURFACE	0.8 1.8 2.5 3.0 3.4 3.8	MILD OIL QUENCH
3/4 RADIUS FROM CENTER	0.5 1.0 1.6 2.0 2.4 2.8 3.2 3.6 4.0	
CENTER	0.2 0.6 1.0 1.4 1.7 2.0 2.4 2.8 3.1	

ROCKWELL HARDNESS C SCALE

DISTANCE FROM QUENCHED END — SIXTEENTHS OF AN INCH

HARDNESS LIMITS FOR SPECIFICATION PURPOSES

"J" DISTANCE SIXTEENTHS OF AN INCH	5132H MAX.	MIN.
1	57	50
2	56	47
3	54	43
4	52	40
5	50	35
6	48	32
7	45	29
8	42	27
9	40	25
10	38	24
11	37	23
12	36	22
13	35	21
14	34	20
15	34	-
16	33	-
18	32	-
20	31	-
22	30	-
24	29	-
26	28	-
28	27	-
30	26	-
32	25	-

HEAT TREATING TEMPERATURES
RECOMMENDED BY SAE
*NORMALIZE   1650 °F
AUSTENITIZE   1800 °F

*For forged or rolled specimens only.

HARDENABILITY BAND ____ 5135 H

C	Mn	Si	Cr
32 /.38	.50 /.90	.20 /.35	.70 /1.15

DIAMETERS OF ROUNDS WITH SAME AS QUENCHED HARDNESS									LOCATION IN ROUND	QUENCH
3.8									SURFACE	MILD
1.1	2.0	2.9	3.8	4.8	5.8	6.7			3/4 RADIUS FROM CENTER	WATER
0.7	1.2	1.6	2.0	2.4	2.8	3.2	3.9		CENTER	QUENCH
0.8	1.8	2.5	3.0	3.4	3.8				SURFACE	MILD
0.5	1.0	1.6	2.0	2.4	2.8	3.2	3.6	4.0	3/4 RADIUS FROM CENTER	OIL
0.2	0.6	1.0	1.4	1.7	2.0	2.4	2.8	3.1	CENTER	QUENCH

ROCKWELL HARDNESS C SCALE

DISTANCE FROM QUENCHED END — SIXTEENTHS OF AN INCH

HARDNESS LIMITS FOR SPECIFICATION PURPOSES

"J" DISTANCE SIXTEENTHS OF AN INCH	5135H MAX.	5135H MIN.
1	58	51
2	57	49
3	56	47
4	55	43
5	54	38
6	52	35
7	50	32
8	47	30
9	45	28
10	43	27
11	41	25
12	40	24
13	39	23
14	38	22
15	37	21
16	37	21
18	36	20
20	35	-
22	34	-
24	33	-
26	32	-
28	32	-
30	31	-
32	30	-

HEAT TREATING TEMPERATURES RECOMMENDED BY SAE

*NORMALIZE 1600 °F
AUSTENITIZE 1550 °F

*For forged or rolled specimens only.

Constructional Steels for Heat Treating — United States (Alloy Bars, Blooms, Billets, and Slabs)

HARDENABILITY BAND — 5140 H

C	Mn	Si	Cr
.37 / .44	.60 / 1.00	.20 / .35	.60 / 1.00

DIAMETERS OF ROUNDS WITH SAME AS QUENCHED HARDNESS

Diameters									LOCATION IN ROUND	QUENCH
3.8									SURFACE	MILD WATER QUENCH
1.1	2.0	2.9	3.8	4.8	5.8	6.7			3/4 RADIUS FROM CENTER	
0.7	1.2	1.6	2.0	2.4	2.8	3.2	3.6	3.9	CENTER	
0.8	1.8	2.5	3.0	3.4	3.8				SURFACE	MILD OIL QUENCH
0.5	1.0	1.6	2.0	2.4	2.8	3.2	3.6	4.0	3/4 RADIUS FROM CENTER	
0.2	0.6	1.0	1.4	1.7	2.0	2.4	2.8	3.1	CENTER	

ROCKWELL HARDNESS C SCALE (20 – 65)

DISTANCE FROM QUENCHED END — SIXTEENTHS OF AN INCH (2 – 32)

HARDNESS LIMITS FOR SPECIFICATION PURPOSES

"J" DISTANCE SIXTEENTHS OF AN INCH	5140H MAX.	5140H MIN.
1	60	53
2	59	52
3	58	50
4	57	48
5	56	43
6	54	38
7	52	35
8	50	33
9	48	31
10	46	30
11	45	29
12	43	28
13	42	27
14	40	27
15	39	26
16	38	25
18	37	24
20	36	23
22	35	21
24	34	20
26	34	–
28	33	–
30	33	–
32	32	–

HEAT TREATING TEMPERATURES RECOMMENDED BY SAE

*NORMALIZE 1600 °F
AUSTENITIZE 1550 °F

*For forged or rolled specimens only.

HARDENABILITY BAND 5145 H

C	Mn	Si	Cr
.42/.49	.60/1.00	.20/.35	.60/1.00

DIAMETERS OF ROUNDS WITH SAME AS QUENCHED HARDNESS

Diameters	LOCATION IN ROUND	QUENCH
3.8	SURFACE	MILD WATER QUENCH
1.1 2.0 2.9 3.8 4.8 5.8 6.7	3/4 RADIUS FROM CENTER	
0.7 1.2 1.6 2.0 2.4 2.8 3.2 3.6 3.9	CENTER	
0.8 1.8 2.5 3.0 3.4 3.8	SURFACE	MILD OIL QUENCH
0.5 1.0 1.6 2.0 2.4 2.8 3.2 3.6 4.0	3/4 RADIUS FROM CENTER	
0.2 0.6 1.0 1.4 1.7 2.0 2.4 2.8 3.1	CENTER	

DISTANCE FROM QUENCHED END – SIXTEENTHS OF AN INCH

ROCKWELL HARDNESS C SCALE

HARDNESS LIMITS FOR SPECIFICATION PURPOSES

"J" DISTANCE SIXTEENTHS OF AN INCH	5145H MAX.	5145H MIN.
1	63	56
2	62	55
3	61	53
4	60	51
5	59	48
6	58	42
7	57	38
8	56	35
9	55	33
10	53	32
11	52	31
12	50	30
13	48	30
14	47	29
15	45	28
16	44	28
18	42	26
20	41	25
22	39	24
24	38	23
26	37	22
28	37	21
30	36	–
32	35	–

HEAT TREATING TEMPERATURES RECOMMENDED BY SAE

*NORMALIZE 1600 °F
AUSTENITIZE 1550 °F

*For forged or rolled specimens only.

Constructional Steels for Heat Treating — United States (Alloy Bars, Blooms, Billets, and Slabs)

HARDENABILITY BAND ___ 5147 H

C	Mn	Si	Cr
.45/.52	.60/1.05	.20/.35	.80/1.25

DIAMETERS OF ROUNDS WITH SAME AS QUENCHED HARDNESS

									LOCATION IN ROUND	QUENCH
3.8									SURFACE	MILD WATER QUENCH
1.1	2.0	2.9	3.8	4.8	5.8	6.7			3/4 RADIUS FROM CENTER	
0.7	1.2	1.6	2.0	2.4	2.8	3.2	3.6	3.9	CENTER	
0.8	1.8	2.5	3.0	3.4	3.8				SURFACE	MILD OIL QUENCH
0.5	1.0	1.6	2.0	2.4	2.8	3.2	3.6	4.0	3/4 RADIUS FROM CENTER	
0.2	0.6	1.0	1.4	1.7	2.0	2.4	2.8	3.1	CENTER	

ROCKWELL HARDNESS C SCALE

DISTANCE FROM QUENCHED END — SIXTEENTHS OF AN INCH

HARDNESS LIMITS FOR SPECIFICATION PURPOSES

"J" DISTANCE SIXTEENTHS OF AN INCH	5147H MAX.	5147H MIN.
1	64	57
2	64	56
3	63	55
4	62	54
5	62	53
6	61	52
7	61	49
8	60	45
9	60	40
10	59	37
11	59	35
12	58	34
13	58	33
14	57	32
15	57	32
16	56	31
18	55	30
20	54	29
22	53	27
24	52	26
26	51	25
28	50	24
30	49	22
32	48	21

HEAT TREATING TEMPERATURES RECOMMENDED BY SAE

*NORMALIZE 1600 °F
AUSTENITIZE 1550 °F

*For forged or rolled specimens only.

HARDENABILITY BAND — 5150 H

C	Mn	Si	Cr
.47 / .54	.60 / 1.00	.20 / .35	.60 / 1.00

DIAMETERS OF ROUNDS WITH SAME AS QUENCHED HARDNESS	LOCATION IN ROUND	QUENCH
3.8	SURFACE	MILD WATER QUENCH
1.1 2.0 2.9 3.8 4.8 5.8 6.7	3/4 RADIUS FROM CENTER	
0.7 1.2 1.6 2.0 2.4 2.8 3.2 3.6 3.9	CENTER	
0.8 1.8 2.5 3.0 3.4 3.8	SURFACE	MILD OIL QUENCH
0.5 1.0 1.6 2.0 2.4 2.8 3.2 3.6 4.0	3/4 RADIUS FROM CENTER	
0.2 0.6 1.0 1.4 1.7 2.0 2.4 2.8 3.1	CENTER	

Chart axes: ROCKWELL HARDNESS C SCALE (vertical, 20–65); DISTANCE FROM QUENCHED END — SIXTEENTHS OF AN INCH (horizontal, 2–32)

HARDNESS LIMITS FOR SPECIFICATION PURPOSES

"J" DISTANCE SIXTEENTHS OF AN INCH	5150H MAX.	MIN.
1	65	59
2	65	58
3	64	57
4	63	56
5	62	53
6	61	49
7	60	42
8	59	38
9	58	36
10	56	34
11	55	33
12	53	32
13	51	31
14	50	31
15	48	30
16	47	30
18	45	29
20	43	28
22	42	27
24	41	26
26	40	25
28	39	24
30	39	23
32	38	22

HEAT TREATING TEMPERATURES RECOMMENDED BY SAE
*NORMALIZE   1600 °F
AUSTENITIZE   1550 °F

*For forged or rolled specimens only.

Constructional Steels for Heat Treating – United States (Alloy Bars, Blooms, Billets, and Slabs)

HARDENABILITY BAND __ 5155 H

C	Mn	Si	Cr
.50 / .60	.60 / 1.00	.20 / .35	.60 / 1.00

DIAMETERS OF ROUNDS WITH SAME AS QUENCHED HARDNESS

DIAMETERS OF ROUNDS									LOCATION IN ROUND	QUENCH
3.8										
1.1	2.0	2.9	3.8	4.8	5.8	6.7			SURFACE	MILD WATER QUENCH
0.7	1.2	1.6	2.0	2.4	2.8	3.2	3.6	3.9	3/4 RADIUS FROM CENTER	
0.8	1.8	2.5	3.0	3.4	3.8				CENTER	
0.5	1.0	1.6	2.0	2.4	2.8	3.2	3.6	4.0	SURFACE	MILD OIL QUENCH
									3/4 RADIUS FROM CENTER	
0.2	0.6	1.0	1.4	1.7	2.0	2.4	2.8	3.1	CENTER	

Chart: ROCKWELL HARDNESS C SCALE (vertical, 20–65) vs. DISTANCE FROM QUENCHED END — SIXTEENTHS OF AN INCH (horizontal, 2–32)

HARDNESS LIMITS FOR SPECIFICATION PURPOSES

"J" DISTANCE SIXTEENTHS OF AN INCH	5155 H MAX.	MIN.
1	-	60
2	65	59
3	64	58
4	64	57
5	63	55
6	63	52
7	62	47
8	62	41
9	61	37
10	60	36
11	59	35
12	57	34
13	55	34
14	52	33
15	51	33
16	49	32
18	47	31
20	45	31
22	44	30
24	43	29
26	42	28
28	41	27
30	41	26
32	40	25

HEAT TREATING TEMPERATURES RECOMMENDED BY SAE

*NORMALIZE 1600 °F
AUSTENITIZE 1550 °F

*For forged or rolled specimens only.

378

HARDENABILITY BAND 5160 H

C	Mn	Si	Cr
.55 / .65	.65 / 1.10	.20 / .35	.60 / 1.00

DIAMETERS OF ROUNDS WITH SAME AS QUENCHED HARDNESS									LOCATION IN ROUND	QUENCH
3.8									SURFACE	MILD
1.1	2.0	2.9	3.8	4.8	5.8	6.7			3/4 RADIUS FROM CENTER	WATER
0.7	1.2	1.8	2.0	2.4	2.8	3.2	3.6	3.9	CENTER	QUENCH
0.8	1.8	2.5	3.0	3.4	3.8				SURFACE	MILD
0.5	1.0	1.6	2.0	2.4	2.8	3.2	3.6	4.0	3/4 RADIUS FROM CENTER	OIL
0.2	0.6	1.0	1.4	1.7	2.0	2.4	2.8	3.1	CENTER	QUENCH

ROCKWELL HARDNESS C SCALE

DISTANCE FROM QUENCHED END —SIXTEENTHS OF AN INCH

HARDNESS LIMITS FOR SPECIFICATION PURPOSES

"J" DISTANCE SIXTEENTHS OF AN INCH	5160H MAX.	MIN.
1	–	80
2	–	80
3	–	80
4	65	59
5	65	58
6	64	56
7	64	52
8	63	47
9	62	42
10	61	39
11	60	37
12	59	36
13	58	35
14	56	35
15	54	34
16	52	34
18	48	33
20	47	32
22	46	31
24	45	30
26	44	29
28	43	28
30	43	28
32	42	27

HEAT TREATING TEMPERATURES RECOMMENDED BY SAE
*NORMALIZE 1600 °F
AUSTENITIZE 1550 °F

*For forged or rolled specimens only.

Constructional Steels for Heat Treating — United States (Alloy Bars, Blooms, Billets, and Slabs)

HARDENABILITY BAND 6118H

C	Mn	Si	Cr	V
.15	.40	.20	.40	.10
.21	.80	.35	.80	.15

DIAMETERS OF ROUNDS WITH SAME AS QUENCHED HARDNESS

LOCATION IN ROUND	QUENCH									
3.8										
SURFACE	MILD WATER QUENCH	1.1	2.0	2.9	3.8	4.8	5.8	6.7		
3/4 RADIUS FROM CENTER		0.7	1.2	1.6	2.0	2.4	2.8	3.2	3.6	3.9
CENTER		0.8	1.8	2.5	3.0	3.4	3.8			
SURFACE	MILD OIL QUENCH	0.5	1.0	1.6	2.0	2.4	2.8	3.2	3.6	4.0
3/4 RADIUS FROM CENTER		0.2	0.6	1.0	1.4	1.7	2.0	2.4	2.8	3.1
CENTER										

ROCKWELL HARDNESS C SCALE

DISTANCE FROM QUENCHED END — SIXTEENTHS OF AN INCH

HARDNESS LIMITS FOR SPECIFICATION PURPOSES

"J" DISTANCE SIXTEENTHS OF AN INCH	6118 H MAX.	MIN.
1	46	39
2	44	36
3	38	28
4	33	24
5	30	22
6	28	20
7	27	-
8	26	-
9	26	-
10	25	-
11	25	-
12	24	-
13	24	-
14	23	-
15	23	-
16	22	-
18	22	-
20	21	-
22	21	-
24	20	-
26	-	-
28	-	-
30	-	-
32	-	-

HEAT TREATING TEMPERATURES RECOMMENDED BY SAE
*NORMALIZE 1700 °F
AUSTENITIZE 1700 °F

*For forged or rolled specimens only.

C	Mn	Si	Cr	V
.47 / .54	.60 / 1.00	.20 / .35	.75 / 1.20	.15 MIN.

DIAMETERS OF ROUNDS WITH SAME AS QUENCHED HARDNESS

Values									LOCATION IN ROUND	QUENCH
3.8									SURFACE	MILD WATER QUENCH
1.1	2.0	2.9	3.8	4.8	5.8	6.7			3/4 RADIUS FROM CENTER	
0.7	1.2	1.6	2.0	2.4	2.8	3.2	3.6	3.9	CENTER	
0.8	1.8	2.5	3.0	3.4	3.8				SURFACE	MILD OIL QUENCH
0.5	1.0	1.6	2.0	2.4	2.8	3.2	3.6	4.0	3/4 RADIUS FROM CENTER	
0.2	0.6	1.0	1.4	1.7	2.0	2.4	2.8	3.1	CENTER	

ROCKWELL HARDNESS C SCALE

DISTANCE FROM QUENCHED END — SIXTEENTHS OF AN INCH

HARDNESS LIMITS FOR SPECIFICATION PURPOSES

"J" DISTANCE SIXTEENTHS OF AN INCH	6150H MAX.	MIN.
1	65	59
2	65	58
3	64	57
4	64	56
5	63	55
6	63	53
7	62	50
8	61	47
9	61	43
10	60	41
11	59	39
12	58	38
13	57	37
14	55	36
15	54	35
16	52	35
18	50	34
20	48	32
22	47	31
24	46	30
26	45	29
28	44	27
30	43	26
32	42	25

HEAT TREATING TEMPERATURES
RECOMMENDED BY SAE
*NORMALIZE 1650 °F
AUSTENITIZE 1800 °F

*For forged or rolled specimens only.

Constructional Steels for Heat Treating – United States (Alloy Bars, Blooms, Billets, and Slabs)

HARDENABILITY BAND 8617H

C	Mn	Si	Ni	Cr	Mo
.14/.20	.60/.95	.20/.35	.35/.75	.35/.65	.15/.25

DIAMETERS OF ROUNDS WITH SAME AS QUENCHED HARDNESS

	LOCATION IN ROUND	QUENCH
3.8	SURFACE	MILD WATER QUENCH
1.1 2.0 2.9 3.8 4.8 5.8 6.7	3/4 RADIUS FROM CENTER	
0.7 1.2 1.6 2.0 2.4 2.8 3.2 3.6 3.9	CENTER	
0.8 1.8 2.5 3.0 3.4 3.8	SURFACE	MILD OIL QUENCH
0.5 1.0 1.6 2.0 2.4 2.8 3.2 3.6 4.0	3/4 RADIUS FROM CENTER	
0.2 0.6 1.0 1.4 1.7 2.0 2.4 2.8 3.1	CENTER	

ROCKWELL HARDNESS C SCALE

DISTANCE FROM QUENCHED END — SIXTEENTHS OF AN INCH

HARDNESS LIMITS FOR SPECIFICATION PURPOSES	8617H	
"J" DISTANCE SIXTEENTHS OF AN INCH	MAX.	MIN.
1	46	39
2	44	33
3	41	27
4	38	24
5	34	20
6	31	-
7	28	-
8	27	-
9	26	-
10	25	-
11	24	-
12	23	-
13	23	-
14	22	-
15	22	-
16	21	-
18	21	-
20	20	-
22	-	-
24	-	-
26	-	-
28	-	-
30	-	-
32	-	-

HEAT TREATING TEMPERATURES RECOMMENDED BY SAE

*NORMALIZE 1700 °F

AUSTENITIZE 1700 °F

*For forged or rolled specimens only.

382

HARDENABILITY BAND 8620 H

C	Mn	Si	Ni	Cr	Mo
.17/.23	.60/.95	.20/.35	.35/.75	.35/.65	.15/.25

DIAMETERS OF ROUNDS WITH SAME AS QUENCHED HARDNESS

(distance)									LOCATION IN ROUND	QUENCH
3.8									SURFACE	MILD WATER QUENCH
1.1	2.0	2.9	3.8	4.8	5.8	6.7			3/4 RADIUS FROM CENTER	
0.7	1.2	1.6	2.0	2.4	2.8	3.2	3.6	3.9	CENTER	
0.8	1.8	2.5	3.0	3.4	3.8				SURFACE	MILD OIL QUENCH
0.5	1.0	1.8	2.0	2.4	2.8	3.2	3.6	4.0	3/4 RADIUS FROM CENTER	
0.2	0.6	1.0	1.4	1.7	2.0	2.4	2.8	3.1	CENTER	

ROCKWELL HARDNESS C SCALE

DISTANCE FROM QUENCHED END — SIXTEENTHS OF AN INCH

HARDNESS LIMITS FOR SPECIFICATION PURPOSES

"J" DISTANCE SIXTEENTHS OF AN INCH	8620H MAX.	MIN.
1	48	41
2	47	37
3	44	32
4	41	27
5	37	23
6	34	21
7	32	–
8	30	–
9	29	–
10	28	–
11	27	–
12	26	–
13	25	–
14	25	–
15	24	–
16	24	–
18	23	–
20	23	–
22	23	–
24	23	–
26	23	–
28	22	–
30	22	–
32	22	–

HEAT TREATING TEMPERATURES
RECOMMENDED BY SAE
*NORMALIZE 1700 °F
AUSTENITIZE 1700 °F

*For forged or rolled specimens only.

Constructional Steels for Heat Treating — United States (Alloy Bars, Blooms, Billets, and Slabs)

383

HARDENABILITY BAND 8622H

C	Mn	Si	Ni	Cr	Mo
.19/.25	.60/.95	.20/.35	.35/.75	.35/.65	.15/.25

DIAMETERS OF ROUNDS WITH SAME AS QUENCHED HARDNESS

	LOCATION IN ROUND	QUENCH								
3.8										
SURFACE	3/4 RADIUS FROM CENTER	MILD WATER QUENCH								
1.1	2.0	2.9	3.8	4.8	5.8	6.7				
0.7	1.2	1.6	2.0	2.4	2.8	3.2	3.9			
0.8	1.8	2.5	3.0	3.4	3.8					
SURFACE	3/4 RADIUS FROM CENTER	CENTER	MILD OIL QUENCH							
0.5	1.0	1.6	2.0	2.4	2.8	3.2	3.6	4.0		
0.2	0.6	1.0	1.4	1.7	2.0	2.4	2.8	3.1		

ROCKWELL HARDNESS C SCALE: 20 25 30 35 40 45 50 55 60 65

DISTANCE FROM QUENCHED END — SIXTEENTHS OF AN INCH: 2 4 6 8 10 12 14 16 18 20 22 24 26 28 30 32

HARDNESS LIMITS FOR SPECIFICATION PURPOSES

"J" DISTANCE SIXTEENTHS OF AN INCH	8622H MAX.	MIN.
1	50	43
2	49	39
3	47	34
4	44	30
5	40	26
6	37	24
7	34	22
8	32	20
9	31	–
10	30	–
11	29	–
12	28	–
13	27	–
14	26	–
15	26	–
16	25	–
18	25	–
20	24	–
22	24	–
24	24	–
26	24	–
28	24	–
30	24	–
32	24	–

HEAT TREATING TEMPERATURES
RECOMMENDED BY SAE
*NORMALIZE 1700 °F
AUSTENITIZE 1700 °F
*For forged or rolled specimens only.

384

HARDENABILITY BAND 8625 H

C	Mn	Si	Ni	Cr	Mo
.22/.28	.60/.95	.20/.35	.35/.75	.35/.65	.15/.25

DIAMETERS OF ROUNDS WITH SAME AS QUENCHED HARDNESS

LOCATION IN ROUND	QUENCH
SURFACE	MILD WATER QUENCH
3/4 RADIUS FROM CENTER	
CENTER	
SURFACE	MILD OIL QUENCH
3/4 RADIUS FROM CENTER	
CENTER	

Diameter values:

3.8									
1.1	2.0	2.9	3.8	4.8	5.8	6.7			
0.7	1.2	1.8	2.0	2.4	2.8	3.2	3.6	3.9	
0.8	1.8	2.5	3.0	3.4	3.8				
0.5	1.0	1.6	2.0	2.4	2.8	3.2	3.6	4.0	
0.2	0.6	1.0	1.4	1.7	2.0	2.4	2.8	3.1	

ROCKWELL HARDNESS C SCALE

DISTANCE FROM QUENCHED END — SIXTEENTHS OF AN INCH

HARDNESS LIMITS FOR SPECIFICATION PURPOSES

"J" DISTANCE SIXTEENTHS OF AN INCH	8625H MAX.	8625H MIN.
1	52	45
2	51	41
3	48	36
4	46	32
5	43	29
6	40	27
7	37	25
8	35	23
9	33	22
10	32	21
11	31	20
12	30	–
13	29	–
14	28	–
15	28	–
16	27	–
18	27	–
20	26	–
22	26	–
24	26	–
26	26	–
28	25	–
30	25	–
32	25	–

HEAT TREATING TEMPERATURES
RECOMMENDED BY SAE
*NORMALIZE   1650 °F
AUSTENITIZE   1600 °F

*For forged or rolled specimens only.

Constructional Steels for Heat Treating – United States (Alloy Bars, Blooms, Billets, and Slabs)

HARDENABILITY BAND 8627 H

C		Mn		Si		Ni		Cr		Mo	
.24	.30	.60	.95	.20	.35	.35	.75	.35	.65	.15	.25

DIAMETERS OF ROUNDS WITH SAME AS QUENCHED HARDNESS									LOCATION IN ROUND	QUENCH
3.8									SURFACE	MILD WATER QUENCH
1.1	2.0	2.9	3.8	4.8	5.8	6.7			3/4 RADIUS FROM CENTER	
0.7	1.2	1.6	2.0	2.4	2.8	3.2	3.6	3.9	CENTER	
0.8	1.8	2.5	3.0	3.4	3.8				SURFACE	MILD OIL QUENCH
0.5	1.0	1.6	2.0	2.4	2.8	3.2	3.6	4.0	3/4 RADIUS FROM CENTER	
0.2	0.6	1.0	1.4	1.7	2.0	2.4	2.8	3.1	CENTER	

ROCKWELL HARDNESS C SCALE

DISTANCE FROM QUENCHED END — SIXTEENTHS OF AN INCH

HARDNESS LIMITS FOR SPECIFICATION PURPOSES

"J" DISTANCE SIXTEENTHS OF AN INCH	8627H MAX.	MIN.
1	54	47
2	52	43
3	50	38
4	48	35
5	45	32
6	43	29
7	40	27
8	38	26
9	36	24
10	34	24
11	33	23
12	32	22
13	31	21
14	30	21
15	30	20
16	29	20
18	28	-
20	28	-
22	28	-
24	27	-
26	27	-
28	27	-
30	27	-
32	27	-

HEAT TREATING TEMPERATURES
RECOMMENDED BY SAE
*NORMALIZE 1650 °F
AUSTENITIZE 1600 °F

*For forged or rolled specimens only.

386

HARDENABILITY BAND 8630 H

C	Mn	Si	Ni	Cr	Mo
.27 / .33	.60 / .95	.20 / .35	.35 / .75	.35 / .65	.15 / .25

DIAMETERS OF ROUNDS WITH SAME AS QUENCHED HARDNESS

LOCATION IN ROUND			QUENCH
SURFACE			MILD WATER QUENCH
3/4 RADIUS FROM CENTER			
CENTER			
SURFACE			MILD OIL QUENCH
3/4 RADIUS FROM CENTER			
CENTER			

3.8								
1.1	2.0	2.9	3.8	4.8	5.8	6.7		
0.7	1.2	1.6	2.0	2.4	2.8	3.2	3.9	
0.8	1.8	2.5	3.0	3.4	3.8			
0.5	1.0	1.6	2.0	2.4	2.8	3.2	3.6	4.0
0.2	0.6	1.0	1.4	1.7	2.0	2.4	2.8	3.1

ROCKWELL HARDNESS C SCALE

DISTANCE FROM QUENCHED END — SIXTEENTHS OF AN INCH

HARDNESS LIMITS FOR SPECIFICATION PURPOSES

"J" DISTANCE SIXTEENTHS OF AN INCH	8630H MAX.	MIN.
1	56	49
2	55	46
3	54	43
4	52	39
5	50	35
6	47	32
7	44	29
8	41	28
9	39	27
10	37	26
11	35	25
12	34	24
13	33	23
14	33	22
15	32	22
16	31	21
18	30	21
20	30	20
22	29	20
24	29	-
26	29	-
28	29	-
30	29	-
32	29	-

HEAT TREATING TEMPERATURES RECOMMENDED BY SAE

*NORMALIZE 1650 °F

AUSTENITIZE 1600 °F

*For forged or rolled specimens only.

Constructional Steels for Heat Treating – United States (Alloy Bars, Blooms, Billets, and Slabs)

HARDENABILITY BAND 863Z H

C	Mn	Si	Ni	Cr	Mo
.34/.41	.70/1.05	.20/.35	.35/.75	.35/.65	.15/.25

DIAMETERS OF ROUNDS WITH SAME AS QUENCHED HARDNESS

							LOCATION IN ROUND	QUENCH
3.8							SURFACE	MILD
1.1	2.0	2.9	3.8	4.8	5.8	6.7	3/4 RADIUS FROM CENTER	WATER
0.7	1.2	1.6	2.0	2.4	2.8	3.2	3.9	CENTER	QUENCH
0.8	1.8	2.5	3.0	3.4	3.8		SURFACE	MILD
0.5	1.0	1.6	2.0	2.4	2.8	3.2	3.6	4.0	3/4 RADIUS FROM CENTER	OIL
0.2	0.6	1.0	1.4	1.7	2.0	2.4	2.8	3.1	CENTER	QUENCH

ROCKWELL HARDNESS C SCALE

DISTANCE FROM QUENCHED END – SIXTEENTHS OF AN INCH

HARDNESS LIMITS FOR SPECIFICATION PURPOSES

"J" DISTANCE SIXTEENTHS OF AN INCH	863H MAX.	MIN.
1	59	52
2	58	51
3	58	50
4	57	48
5	56	45
6	55	42
7	54	39
8	53	36
9	51	34
10	49	32
11	47	31
12	46	30
13	44	29
14	43	28
15	41	27
16	40	26
18	39	25
20	37	25
22	36	24
24	36	24
26	35	24
28	35	24
30	35	23
32	35	23

HEAT TREATING TEMPERATURES
RECOMMENDED BY SAE
*NORMALIZE 1600 °F
AUSTENITIZE 1550 °F

*For forged or rolled specimens only.

388

HARDENABILITY BAND 8640 H

C	Mn	Si	Ni	Cr	Mo
.37/.44	.70/1.05	.20/.35	.35/.75	.35/.65	.15/.25

DIAMETERS OF ROUNDS WITH SAME AS QUENCHED HARDNESS									LOCATION IN ROUND	QUENCH
3.8									SURFACE	MILD WATER QUENCH
1.1	2.0	2.9	3.8	4.8	5.8	6.7			3/4 RADIUS FROM CENTER	
0.7	1.2	1.6	2.0	2.4	2.8	3.2	3.6	3.9	CENTER	
0.8	1.8	2.5	3.0	3.4	3.8				SURFACE	MILD OIL QUENCH
0.5	1.0	1.6	2.0	2.4	2.8	3.2	3.6	4.0	3/4 RADIUS FROM CENTER	
0.2	0.6	1.0	1.4	1.7	2.0	2.4	2.8	3.1	CENTER	

DISTANCE FROM QUENCHED END — SIXTEENTHS OF AN INCH
2 4 6 8 10 12 14 16 18 20 22 24 26 28 30 32

ROCKWELL HARDNESS C SCALE
65 60 55 50 45 40 35 30 25 20

HARDNESS LIMITS FOR SPECIFICATION PURPOSES

"J" DISTANCE SIXTEENTHS OF AN INCH	8640H MAX.	MIN.
1	60	53
2	60	53
3	60	52
4	59	51
5	59	49
6	58	46
7	57	42
8	55	39
9	54	36
10	52	34
11	50	32
12	49	31
13	47	30
14	45	29
15	44	28
16	42	28
18	41	26
20	39	26
22	38	25
24	38	25
26	37	24
28	37	24
30	37	24
32	37	24

HEAT TREATING TEMPERATURES
RECOMMENDED BY SAE
*NORMALIZE   1600 °F
AUSTENITIZE 1550 °F

*For forged or rolled specimens only.

Constructional Steels for Heat Treating — United States (Alloy Bars, Blooms, Billets, and Slabs)

389

HARDENABILITY BAND 8642H

C	Mn	Si	Ni	Cr	Mo
.39/.46	.70/1.05	.20/.35	.35/.75	.35/.65	.15/.25

HARDNESS LIMITS FOR SPECIFICATION PURPOSES

"J" DISTANCE SIXTEENTHS OF AN INCH	8642H MAX.	MIN.
1	62	55
2	62	54
3	62	53
4	61	52
5	61	50
6	60	48
7	59	45
8	58	42
9	57	39
10	55	37
11	54	34
12	52	33
13	50	32
14	49	31
15	48	30
16	46	29
18	44	28
20	42	28
22	41	27
24	40	27
26	40	26
28	39	26
30	39	26
32	39	26

HEAT TREATING TEMPERATURES
RECOMMENDED BY SAE
*NORMALIZE 1600 °F
AUSTENITIZE 1550 °F

*For forged or rolled specimens only.

DIAMETERS OF ROUNDS WITH SAME AS QUENCHED HARDNESS

DIAMETERS									LOCATION IN ROUND	QUENCH
3.8									SURFACE	MILD WATER QUENCH
1.1	2.0	2.9	3.8	4.8	5.8	6.7			3/4 RADIUS FROM CENTER	
0.7	1.2	1.6	2.0	2.4	2.8	3.2	3.9		CENTER	
0.8	1.8	2.5	3.0	3.4	3.8				SURFACE	MILD OIL QUENCH
0.5	1.0	1.6	2.0	2.4	2.8	3.2	3.6	4.0	3/4 RADIUS FROM CENTER	
0.2	0.6	1.0	1.4	1.7	2.0	2.4	2.8	3.1	CENTER	

ROCKWELL HARDNESS C SCALE

DISTANCE FROM QUENCHED END — SIXTEENTHS OF AN INCH

2 4 6 8 10 12 14 16 18 20 22 24 26 28 30 32

65 60 55 50 45 40 35 30 25 20

HARDENABILITY BAND 8645 H

C	Mn	Si	Ni	Cr	Mo
.42 / .49	.70 / 1.05	.20 / .35	.35 / .75	.35 / .65	.15 / .25

DIAMETERS OF ROUNDS WITH SAME AS QUENCHED HARDNESS

Values	LOCATION IN ROUND	QUENCH
3.8	SURFACE	MILD WATER QUENCH
1.1 2.0 2.9 3.8 4.8 5.8 6.7	3/4 RADIUS FROM CENTER	
0.7 1.2 1.6 2.0 2.4 2.8 3.2 3.6 3.9	CENTER	
0.8 1.8 2.5 3.0 3.4 3.8	SURFACE	MILD OIL QUENCH
0.5 1.0 1.6 2.0 2.4 2.8 3.2 3.6 4.0	3/4 RADIUS FROM CENTER	
0.2 0.6 1.0 1.4 1.7 2.0 2.4 2.8 3.1	CENTER	

ROCKWELL HARDNESS C SCALE

DISTANCE FROM QUENCHED END — SIXTEENTHS OF AN INCH

HARDNESS LIMITS FOR SPECIFICATION PURPOSES

"J" DISTANCE SIXTEENTHS OF AN INCH	8645H MAX.	8645H MIN.
1	63	56
2	63	56
3	63	55
4	63	54
5	62	52
6	61	50
7	61	48
8	60	45
9	59	41
10	58	39
11	56	37
12	55	35
13	54	34
14	52	33
15	51	32
16	49	31
18	47	30
20	45	29
22	43	28
24	42	28
26	42	27
28	41	27
30	41	27
32	41	27

HEAT TREATING TEMPERATURES RECOMMENDED BY SAE

*NORMALIZE 1600 °F

AUSTENITIZE 1550 °F

*For forged or rolled specimens only.

Constructional Steels for Heat Treating – United States (Alloy Bars, Blooms, Billets, and Slabs)

391

HARDENABILITY BAND 8655H

C	Mn	Si	Ni	Cr	Mo
.50/.60	.70/1.05	.20/.35	.35/.75	.35/.65	.15/.25

DIAMETERS OF ROUNDS WITH SAME AS QUENCHED HARDNESS

LOCATION IN ROUND									QUENCH
SURFACE	3.8								MILD WATER QUENCH
3/4 RADIUS FROM CENTER	1.1	2.0	2.9	3.8	4.8	5.8	6.7		
CENTER	0.7	1.2	1.6	2.0	2.4	2.8	3.2	3.9	
SURFACE	0.8	1.8	2.5	3.0	3.4	3.8			MILD OIL QUENCH
3/4 RADIUS FROM CENTER	0.5	1.0	1.6	2.0	2.4	2.8	3.2	3.6	4.0
CENTER	0.2	0.6	1.0	1.4	1.7	2.0	2.4	2.8	3.1

ROCKWELL HARDNESS C SCALE

DISTANCE FROM QUENCHED END — SIXTEENTHS OF AN INCH

HARDNESS LIMITS FOR SPECIFICATION PURPOSES

"J" DISTANCE SIXTEENTHS OF AN INCH	8655H MAX.	MIN.
1	-	60
2	-	59
3	-	59
4	-	58
5	-	57
6	-	56
7	-	55
8	-	54
9	-	52
10	65	49
11	65	46
12	64	43
13	64	41
14	63	40
15	63	39
16	62	38
18	61	37
20	60	35
22	59	34
24	58	34
26	57	33
28	56	33
30	55	32
32	53	32

HEAT TREATING TEMPERATURES RECOMMENDED BY SAE

*NORMALIZE 1600 °F
AUSTENITIZE 1550 °F

*For forged or rolled specimens only.

HARDENABILITY BAND 8720H

C	Mn	Si	Ni	Cr	Mo
.17/.23	.60/.95	.20/.35	.35/.75	.35/.65	.20/.30

HARDNESS LIMITS FOR SPECIFICATION PURPOSES

"J" DISTANCE SIXTEENTHS OF AN INCH	8720H MAX.	8720H MIN.
1	48	41
2	47	38
3	45	35
4	42	30
5	38	26
6	35	24
7	33	22
8	31	21
9	30	20
10	29	–
11	28	–
12	27	–
13	26	–
14	26	–
15	25	–
16	25	–
18	24	–
20	24	–
22	23	–
24	23	–
26	23	–
28	23	–
30	22	–
32	22	–

HEAT TREATING TEMPERATURES RECOMMENDED BY SAE
*NORMALIZE 1700 °F
AUSTENITIZE 1700 °F
*For forged or rolled specimens only.

DIAMETERS OF ROUNDS WITH SAME AS QUENCHED HARDNESS

LOCATION IN ROUND	QUENCH								
SURFACE	MILD WATER QUENCH	1.1	2.0	2.9	3.8	4.8	5.8	6.7	
3/4 RADIUS FROM CENTER	MILD WATER QUENCH	0.7	1.2	1.6	2.0	2.4	2.8	3.2	3.6 3.9
CENTER	MILD WATER QUENCH	0.8	1.8	2.5	3.0	3.4	3.8		
SURFACE	MILD OIL QUENCH	0.5	1.0	1.6	2.0	2.4	2.8	3.2	3.6 4.0
3/4 RADIUS FROM CENTER	MILD OIL QUENCH	0.2	0.6	1.0	1.4	1.7	2.0	2.4	2.8 3.1
CENTER	MILD OIL QUENCH	3.8							

ROCKWELL HARDNESS C SCALE (20–65)

DISTANCE FROM QUENCHED END — SIXTEENTHS OF AN INCH (2 4 6 8 10 12 14 16 18 20 22 24 26 28 30 32)

Constructional Steels for Heat Treating – United States (Alloy Bars, Blooms, Billets, and Slabs)

393

HARDENABILITY BAND 8740 H

C	Mn	Si	Ni	Cr	Mo
.37/.44	.70/1.05	.20/.35	.35/.75	.35/.65	.20/.30

DIAMETERS OF ROUNDS WITH SAME AS QUENCHED HARDNESS

									LOCATION IN ROUND	QUENCH
3.8									SURFACE	MILD WATER QUENCH
1.1	2.0	2.9	3.8	4.8	5.8	6.7			3/4 RADIUS FROM CENTER	
0.7	1.2	1.6	2.0	2.4	2.8	3.2	3.6	3.9	CENTER	
0.8	1.8	2.5	3.0	3.4	3.8				SURFACE	MILD OIL QUENCH
0.5	1.0	1.6	2.0	2.4	2.8	3.2	3.6	4.0	3/4 RADIUS FROM CENTER	
0.2	0.6	1.0	1.4	1.7	2.0	2.4	2.8	3.1	CENTER	

ROCKWELL HARDNESS C SCALE

DISTANCE FROM QUENCHED END — SIXTEENTHS OF AN INCH

HARDNESS LIMITS FOR SPECIFICATION PURPOSES

"J" DISTANCE SIXTEENTHS OF AN INCH	8740H MAX.	8740H MIN.
1	60	53
2	60	53
3	60	52
4	60	51
5	59	49
6	58	46
7	57	43
8	56	40
9	55	37
10	53	35
11	52	34
12	50	32
13	49	31
14	48	31
15	46	30
16	45	29
18	43	28
20	42	28
22	41	27
24	40	27
26	39	27
28	39	27
30	38	26
32	38	26

HEAT TREATING TEMPERATURES RECOMMENDED BY SAE
*NORMALIZE 1600 °F
AUSTENITIZE 1550 °F

*For forged or rolled specimens only.

HARDENABILITY BAND ___ 8822 H

C	Mn	Si	Ni	Cr	Mo
.19 / .25	.70 / 1.05	.20 / .35	.35 / .75	.35 / .65	.30 / .40

DIAMETERS OF ROUNDS WITH SAME AS QUENCHED HARDNESS

Diameters									LOCATION IN ROUND	QUENCH
3.8									SURFACE	MILD WATER QUENCH
1.1	2.0	2.9	3.8	4.8	5.8	6.7			3/4 RADIUS FROM CENTER	
0.7	1.2	1.6	2.0	2.4	2.8	3.2	3.6	3.9	CENTER	
0.8	1.8	2.5	3.0	3.4	3.8				SURFACE	MILD OIL QUENCH
0.5	1.0	1.6	2.0	2.4	2.8	3.2	3.6	4.0	3/4 RADIUS FROM CENTER	
0.2	0.6	1.0	1.4	1.7	2.0	2.4	2.8	3.1	CENTER	

Chart: ROCKWELL HARDNESS C SCALE (vertical, 20–65) vs. DISTANCE FROM QUENCHED END — SIXTEENTHS OF AN INCH (horizontal, 2–32)

HARDNESS LIMITS FOR SPECIFICATION PURPOSES

"J" DISTANCE SIXTEENTHS OF AN INCH	8822 H MAX.	8822 H MIN.
1	50	43
2	49	42
3	48	39
4	46	33
5	43	29
6	40	27
7	37	25
8	35	24
9	34	24
10	33	23
11	32	23
12	31	22
13	31	22
14	30	22
15	30	21
16	29	21
18	29	20
20	28	–
22	27	–
24	27	–
26	27	–
28	27	–
30	27	–
32	27	–

HEAT TREATING TEMPERATURES RECOMMENDED BY SAE
*NORMALIZE 1700 °F
AUSTENITIZE 1700 °F

*For forged or rolled specimens only.

Constructional Steels for Heat Treating — United States (Alloy Bars, Blooms, Billets, and Slabs)

HARDENABILITY BAND 9260H

C	Mn	Si
.55 / .65	.65 / 1.10	1.70 / 2.20

DIAMETERS OF ROUNDS WITH SAME AS QUENCHED HARDNESS

									LOCATION IN ROUND	QUENCH
3.8										
1.1	2.0	2.9	3.8	4.8	5.8	6.7			SURFACE	MILD WATER QUENCH
0.7	1.2	1.6	2.0	2.4	2.8	3.2	3.6	3.9	3/4 RADIUS FROM CENTER	
0.8	1.8	2.5	3.0	3.4	3.8				CENTER	
0.5	1.0	1.6	2.0	2.4	2.8	3.2	3.6	4.0	SURFACE	MILD OIL QUENCH
0.2	0.6	1.0	1.4	1.7	2.0	2.4	2.8	3.1	3/4 RADIUS FROM CENTER	
									CENTER	

ROCKWELL HARDNESS C SCALE

DISTANCE FROM QUENCHED END – SIXTEENTHS OF AN INCH

HARDNESS LIMITS FOR SPECIFICATION PURPOSES

"J" DISTANCE SIXTEENTHS OF AN INCH	9260H MAX.	9260H MIN.
1	–	60
2	–	60
3	65	57
4	64	53
5	63	46
6	62	41
7	60	38
8	58	36
9	55	36
10	52	35
11	49	34
12	47	34
13	45	33
14	43	33
15	42	32
16	40	32
18	38	31
20	37	31
22	36	30
24	36	30
26	35	29
28	35	29
30	35	28
32	34	28

HEAT TREATING TEMPERATURES RECOMMENDED BY SAE
*NORMALIZE 1650 °F
AUSTENITIZE 1600 °F

*For forged or rolled specimens only.

396

HARDENABILITY BAND 50B44H

C	Mn	Si	Cr	B
.42/.49	.65/1.10	.20/.35	.30/.70	*

* Can be expected to have 0.0005 per cent minimum boron content.

DIAMETERS OF ROUNDS WITH SAME AS QUENCHED HARDNESS

DIAMETERS	LOCATION IN ROUND	QUENCH
3.8		
1.1 2.0 2.9 3.8 4.8 5.8 6.7	SURFACE	MILD WATER QUENCH
0.7 1.2 1.6 2.0 2.4 2.8 3.2 3.6 3.9	3/4 RADIUS FROM CENTER	
0.8 1.8 2.5 3.0 3.4 3.8	CENTER	
0.8 1.0 1.6 2.0 2.4 2.8 3.2 3.6 4.0	SURFACE	MILD OIL QUENCH
0.5 1.0 1.6 2.0 2.4 2.8 3.2 3.6 4.0	3/4 RADIUS FROM CENTER	
0.2 0.6 1.0 1.4 1.7 2.0 2.4 2.8 3.1	CENTER	

ROCKWELL HARDNESS C SCALE

DISTANCE FROM QUENCHED END — SIXTEENTHS OF AN INCH

HARDNESS LIMITS FOR SPECIFICATION PURPOSES

"J" DISTANCE SIXTEENTHS OF AN INCH	50B44 H MAX.	MIN.
1	63	56
2	63	56
3	62	55
4	62	55
5	61	54
6	61	52
7	60	48
8	60	43
9	59	38
10	58	34
11	57	31
12	56	30
13	54	29
14	52	29
15	50	28
16	48	27
18	44	26
20	40	24
22	38	23
24	37	21
26	36	20
28	35	—
30	34	—
32	33	—

HEAT TREATING TEMPERATURES
RECOMMENDED BY SAE
*NORMALIZE   1600°F
AUSTENITIZE   1550°F

*For forged or rolled specimens only.

Constructional Steels for Heat Treating — United States (Alloy Bars, Blooms, Billets, and Slabs)

HARDENABILITY BAND 50B46 H

C	Mn	Si	Cr	B
.43/.50	.65/1.10	.20/.35	.13/.43	*

* Can be expected to have 0.0005 per cent minimum boron content.

DIAMETERS OF ROUNDS WITH SAME AS QUENCHED HARDNESS									LOCATION IN ROUND	QUENCH
3.8										
1.1	2.0	2.9	3.8	4.8	5.8	6.7			SURFACE	MILD WATER QUENCH
0.7	1.2	1.6	2.0	2.4	2.8	3.2	3.6	3.9	3/4 RADIUS FROM CENTER	
0.8	1.8	2.5	3.0	3.4	3.8				CENTER	
0.5	1.0	1.6	2.0	2.4	2.8	3.2	3.6	4.0	SURFACE	MILD OIL QUENCH
0.2	0.6	1.0	1.4	1.7	2.0	2.4	2.8	3.1	3/4 RADIUS FROM CENTER	
									CENTER	

ROCKWELL HARDNESS C SCALE

DISTANCE FROM QUENCHED END — SIXTEENTHS OF AN INCH

HARDNESS LIMITS FOR SPECIFICATION PURPOSES

"J" DISTANCE SIXTEENTHS OF AN INCH	50B46 H MAX.	MIN.
1	63	56
2	62	54
3	61	52
4	60	50
5	59	41
6	58	32
7	57	31
8	56	30
9	54	29
10	51	28
11	47	27
12	43	26
13	40	26
14	38	25
15	37	25
16	36	24
18	35	23
20	34	22
22	33	21
24	32	20
26	31	–
28	30	–
30	29	–
32	28	–

HEAT TREATING TEMPERATURES RECOMMENDED BY SAE
*NORMALIZE 1600 °F
AUSTENITIZE 1550 °F

*For forged or rolled specimens only.

398

HARDENABILITY BAND 50B50H

C	Mn	Si	Cr	B
.47 / .54	.65 / 1.10	.20 / .35	.30 / .70	*

* Can be expected to have 0.0005 per cent minimum boron content.

DIAMETERS OF ROUNDS WITH SAME AS QUENCHED HARDNESS									LOCATION IN ROUND	QUENCH
3.8									SURFACE	MILD WATER QUENCH
1.1	2.0	2.9	3.8	4.8	5.8	6.7			3/4 RADIUS FROM CENTER	
0.7	1.2	1.6	2.0	2.4	2.8	3.2	3.6	3.9	CENTER	MILD QUENCH
0.8	1.8	2.5	3.0	3.4	3.8				SURFACE	MILD OIL QUENCH
0.5	1.0	1.6	2.0	2.4	2.8	3.2	3.6	4.0	3/4 RADIUS FROM CENTER	
0.2	0.6	1.0	1.4	1.7	2.0	2.4	2.8	3.1	CENTER	OIL QUENCH

ROCKWELL HARDNESS C SCALE

DISTANCE FROM QUENCHED END – SIXTEENTHS OF AN INCH

HARDNESS LIMITS FOR SPECIFICATION PURPOSES

"J" DISTANCE SIXTEENTHS OF AN INCH	50B50H MAX.	MIN.
1	65	59
2	65	59
3	64	58
4	64	57
5	63	56
6	63	55
7	62	52
8	62	47
9	61	42
10	60	37
11	60	35
12	59	33
13	58	32
14	57	31
15	56	30
16	54	29
18	50	28
20	47	27
22	44	26
24	41	25
26	39	24
28	38	22
30	37	21
32	36	20

HEAT TREATING TEMPERATURES RECOMMENDED BY SAE
*NORMALIZE 1600 °F
AUSTENITIZE 1550 °F

*For forged or rolled specimens only.

Constructional Steels for Heat Treating – United States (Alloy Bars, Blooms, Billets, and Slabs)

HARDENABILITY BAND 50B60 H

C	Mn	Si	Cr	B
.55/.65	.65/1.10	.20/.35	.30/.70	*

* Can be expected to have 0.0005 per cent minimum boron content.

DIAMETERS OF ROUNDS WITH SAME AS QUENCHED HARDNESS										LOCATION IN ROUND	QUENCH
3.8										SURFACE	MILD
1.1	2.0	2.9	3.8	4.8	5.8	6.7				3/4 RADIUS FROM CENTER	WATER
0.7	1.2	1.6	2.0	2.4	2.8	3.2	3.6	3.9		CENTER	QUENCH
0.8	1.8	2.5	3.0	3.4	3.8					SURFACE	MILD
0.5	1.0	1.6	2.0	2.4	2.8	3.2	3.6	4.0		3/4 RADIUS FROM CENTER	OIL
0.2	0.6	1.0	1.4	1.7	2.0	2.4	2.8	3.1		CENTER	QUENCH

Graph: ROCKWELL HARDNESS C SCALE vs DISTANCE FROM QUENCHED END — SIXTEENTHS OF AN INCH

HARDNESS LIMITS FOR SPECIFICATION PURPOSES

"J" DISTANCE SIXTEENTHS OF AN INCH	50B60 H MAX.	MIN.
1	–	60
2	–	60
3	–	60
4	–	60
5	–	60
6	–	59
7	–	57
8	65	53
9	65	47
10	64	42
11	64	39
12	64	37
13	63	36
14	63	35
15	63	34
16	62	34
18	60	33
20	58	31
22	55	30
24	53	29
26	51	28
28	49	27
30	47	26
32	44	25

HEAT TREATING TEMPERATURES RECOMMENDED BY SAE
*NORMALIZE 1600 °F
AUSTENITIZE 1550 °F

*For forged or rolled specimens only.

400

HARDENABILITY BAND 51B60 H

C	Mn	Si	Cr	B
.55/.65	.65/1.10	.20/.35	.60/1.00	*

* Can be expected to have 0.0005 per cent minimum boron content.

DIAMETERS OF ROUNDS WITH SAME AS QUENCHED HARDNESS

DIAMETERS OF ROUNDS WITH SAME AS QUENCHED HARDNESS									LOCATION IN ROUND	QUENCH
3.8										
1.1	2.0	2.9	3.8	4.8	5.8	6.7			SURFACE	MILD WATER QUENCH
0.7	1.2	1.6	2.0	2.4	2.8	3.2	3.6	3.9	3/4 RADIUS FROM CENTER	
0.8	1.8	2.5	3.0	3.4	3.8				CENTER	
0.5	1.0	1.6	2.0	2.4	2.8	3.2	3.6	4.0	SURFACE	MILD OIL QUENCH
0.2	0.6	1.0	1.4	1.7	2.0	2.4	2.8	3.1	3/4 RADIUS FROM CENTER / CENTER	

ROCKWELL HARDNESS C SCALE

DISTANCE FROM QUENCHED END —SIXTEENTHS OF AN INCH

HARDNESS LIMITS FOR SPECIFICATION PURPOSES

"J" DISTANCE SIXTEENTHS OF AN INCH	51B60 H MAX.	MIN.
1	-	60
2	-	60
3	-	60
4	-	60
5	-	60
6	-	59
7	-	58
8	-	57
9	-	54
10	-	50
11	-	44
12	65	41
13	65	40
14	64	39
15	64	38
16	63	37
18	61	36
20	59	34
22	57	33
24	55	31
26	53	30
28	51	28
30	49	27
32	47	25

HEAT TREATING TEMPERATURES RECOMMENDED BY SAE
*NORMALIZE 1600 °F
AUSTENITIZE 1550 °F

*For forged or rolled specimens only.

Constructional Steels for Heat Treating – United States (Alloy Bars, Blooms, Billets, and Slabs)

401

HARDENABILITY BAND 81B45H

C	Mn	Si	Ni	Cr	Mo	B
.42/.49	.70/1.05	.20/.35	.15/.45	.30/.60	.08/.15	*

* Can be expected to have 0.0005 per cent minimum boron content.

DIAMETERS OF ROUNDS WITH SAME AS QUENCHED HARDNESS									LOCATION IN ROUND	QUENCH
3.8									SURFACE	MILD WATER QUENCH
1.1	2.0	2.9	3.8	4.8	5.8	6.7			3/4 RADIUS FROM CENTER	
0.7	1.2	1.6	2.0	2.4	2.8	3.2	3.6	3.9	CENTER	
0.8	1.8	2.5	3.0	3.4	3.8				SURFACE	MILD OIL QUENCH
0.5	1.0	1.6	2.0	2.4	2.8	3.2	3.6	4.0	3/4 RADIUS FROM CENTER	
0.2	0.6	1.0	1.4	1.7	2.0	2.4	2.8	3.1	CENTER	OIL QUENCH

ROCKWELL HARDNESS C SCALE

DISTANCE FROM QUENCHED END — SIXTEENTHS OF AN INCH

HARDNESS LIMITS FOR SPECIFICATION PURPOSES

"J" DISTANCE SIXTEENTHS OF AN INCH	81B45H MAX.	MIN.
1	63	56
2	63	56
3	63	56
4	63	56
5	63	55
6	63	54
7	62	53
8	62	51
9	61	48
10	60	44
11	60	41
12	59	39
13	58	38
14	57	37
15	57	36
16	56	35
18	55	34
20	53	32
22	52	31
24	50	30
26	49	29
28	47	28
30	45	28
32	43	27

HEAT TREATING TEMPERATURES
RECOMMENDED BY SAE
*NORMALIZE 1600 °F
AUSTENITIZE 1550 °F

*For forged or rolled specimens only.

402

HARDENABILITY BAND 94B17H

C	Mn	Si	Ni	Cr	Mo	B
.14 / .20	.70 / 1.05	.20 / .35	.25 / .65	.25 / .55	.08 / .15	*

Can be expected to have 0.0005 per cent minimum boron content.

DIAMETERS OF ROUNDS WITH SAME AS QUENCHED HARDNESS

Diameters	LOCATION IN ROUND	QUENCH
3.8	SURFACE	MILD WATER QUENCH
1.1 2.0 2.9 3.8 4.8 5.8 6.7	3/4 RADIUS FROM CENTER	MILD WATER QUENCH
0.7 1.2 1.6 2.0 2.4 2.8 3.2 3.6 3.9	CENTER	MILD WATER QUENCH
0.8 1.8 2.5 3.0 3.4 3.8	SURFACE	MILD OIL QUENCH
0.5 1.0 1.6 2.0 2.4 2.8 3.2 3.6 4.0	3/4 RADIUS FROM CENTER	MILD OIL QUENCH
0.2 0.6 1.0 1.4 1.7 2.0 2.4 2.8 3.1	CENTER	MILD OIL QUENCH

ROCKWELL HARDNESS C SCALE (65 – 20)

DISTANCE FROM QUENCHED END — SIXTEENTHS OF AN INCH (2 – 32)

HARDNESS LIMITS FOR SPECIFICATION PURPOSES

"J" DISTANCE SIXTEENTHS OF AN INCH	94B17H MAX.	MIN.
1	46	39
2	46	39
3	45	38
4	45	37
5	44	34
6	43	29
7	42	26
8	41	24
9	40	23
10	38	21
11	36	20
12	34	–
13	33	–
14	32	–
15	31	–
16	30	–
18	28	–
20	27	–
22	26	–
24	25	–
26	24	–
28	24	–
30	23	–
32	23	–

HEAT TREATING TEMPERATURES RECOMMENDED BY SAE

*NORMALIZE 1700 °F
AUSTENITIZE 1700 °F

*For forged or rolled specimens only.

Constructional Steels for Heat Treating — United States (Alloy Bars, Blooms, Billets, and Slabs)

403

HARDENABILITY BAND 94B30H

C	Mn	Si	Ni	Cr	Mo	B
.27 / .33	.70 / 1.05	.20 / .35	.25 / .65	.25 / .55	.08 / .15	*

* Can be expected to have 0.0005 per cent minimum boron content.

DIAMETERS OF ROUNDS WITH SAME AS QUENCHED HARDNESS

LOCATION IN ROUND									QUENCH
SURFACE	3.8								MILD WATER QUENCH
3/4 RADIUS FROM CENTER	1.1	2.0	2.9	3.8	4.8	5.8	6.7		
CENTER	0.7	1.2	1.6	2.0	2.4	2.8	3.2	3.6	3.9
SURFACE	0.8	1.8	2.5	3.0	3.4	3.8			MILD OIL QUENCH
3/4 RADIUS FROM CENTER	0.5	1.0	1.6	2.0	2.4	2.8	3.2	3.6	4.0
CENTER	0.2	0.6	1.0	1.4	1.7	2.0	2.4	2.8	3.1

ROCKWELL HARDNESS C SCALE

DISTANCE FROM QUENCHED END — SIXTEENTHS OF AN INCH

HARDNESS LIMITS FOR SPECIFICATION PURPOSES

"J" DISTANCE SIXTEENTHS OF AN INCH	94B30 H MAX.	MIN.
1	56	49
2	56	49
3	55	48
4	55	48
5	54	47
6	54	46
7	53	44
8	53	42
9	52	39
10	52	37
11	51	34
12	51	32
13	50	30
14	49	29
15	48	28
16	46	27
18	44	25
20	42	24
22	40	23
24	38	23
26	37	22
28	35	21
30	34	21
32	34	20

HEAT TREATING TEMPERATURES RECOMMENDED BY SAE
*NORMALIZE 1650 °F
AUSTENITIZE 1600 °F

*For forged or rolled specimens only.

VII Methods of Calculating the Hardenability of Heat Treating Grades of Steel

Four methods are described: A. The Grossmann Method. B. The Climax Molybdenum Method. C. The Jatczak and Girardi Method. D. The R. A. Grange Method.

A The Grossmann Method

In 1942 M. A. Grossmann published a method for deriving the hardenability of steel from chemical composition and grain size by the use of hardenability multiplying factors (Ref. 6.9). Further work by J. Field (Ref. 6.10) made it possible to derive the familiar end-quench hardenability curve. By the use of the following tables it is possible to predict the end-quench hardenability curve of a steel of known chemical composition and grain size with a reasonable degree of accuracy.

Because most of Grossmann's work was done with direct-hardening grades of steel, its accuracy is at its best, approximately ±10%. This method of calculating hardenability is also useful when working in reverse; that is, by knowing the hardenability desired it is possible to work out the various combinations of chemical composition and grain size that will produce the desired result.

The Grossmann method consists of the following steps:

1 Determine ideal critical diameter (D_I)* from composition by using the multiplying factors for each element of the composition. These factors are taken from Table VII.1.

2 To predict the Rockwell hardness at various distances from the quenched end of the Jominy test bar the hardness at J_1 (1/16th from the quenched end) is divided by the appropriate factor from Table VII.2. The J_1 hardness is a function of carbon content only and is also listed in Table VII.1.

Example. Assume that the following chemical analysis and grain size = ASTM 8:

C	Mn	Si	S	P	Ni	Cr	Mo
0.35	0.60	0.20	0.020	0.025	0.35	0.50	0.10

---

*D_I is defined as the diameter round bar that will harden to 50% martensite in the center when quenched with a severity that equals the thermal conductivity of steel (ideal quench).

Table VII.1 Multiplying Factors and 1/16 In. Hardness Values

%	Carbon-Grain Size				Mn	Si	Ni	Cr	Mo	Rc (1/16")[a]	
	#5	#6	#7	#8							
0.01	0.0380	0.0340	0.0319	0.0300	1.033	1.007	1.004	1.0216	1.03	· · ·	1.02
0.02	0.0550	0.0510	0.0460	0.0430	1.067	1.014	1.007	1.0432	1.06	· · ·	1.03
0.03	0.0680	0.0629	0.0578	0.0538	1.100	1.021	1.011	1.0648	1.09	· · ·	1.05
0.04	0.0791	0.0727	0.0672	0.0619	1.133	1.028	1.015	1.0864	1.12	· · ·	1.07
0.05	0.0883	0.0814	0.0750	0.0697	1.167	1.035	1.018	1.1080	1.15	· · ·	1.09
0.06	0.0949	0.0888	0.0821	0.0762	1.200	1.042	1.022	1.1296	1.18	· · ·	1.10
0.07	0.1050	0.0960	0.0891	0.0833	1.233	1.049	1.026	1.1512	1.21	· · ·	1.12
0.08	0.1122	0.1029	0.0957	0.0888	1.267	1.056	1.029	1.1728	1.24	· · ·	1.14
0.09	0.1191	0.1090	0.1014	0.0941	1.300	1.063	1.033	1.1944	1.27	· · ·	
0.10	0.1259	0.1153	0.1065	0.0995	1.333	1.070	1.036	1.2160	1.30	39.0	
0.11	0.1319	0.1217	0.1121	0.1041	1.367	1.077	1.040	1.2376	1.33	39.5	
0.12	0.1373	0.1268	0.1177	0.1088	1.400	1.084	1.044	1.2592	1.36	40.0	
0.13	0.1430	0.1320	0.1221	0.1137	1.433	1.091	1.047	1.2808	1.39	40.5	
0.14	0.1480	0.1368	0.1270	0.1175	1.467	1.098	1.051	1.3024	1.42	41.0	
0.15	0.1528	0.1413	0.1315	0.1212	1.500	1.105	1.055	1.3240	1.45	41.5	
0.16	0.1579	0.1460	0.1355	0.1250	1.533	1.112	1.058	1.3456	1.48	42.0	
0.17	0.1622	0.1500	0.1399	0.1287	1.567	1.119	1.062	1.3672	1.51	42.5	
0.18	0.1670	0.1542	0.1438	0.1322	1.600	1.126	1.066	1.3888	1.54	43.0	
0.19	0.1720	0.1583	0.1474	0.1361	1.633	1.133	1.069	1.4104	1.57	44.0	
0.20	0.1761	0.1623	0.1509	0.1400	1.667	1.140	1.073	1.4320	1.60	44.5	

45.0	1.63	1.4536	1.077	1.147	1.700	0.1436	0.1542	0.1662	0.1810	0.21
45.5	1.66	1.4752	1.080	1.154	1.733	0.1462	0.1579	0.1701	0.1855	0.22
46.0	1.69	1.4968	1.084	1.161	1.767	0.1499	0.1614	0.1741	0.1897	0.23
47.0	1.72	1.5184	1.088	1.168	1.800	0.1528	0.1642	0.1780	0.1939	0.24
47.5	1.75	1.54	1.091	1.175	1.833	0.1560	0.1678	0.1820	0.1980	0.25
48.0	1.78	1.5616	1.095	1.182	1.867	0.1589	0.1712	0.1857	0.2021	0.26
48.5	1.81	1.5832	1.098	1.189	1.900	0.1619	0.1743	0.1891	0.2060	0.27
49.0	1.84	1.6048	1.102	1.196	1.933	0.1642	0.1779	0.1923	0.2099	0.28
49.5	1.87	1.6264	1.106	1.203	1.967	0.1672	0.1814	0.1960	0.2138	0.29
50.0	1.90	1.6480	1.109	1.210	2.000	0.1700	0.1849	0.1991	0.2172	0.30
50.5	1.93	1.6696	1.113	1.217	2.033	0.1733	0.1879	0.2022	0.2205	0.31
51.5	1.96	1.6912	1.117	1.224	2.067	0.1760	0.1911	0.2059	0.2240	0.32
52.0	1.99	1.7128	1.120	1.231	2.100	0.1790	0.1941	0.2088	0.2272	0.33
53.0	2.02	1.7344	1.124	1.238	2.133	0.1819	0.1969	0.2120	0.2300	0.34
53.5	2.05	1.7560	1.128	1.245	2.167	0.1842	0.2000	0.2154	0.2339	0.35
54.0	2.08	1.7760	1.131	1.252	2.200	0.1869	0.2023	0.2181	0.2368	0.36
54.5	2.11	1.7992	1.135	1.259	2.233	0.1900	0.2052	0.2214	0.2401	0.37
55.0	2.14	1.8208	1.139	1.266	2.267	0.1922	0.2079	0.2241	0.2431	0.38
55.5	2.17	1.8424	1.142	1.273	2.300	0.1950	0.2107	0.2273	0.2460	0.39
56.0	2.20	1.8640	1.146	1.280	2.333	0.1976	0.2130	0.2300	0.2492	0.40
56.5	2.23	1.8856	1.150	1.287	2.367	0.2000	0.2159	0.2331	0.2522	0.41
57.0	2.26	1.9072	1.153	1.294	2.400	0.2181	0.2181	0.2360	0.2559	0.42
58.0	2.29	1.9288	1.157	1.301	2.433	0.2042	0.2209	0.2390	0.2586	0.43
58.5	2.32	1.9504	1.160	1.308	2.467	0.2065	0.2233	0.2420	0.2618	0.44
59.0	2.35	1.9720	1.164	1.315	2.500	0.2090	0.2259	0.2440	0.2642	0.45

Continued

Table VII.1 Continued

	Carbon–Grain Size				Mn	Si	Ni	Cr	Mo	Rc (1/16")[a]
%	#5	#6	#7	#8						
0.46	0.2672	0.2465	0.2280	0.2117	2.533	1.322	1.168	1.9936	2.38	60.0
0.47	0.2701	0.2495	0.2301	0.2139	2.567	1.329	1.171	2.0152	2.41	60.5
0.48	0.2733	0.2521	0.2325	0.2159	2.600	1.336	1.175	2.0368	2.44	61.0
0.49	0.2762	0.2549	0.2352	0.2180	2.633	1.343	1.179	2.0584	2.47	61.5
0.50	0.2800	0.2580	0.2380	0.2200	2.667	1.350	1.182	2.0800	Factor = % Mo × 3 + 1	62.0
0.51	0.283	0.263	0.242	0.223	2.700	1.357	1.186	2.1016		62.5
0.52	0.286	0.266	0.244	0.225	2.733	1.364	1.190	2.1232		63.0
0.53	0.289	0.268	0.246	0.227	2.767	1.371	1.193	2.1448		63.5
0.54	0.292	0.271	0.249	0.229	2.800	1.378	1.197	2.1664		63.5
0.55	0.294	0.273	0.251	0.231	2.833	1.385	1.201	2.1880		64.0
0.56	0.297	0.275	0.253	0.233	2.867	1.392	1.204	2.2096		64.0
0.57	0.300	0.277	0.256	0.235	2.900	1.399	1.208	2.2312		64.5
0.58	0.303	0.280	0.258	0.237	2.933	1.406	1.212	2.2528		64.5
0.59	0.305	0.282	0.260	0.239	2.967	1.413	1.215	2.2744		64.5
0.60	0.308	0.284	0.262	0.241	3.000	1.420	1.219	2.2960		65.0
0.61	0.311	0.287	0.264	0.243	3.033	1.427	1.222	2.3176		65
0.62	0.314	0.289	0.267	0.245	3.067	1.434	1.226	2.3392		65
0.63	0.316	0.291	0.269	0.247	3.100	1.441	1.230	2.3608		65
0.64	0.319	0.293	0.271	0.249	3.133	1.448	1.233	2.3824		65
0.65	0.321	0.295	0.273	0.251	3.167	1.455	1.237	2.4040		65

0.66	0.324	0.297	0.275	0.253	3.200	1.462	1.241	2.4256	65
0.67	0.326	0.299	0.277	0.255	3.233	1.469	1.244	2.4472	65
0.68	0.329	0.301	0.279	0.257	3.267	1.476	1.248	2.4688	65
0.69	0.331	0.303	0.281	0.259	3.300	1.483	1.252	2.4904	65
0.70	0.334	0.306	0.283	0.260	3.333	1.490	1.255	2.5120	65
0.71	0.336	0.308	0.285	0.262	3.367	1.497	1.259	2.5336	65
0.72	0.339	0.310	0.287	0.264	3.400	1.504	1.262	2.5552	65
0.73	0.341	0.312	0.289	0.266	3.433	1.511	1.266	2.5768	65
0.74	0.343	0.314	0.291	0.268	3.467	1.518	1.270	2.5984	65
0.75	0.346	0.316	0.293	0.270	3.500	1.525	1.273	2.62	65
0.76	0.348	0.318	0.295	0.271	3.533	1.532	1.276	2.6416	65
0.77	0.350	0.320	0.297	0.273	3.567	1.539	1.280	2.6632	65
0.78	0.352	0.322	0.299	0.275	3.600	1.546	1.284	2.6848	65
0.79	0.354	0.324	0.301	0.276	3.633	1.553	1.287	2.7064	65
0.80	0.356	0.326	0.303	0.278	3.667	1.560	1.291	2.7280	65
0.81	0.358	0.328	0.305	0.280	3.700	1.567	1.294	2.7496	65
0.82	0.360	0.330	0.307	0.282	3.733	1.574	1.298	2.7712	65
0.83	0.362	0.332	0.309	0.284	3.767	1.581	1.301	2.7928	65
0.84	0.364	0.334	0.310	0.286	3.800	1.588	1.306	2.8144	65
0.85	0.368	0.336	0.312	0.287	3.833	1.595	1.309	2.8360	65
0.86	0.368	0.338	0.314	0.289	3.867	1.602	1.313	2.8576	65
0.87	0.370	0.340	0.316	0.291	3.900	1.609	1.317	2.8792	65
0.88	0.372	0.342	0.318	0.293	3.933	1.616	1.320	2.9002	65
0.89	0.374	0.344	0.319	0.294	3.967	1.623	1.324	2.9224	65
0.90	0.375	0.346	0.321	0.296	4.000	1.630	1.321	2.9440	65

Factor = % Mo X 3 + 1.00

Continued

Table VII.1 Continued

%	Carbon-Grain Size				Mn	Si	Ni	Cr	Mo	Rc (1/16")[a]
	#5	#6	#7	#8						
0.91					4.033	1.637	1.331	2.9656	Factor	
0.92					4.067	1.644	1.334	2.9872	= % Mo	
0.93					4.100	1.651	1.338	3.0088	× 3	
0.94					4.133	1.658	1.343	3.0304	+ 1.00	
0.95					4.167	1.665	1.345	3.0520		
0.96					4.200	1.672	1.349	3.0736		
0.97					4.233	1.679	1.352	3.0952		
0.98					4.267	1.686	1.356	3.1168		
0.99					4.300	1.693	1.360	3.1384		
1.00					4.333	1.700	1.364	3.1600		
1.01					4.367	1.707	1.367	3.1816		
1.02					4.400	1.714	1.370	3.2032		
1.03					4.433	1.721	1.375	3.2248		
1.04					4.467	1.728	1.378	3.2464		
1.05					4.500	1.735	1.382	3.2680		
1.06					4.533	1.742	1.386	3.2896		
1.07					4.567	1.749	1.389	3.3112		
1.08					4.600	1.756	1.393	3.3328		
1.09					4.633	1.763	1.396	3.3544		
1.10					4.667	1.770	1.400	3.3760		

Continued

1.11	4.700	1.777	1.403	3.3976
1.12	4.733	1.784	1.406	3.4192
1.13	4.767	1.791	1.411	3.4408
1.14	4.800	1.798	1.414	3.4624
1.15	4.833	1.805	1.418	3.4840
1.16	4.867	1.812	1.422	3.5056
1.17	4.900	1.819	1.426	3.5272
1.18	4.933	1.826	1.429	3.5488
1.19	4.967	1.833	1.422	3.5704
1.20	5.000	1.840	1.437	3.5920
1.21	5.051	1.847	1.440	3.6136
1.22	5.102	1.854	1.444	3.6352
1.23	5.153	1.861	1.447	3.6568
1.24	5.204	1.868	1.450	3.6784
1.25	5.255	1.875	1.454	3.700
1.26	5.306	1.882	1.458	3.7216
1.27	5.357	1.889	1.461	3.7432
1.28	5.408	1.896	1.465	3.7648
1.29	5.459	1.903	1.470	3.7864
1.30	5.510	1.910	1.473	3.8080
1.31	5.561	1.917	1.476	3.8296
1.32	5.612	1.924	1.481	3.8512
1.33	5.663	1.931	1.484	3.8728
1.34	5.714	1.938	1.487	3.8944
1.35	5.765	1.945	1.491	3.9160

Table VII.1 Continued

	Carbon-Grain Size									
%	#5	#6	#7	#8	Mn	Si	Ni	Cr	Mo	Rc (1/16″)[a]
1.36					5.816	1.952	1.495	3.9376		
1.37					5.867	1.959	1.498	3.9592		
1.38					5.918	1.966	1.501	3.9808		
1.39					5.969	1.973	1.506	4.0024		
1.40					6.020	1.980	1.509	4.0240		
1.41					6.071	1.987	1.512	4.0456		
1.42					6.122	1.994	1.517	4.0672		
1.43					6.173	2.001	1.520	4.0888		
1.44					6.224	2.008	1.523	4.1104		
1.45					6.275	2.015	1.527	4.1320		
1.46					6.326	2.022	1.531	4.1536		
1.47					6.377	2.029	1.535	4.1952		
1.48					6.428	2.036	1.538	4.1968		
1.49					6.479	2.043	1.541	4.2174		
1.50					6.530	2.050	1.545	4.2390		
1.51					6.581	2.057	1.5563	4.2616		
1.52					6.632	2.064	1.5606	4.2832		
1.53					6.683	2.071	1.5649	4.3048		
1.54					6.734	2.078	1.5692	4.3264		
1.55					6.785	2.085	1.5735	4.3480		

1.56	6.836	2.092	1.5778	4.3696
1.57	6.887	2.099	1.5821	4.3912
1.58	6.938	2.106	1.5864	4.4128
1.59	6.989	2.113	1.5907	4.4344
1.60	7.040	2.120	1.5950	4.4560
1.61	7.091	2.127	1.5995	4.4776
1.62	7.142	2.134	1.6040	4.4990
1.63	7.193	2.141	1.6085	4.5208
1.64	7.224	2.148	1.6230	4.5424
1.65	7.295	2.155	1.6175	4.5640
1.66	7.346	2.162	1.6220	4.5856
1.67	7.397	2.169	1.6265	4.6072
1.68	7.448	2.176	1.6310	4.6288
1.69	7.499	2.183	1.6355	4.6504
1.70	7.550	2.190	1.640	4.6720
1.71	7.601	2.197	1.644	4.6936
1.72	7.652	2.204	1.648	4.7152
1.73	7.703	2.211	1.652	4.7368
1.74	7.754	2.218	1.656	4.7544
1.75	7.805	2.225	1.660	4.7800
1.76	7.856	2.232	1.664	
1.77	7.907	2.239	1.668	
1.78	7.958	2.246	1.672	
1.79	8.009	2.253	1.670	
1.80	8.060	2.280	1.680	

Table VII.1 Continued

| | Carbon-Grain Size | | | | | | | | | |
%	#5	#6	#7	#8	Mn	Si	Ni	Cr	Mo	Rc (1/16")[a]
1.81					8.111	2.267	1.687			
1.82					8.162	2.274	1.694			
1.83					8.213	2.281	1.701			
1.84					8.315	2.288	1.708			
1.85					8.366	2.295	1.715			
1.86					8.417	2.302	1.722			
1.87					8.468	2.309	1.729			
1.88					8.519	2.316	1.736			
1.89					8.570	2.323	1.743			
1.90					8.621	2.330	1.750			
1.91					8.672	2.337	1.7529			
1.92					8.723	2.344	1.7558			
1.93					8.774	2.351	1.7587			
1.94					8.825	2.358	1.7616			
1.95					8.876	2.364	1.7645			
1.96						2.372	1.7674			
1.97						2.379	1.7703			
1.98						2.386	1.7732			
1.99						2.393	1.7761			
2.00						2.400	1.7790			

%	Ni	%	Ni	%	Ni	%	Ni	%	Ni
2.01	1.7841	2.41	2.015	2.81	2.3408	3.20	2.80	3.60	
2.02	1.7892	2.42	2.022	2.82	2.3516	3.21	2.815	3.61	
2.03	1.7943	2.43	2.029	2.83	2.3624	3.22	2.830	3.62	
2.04	1.7994	2.44	2.036	2.84	2.3732	3.23	2.845	3.63	
2.05	1.8045	2.45	2.043	2.85	2.3840	3.24	2.860	3.64	
2.06	1.8096	2.46	2.050	2.86	2.3948	3.25	2.875	3.65	
2.07	1.8147	2.47	2.057	2.87	2.4056	3.26	2.890	3.66	
2.08	1.8198	2.48	2.064	2.88	2.4164	3.27	2.905	3.67	
2.09	1.8249	2.49	2.071	2.89	2.4272	3.28	2.920	3.68	
2.10	1.830	2.50	2.078	2.90	2.4380	3.29	2.935	3.69	
2.11	1.8352	2.51	2.0857	2.91	2.4487	3.30	2.950	3.70	
2.12	1.8402	2.52	2.0934	2.92	2.4594	3.31	2.965	3.71	
2.13	1.8454	2.53	2.1011	2.93	2.4701	3.32	2.980	3.72	
2.14	1.8506	2.54	2.1088	2.94	2.4808	3.33	2.995	3.73	
2.15	1.8558	2.55	2.1165	2.95	2.4915	3.34	3.010	3.74	
2.16	1.8610	2.56	2.1242	2.96	2.5022	3.35	3.025	3.75	
2.17	1.8662	2.57	2.1319	2.97	2.5129	3.36	3.040	3.76	
2.18	1.8714	2.58	2.1396	2.98	2.5236	3.37	3.055	3.77	
2.19	1.8766	2.59	2.1473	2.99	2.5343	3.38	3.070	3.78	
2.20	1.8820	2.60	2.1550	3.00	2.5450	3.39	3.085	3.79	
2.21	1.888	2.61	2.1635	3.01	2.5575	3.40	3.10	3.80	
2.22	1.894	2.62	2.1720	3.02	2.5700	3.41	3.116	3.81	
2.23	1.900	2.63	2.1805	3.03	2.5825	3.42	3.132	3.82	

Continued

415

Table VII.1 Continued

%	Ni	%	Ni	%	Ni	%	Ni	%	Ni
2.24	1.906	2.64	2.1890	3.04	2.5950	3.43	3.148	3.83	
2.25	1.912	2.65	2.1975	3.05	2.6075	3.44	3.164	3.84	
2.26	1.918	2.66	2.2060	3.06	2.6200	3.45	3.180	3.85	
2.27	1.924	2.67	2.2145	3.07	2.6325	3.46	3.196	3.86	
2.28	1.930	2.68	2.2230	3.08	2.6450	3.47	3.212		
2.29	1.936	2.69	2.2315	3.09	2.6575	3.48	3.228		
2.30	1.942	2.70	2.240	3.10	2.67	3.49	3.244		
2.31	1.9486	2.71	2.249	3.11	2.696	3.50	3.26		
2.32	1.9552	2.72	2.258	3.12	2.709	3.51			
2.33	1.9618	2.73	2.267	3.13	2.722	3.52			
2.34	1.9684	2.74	2.276	3.14		3.53			
2.35	1.9750	2.75	2.285			3.54			
2.36	1.9816	2.76	2.294	3.15	2.735	3.55			
2.37	1.9882	2.77	2.303	3.16	2.748	3.56			
2.38	1.9948	2.78	2.312	3.17	2.761	3.57			
2.39	2.0014	2.79	2.321	3.18	2.774	3.58			
2.40	2.0080	2.80	2.330	3.19	2.787	3.59			

Sulfur Factors

Sulfur (%)	Factor	Sulfur (%)	Factor
0.020	0.985	0.060	0.955
0.030	0.977	0.070	0.947
0.040	0.970	0.080	0.940
0.050	0.962	0.090	0.932

Phosphorus Factors

Phosphorus (%)	Factor	Phosphorus (%)	Factor
0.020	1.052	0.050	1.130
0.030	1.078	0.060	1.157
0.040	1.104	0.070	1.183

Copper Factors

Copper (%)	Factor	Copper (%)	Factor
0.15	1.00	1.00	1.35
0.20	1.07	1.25	1.40
0.25	1.09	1.50	1.55
0.50	1.18		
0.75	1.30		

Zirconium Factors

Zr (%)	Factor	Zr (%)	Factor
.03	1.070	.20	1.500
.05	1.120	.25	1.625
.10	1.250	.30	1.750
.15	1.370		

[a] From Crafts and Lamont, Ref. 6.14.

Table VII.2 Dividing Factors (Ratio of 1/16″ Hardness to Distance Hardness)

D_I	J/4	J/8	J/12	J/16	J/20	J/24	J/28	J/32
1.50	1.50	2.345	2.88	3.26	3.53	3.71	3.89	4.07
1.55	1.45	2.27	2.82	3.17	3.45	3.62	3.79	3.96
1.60	1.41	2.185	2.735	3.08	3.36	3.53	3.70	3.85
1.65	1.37	2.12	2.665	3.015	3.285	3.45	3.60	3.725
1.70	1.35	2.055	2.60	2.94	3.20	3.36	3.50	3.615
1.75	1.33	2.00	2.53	2.875	3.125	3.275	3.42	3.52
1.80	1.305	1.945	2.46	2.81	3.05	3.195	3.33	3.42
1.85	1.29	1.90	2.405	2.75	2.975	3.12	3.25	3.33
1.90	1.27	1.855	2.345	2.69	2.91	3.05	3.175	3.255
1.95	1.255	1.815	2.295	2.625	2.85	2.985	3.11	3.19
2.00	1.235	1.78	2.24	2.56	2.78	2.92	3.04	3.125
2.05	1.225	1.745	2.195	2.51	2.73	2.87	2.975	3.065
2.10	1.21	1.71	2.15	2.465	2.685	2.815	2.92	3.01
2.15	1.20	1.685	2.105	2.42	2.64	2.765	2.87	2.96
2.20	1.185	1.66	2.06	2.38	2.595	2.715	2.82	2.91
2.25	1.175	1.635	2.025	2.34	2.56	2.675	2.775	2.865
2.30	1.165	1.615	1.99	2.305	2.53	2.635	2.735	2.825
2.35	1.16	1.595	1.955	2.27	2.485	2.60	2.695	2.78
2.40	1.145	1.575	1.92	2.235	2.445	2.565	2.66	2.74
2.45	1.14	1.555	1.89	2.20	2.415	2.53	2.625	2.705
2.50	1.13	1.535	1.86	2.165	2.39	2.50	2.59	2.67
2.60	1.115	1.50	1.81	2.10	2.325	2.44	2.53	2.60

2.70	1.10	1.47	1.77	2.04	2.27	2.38	2.47	2.54
2.80	1.09	1.445	1.73	1.995	2.215	2.325	2.415	2.48
2.90	1.08	1.415	1.695	1.945	2.165	2.275	2.365	2.425
3.00	1.07	1.39	1.66	1.905	2.11	2.225	2.31	2.365
3.10	1.06	1.365	1.63	1.87	2.07	2.175	2.26	2.315
3.20	1.055	1.345	1.595	1.83	2.025	2.13	2.215	2.265
3.30	1.05	1.325	1.57	1.80	1.98	2.085	2.17	2.22
3.40	1.045	1.30	1.54	1.76	1.94	2.04	2.125	2.17
3.50	1.04	1.28	1.51	1.73	1.895	2.00	2.085	2.125
3.60	1.035	1.26	1.485	1.695	1.86	1.955	2.05	2.085
3.70	1.03	1.245	1.46	1.665	1.82	1.915	2.00	2.04
3.80	1.03	1.225	1.435	1.635	1.785	1.875	1.955	2.00
3.80	1.025	1.21	1.41	1.605	1.75	1.84	1.915	1.96
4.00	1.02	1.195	1.385	1.575	1.71	1.80	1.875	1.92
4.10	1.02	1.18	1.36	1.545	1.68	1.76	1.835	1.88
4.20	1.02	1.165	1.34	1.515	1.645	1.725	1.80	1.845
4.30	1.015	1.155	1.315	1.485	1.61	1.69	1.76	1.81
4.40	1.01	1.45	1.29	1.455	1.58	1.655	1.725	1.775
4.50	1.01	1.13	1.27	1.43	1.55	1.62	1.69	1.735
4.60	1.01	1.12	1.25	1.40	1.515	1.585	1.65	1.70
4.70	1.005	1.11	1.23	1.375	1.485	1.555	1.62	1.665
4.80	1.005	1.10	1.21	1.35	1.455	1.525	1.585	1.63
4.90	1.005	1.09	1.195	1.325	1.425	1.495	1.55	1.595
5.00	1.00	1.08	1.18	1.305	1.40	1.46	1.52	1.56

Continued

Table VII.2 Continued

D_I	J/4	J/8	J/12	J/16	J/20	J/24	J/28	J/32
5.10	1.00	1.07	1.16	1.28	1.37	1.43	1.49	1.53
5.20	1.00	1.065	1.145	1.26	1.345	1.40	1.455	1.495
5.30	1.00	1.06	1.13	1.235	1.32	1.375	1.425	1.46
5.40	1.00	1.05	1.115	1.22	1.295	1.345	1.395	1.43
5.50	1.00	1.04	1.10	1.20	1.275	1.32	1.365	1.40
5.60	1.00	1.035	1.085	1.18	1.25	1.29	1.33	1.365
5.70	1.00	1.03	1.075	1.16	1.225	1.265	1.30	1.335
5.80	1.00	1.025	1.065	1.145	1.20	1.24	1.27	1.305
5.90	1.00	1.02	1.055	1.125	1.18	1.215	1.24	1.275
6.00	1.00	1.015	1.05	1.11	1.155	1.19	1.215	1.24
6.10	1.00	1.01	1.04	1.09	1.135	1.165	1.185	1.215
6.20	1.00	1.01	1.035	1.08	1.115	1.14	1.16	1.185
6.30	1.00	1.005	1.025	1.065	1.095	1.12	1.14	1.16
6.40	1.00	1.005	1.02	1.05	1.075	1.10	1.115	1.14
6.50	1.00	1.00	1.015	1.04	1.06	1.08	1.095	1.115
6.60	1.00	1.00	1.01	1.03	1.05	1.06	1.08	1.095
6.70	1.00	1.00	1.005	1.02	1.035	1.05	1.06	1.075
6.80	1.00	1.00	1.00	1.01	1.02	1.035	1.045	1.055
6.90	1.00	1.00	1.00	1.00	1.01	1.02	1.03	1.04
7.00	1.00	1.00	1.00	1.00	1.00	1.015	1.02	1.025
7.10	1.00	1.00	1.00	1.00	1.00	1.00	1.015	1.02
7.20	1.00	1.00	1.00	1.00	1.00	1.00	1.00	1.01
7.30	1.00	1.00	1.00	1.00	1.00	1.00	1.00	1.00

Calculate D_I from factors in Table VII.1.

$D_I = 0.1842 \times 3.000 \times 1.140 \times 0.985 \times 1.065 \times 1.128 \times 2.080 \times 1.30$

$D_I = 2.01$ in.

Hardness at J1 from Table VII.1 = Rc 53.5.

Using dividing factors from Table VII.2, calculate hardness for various Jominy positions:

Jominy Distance	Dividing Factor	Calculated Hardness (Rc)
1	. . .	53.5
4	1.235	43.3
8	1.78	30.0
12	2.24	23.9
16	2.56	20.1
20	2.78	19.2
24	2.92	18.3
28	3.04	17.6
32	3.125	17.1

The Grossmann method is also useful for determining the hardenability of boron-heated steel. When steel is properly made, the boron factor, or its contribution to increased hardenability, is an inverse function of the carbon content. The higher the carbon, the lower the boron factor and the less the contribution to increasing hardenability. The *approximate* effect can be expressed by the following formula to obtain a multiplying factor

$$B_f = 1 + 1.6(1.01 - \% \text{ carbon})$$

for example, if the steel in the preceding example were properly boron-treated, the hardenability (D_I) would be increased by a factor of

$$B_f = 1.1 + 1.6(1.01 - 0.35)$$
$$B_f = 2.16$$

The ideal critical diameter would be increased to

$$D_I = (2.01)(2.16) = 4.34 \text{ in.}$$

The calculated hardness values at various J distances would then increase as follows:

Jominy Distance	Dividing Factor	Calculated Hardness (Rc)
J_1	. . .	53.5
J_4	1.01	52.9
J_8	1.151	46.5
J_{12}	1.30	41.1
J_{16}	1.47	36.4
J_{20}	1.60	33.4
J_{24}	1.67	32.0
J_{28}	1.74	30.7
J_{32}	1.79	29.8

Calculated hardenability is useful for determining the effectiveness of the boron addition by comparing the actual D_I with its calculated D_I. The actual boron factor is expressed by the following equation:

$$\text{Actual } B_f = \frac{\text{actual } D_I}{\text{calculated } D_I \text{ (without boron)}}$$

Table VII.3 "As-Quenched" Hardness of 50% Martensite[a]

Carbon (%)	Hardness (Rc)	Carbon (%)	Hardness (Rc)	Carbon (%)	Hardness (Rc)	Carbon (%)	Hardness (Rc)	Carbon (%)	Hardness (Rc)
0.11	26.0	0.21	31.9	0.31	35.8	0.41	39.4	0.51	43.4
0.12	26.7	0.22	32.3	0.32	26.2	0.42	39.8	0.52	43.8
0.13	27.3	0.23	32.7	0.33	36.5	0.43	40.2	0.53	44.2
0.14	27.9	0.24	33.1	0.34	36.9	0.44	40.6	0.54	44.6
0.15	28.5	0.25	33.5	0.35	37.2	0.45	41.0	0.55	45.0
0.16	29.1	0.26	33.9	0.36	37.8	0.46	41.4	0.56	45.5
0.17	29.7	0.27	34.3	0.37	37.9	0.47	41.8	0.57	45.8
0.18	30.3	0.28	34.7	0.38	38.3	0.48	42.2	0.58	46.2
0.19	30.9	0.29	35.1	0.39	38.6	0.49	42.6	0.59	46.6
0.20	31.5	0.30	35.5	0.40	39.0	0.50	43.0	0.60	47.0

[a]From Hodge and Orehoski Ref. 6.17.

To determine the actual D_I, a Jominy test must be made. The Jominy distance to 50% martensite is determined from the results Table VII.3. Once this is established Table VII.4 is used to determine actual D_I based on Jominy distance to hardness corresponding to 50% martensite; for example, consider the following Jominy test results for this boron-treated steel:

Analysis:	C	Mn	Si	S	P	Ni	Cr	Mo	B
	0.35	0.60	0.20	0.020	0.025	0.35	0.50	0.10	0.001

Jominy Results:

Jominy Distance (in./16)	Rc
1	53
2	53
3	52.5
4	52
6	48.3
8	44.3
10	41.5
12	38.3
14	36.0
16	33.6
20	31.0
24	29.4
28	28
32	26

Table VII.3 indicates that the hardness of 50% martensite for a 0.35% C steel is Rockwell C 37.2. This occurs at $J = 12.5$. Referring to Table VII.4, a Jominy distance of 12.5 sixteenths to 50% martensite hardness corresponds to an actual $D_I = 3.94$ in. Therefore

$$\text{Actual boron factor} = \frac{3.94}{2.01}$$

$$B_f = 1.96$$

This corresponds well with the previously calculated boron factor of 2.16 and the boron addition was effective.

Table VII.4   Relationship Between Jominy Distance to 50% Martensite Hardness and D_I

Jominy Distance	D_I	Jominy Distance	D_I	Jominy Distance	D_I
0.5	0.27	11.5	3.74	22.5	5.46
1.0	0.50	12.0	3.83	23.0	5.51
1.5	0.63	12.5	3.94	23.5	5.57
2.0	0.95	13.0	4.04	24.0	5.63
2.5	1.16	13.5	4.13	24.5	5.69
3.0	1.37	14.0	4.22	25.0	5.74
3.5	1.57	14.5	4.32	25.5	5.80
4.0	1.75	15.0	4.40	26.0	5.86
4.5	1.93	15.5	4.48	26.5	5.91
5.0	2.12	16.0	4.57	27.0	5.96
5.5	2.29	16.5	4.64	27.5	6.02
6.0	2.45	17.0	4.72	28.0	6.06
6.5	2.58	17.5	4.80	28.5	6.12
7.0	2.72	18.0	4.87	29.0	6.16
7.5	2.86	18.5	4.94	29.5	6.20
8.0	2.97	19.0	5.02	30.0	6.25
8.5	3.07	19.5	5.08	30.5	6.29
9.0	3.20	20.0	5.15	31.0	6.33
9.5	3.32	20.5	5.22	31.5	6.37
10.0	3.43	21.0	5.28	32.0	6.42
10.5	3.54	21.5	5.33		
11.0	3.64	22.0	5.39		

.From Hodge and Orehoski, Ref. 6.17.

B The Climax Molybdenum Method (Ref 6.13 and 6.14)

This method of calculating the end-quench hardenability of steel follows the same scheme proposed by Grossmann; however, it is especially useful for carburizing grades of alloy steel. It has different multiplying factors but the method of determining a calculated D_I and constructing a theoretical Jominy curve is identical to Grossman's. The recommended carbon and grain-size factors are those extrapolated from the work of Kramer (Ref. 6.14).

Table VII.5 Multiplying Factors — Climax Molybdenum Method

Carbon				Manganese			
	Grain Size						
% C	ASTM 5	ASTM 7	ASTM 9	% Mn	Factor	% Mn	Factor
0.15	0.35	0.300	0.260	0.1	1.07	1.1	2.91
0.20	0.43	0.37	0.33	0.2	1.18	1.2	3.13
0.25	0.50	0.44	0.39	0.3	1.33	1.3	3.35
0.30	0.57	0.51	0.45	0.4	1.49	1.4	3.56
				0.5	1.67		
				0.6	1.87		
				0.7	2.07		
				0.8	2.28		
				0.9	2.48		
				1.0	2.70		

Nickel		Chromium			
% Ni	Factor	% Cr	Factor	% Cr	Factor
0.5	1.15	0.1	1.04	1.1	2.25
1.0	1.31	0.2	1.09	1.2	2.47
1.5	1.44	0.3	1.18	1.3	2.75
2.0	1.60	0.4	1.28	1.4	3.14
2.25	1.73	0.5	1.41		
2.5	1.88	0.6	1.54		
2.75	2.06	0.7	1.67		
3.0	2.26	0.8	1.81		
3.2	2.50	0.9	1.95		
3.4	2.76	1.0	2.09		
3.6	3.08				

Continued

Table VII.5   Continued

Molybdenum (with Ni < 0.75%)		Molybdenum (with Ni > 0.75%)	
% Mo	Factor	% Mo	Factor
0.1	1.11	0.1	1.22
0.2	1.25	0.2	1.41
0.3	1.43	0.3	1.68
0.4	1.61	0.4	1.98
0.5	1.82	0.5	2.29
0.6	2.03	0.6	2.59
0.7	2.27	0.7	2.90
0.8	2.51	0.8	3.20
0.9	2.75	0.9	3.51
1.0	3.02	1.0	3.82

C   The Jatczak and Girardi Method

This method of calculating hardenability was first announced in 1954 (Ref. 6.15). Not only was it a technical breakthrough that provided a means of calculating the hardenability of steels with more than 0.75% carbon but it gave the results in terms of microstructure (originally as 10% transformation), balance martensite, and austenite. In 1972 C. F. Jatczak announced that it had been updated to include more alloying elements, synergestic effects between them, and the effect of different austenitizing temperatures (Ref. 6.16). A means was also provided to obtain results for expressing the hardenability not only in terms of 10% transformation, but 1% as well.

The method employs multiplying factors (Tables VII.6 to VII.10) as in preceding methods, but in using this procedure for calculating hardenability there are several guidelines to keep in mind:

1 The multiplying factors were developed for steels that had been normalized or quenched before reaustenitization for hardening. If the steel is spheroidize-annealed before hardening, its hardenability will be slightly higher as shown in Figure VII.1.

2 For calculation of case hardenability of carburizing steels hardened by direct quenching from the carburizer (1700 F) use the factors in the tables as follows:

   (a) If the composition is pearlitic (i.e., contains less than 1.0% nickel and 0.15% molybdenum), use the factors Mn,* Si,* Ni,* Mo,* and

Table VII.6 Multiplying Factors for Carbon (Grain Size Range 5/9 ASTM)

Percent	1475 F	1525 F	1575 F	1700 F
0.60	0.77	0.79	0.79	0.79
0.65	0.795	0.81	0.82	0.82
0.70	0.82	0.83	0.85	0.85
0.75	0.83	0.845	0.875	0.875
0.80	0.83	0.86	0.90	0.90
0.85	0.80	0.85	0.91	0.93
0.90	0.73	0.81	0.90	0.935
0.91	0.715	0.785	0.89	0.935
0.92	0.70	0.765	0.88	0.93
0.93	0.685	0.745	0.87	0.92
0.94	0.675	0.73	0.86	0.91
0.95	0.66	0.71	0.85	0.90
0.96	0.65	0.70	0.835	0.89
0.97	0.64	0.69	0.825	0.875
0.98	0.625	0.675	0.81	0.86
0.99	0.62	0.665	0.795	0.845
1.00	0.61	0.655	0.78	0.83
1.01	0.60	0.645	0.76	0.815
1.02	0.595	0.64	0.74	0.80
1.03	0.59	0.63	0.725	0.79
1.04	0.58	0.625	0.71	0.78
1.05	0.575	0.62	0.695	0.77
1.06	0.57	0.61	0.68	0.76
1.07	0.565	0.605	0.67	0.75
1.08	0.56	0.60	0.655	0.74
1.09	0.557	0.595	0.645	0.735
1.10	0.555	0.59	0.64	0.73
Av G.S.	8/9	7/8	6/7	5/7

Al with the Cr factor labeled "Carburized Steels." Multiply them together with the proper factors for carbon and boron.

(b) If composition is bainitic (i.e., contains more than 1.0% nickel and 0.15% molybdenum), again use the factors Mn,* Ni,* Mo,* Al, and Cr "Carburized Steels," but for silicon use the factor labeled Si.† Also use the proper carbon and boron factors.

3 For the calculation of case hardenability of carburizing steels, which are normalized or quenched from carburizing and reheated to 1700 F for

Table VII.7 Multiplying Factors for the Calculation of Case Hardenability of Carburizing Steels and the Hardenability of High Carbon Steels Hardened After a Prior Normalize or Quench Treatment

Per-cent	Mn*				Si*				Cr*				Cr Car-burized† Steels
	1475	1525	1575	1700	1475	1525	1575	1700	1475	1525	1575	1700	1700
0.05	1.02	1.04	1.04	1.04	1.02	1.04	1.04	1.04	1.02	1.04	1.04	1.04	1.04
0.10	1.06	1.06	1.06	1.06	1.06	1.06	1.06	1.06	1.06	1.06	1.06	1.06	1.06
0.15	1.10	1.10	1.10	1.10	1.10	1.10	1.10	1.10	1.10	1.10	1.10	1.10	1.10
0.20	1.14	1.14	1.14	1.14	1.14	1.14	1.14	1.14	1.14	1.14	1.14	1.14	1.14
0.25	1.18	1.18	1.18	1.18	1.18	1.18	1.18	1.18	1.18	1.18	1.18	1.18	1.18
0.30	1.26	1.26	1.27	1.26	1.19	1.19	1.20	1.24	1.22	1.22	1.24	1.23	1.30
0.35	1.31	1.32	1.33	1.33	1.20	1.20	1.21	1.33	1.26	1.27	1.28	1.31	1.41
0.40	1.35	1.38	1.39	1.39	1.21	1.21	1.21	1.40	1.28	1.32	1.33	1.29	1.49
0.45	1.41	1.44	1.45	1.44	1.22	1.22	1.22	1.48	1.31	1.35	1.36	1.44	1.59
0.50	1.45	1.47	1.48	1.47	1.23	1.23	1.24	1.54	1.33	1.38	1.39	1.47	1.67
0.55	1.48	1.53	1.53	1.53	1.24	1.24	1.25	1.61	1.34	1.41	1.41	1.52	1.75
0.60	1.52	1.58	1.58	1.56	1.25	1.25	1.26	1.67	1.35	1.43	1.44	1.56	1.84
0.65	1.55	1.61	1.62	1.59	1.26	1.26	1.27	1.72	1.36	1.46	1.46	1.61	1.89
0.70	1.59	1.65	1.67	1.61	1.27	1.27	1.28	1.78	1.38	1.47	1.47	1.66	1.95
0.75	1.62	1.69	1.72	1.66	1.28	1.28	1.29	1.84	1.39	1.49	1.48	1.73	2.00
0.80	1.65	1.73	1.76	1.71	1.29	1.29	1.30	1.88	1.39	1.51	1.50	1.76	2.05
0.85	1.67	1.76	1.81	1.75	1.31	1.31	1.31	1.93	1.40	1.51	1.51	1.81	2.09
0.90	1.69	1.81	1.86	1.81	1.32	1.32	1.32	1.96	1.41	1.52	1.52	1.86	2.12
0.95	1.73	1.85	1.89	1.91	1.33	1.33	1.32	1.99	1.41	1.53	1.53	1.92	2.15
1.00	1.75	1.88	1.93	2.00	1.34	1.34	1.33	2.00	1.41	1.54	1.54	1.96	2.18
1.05	1.78	1.92	1.98	2.09	1.35	1.34	1.34	2.01	1.41	1.55	1.54	2.00	2.21
1.10	1.80	1.96	2.00	2.19	1.35	1.35	1.34	2.01	1.41	1.56	1.55	2.06	2.24

Table (continued). Multiplying factors for hardenability.

Per-cent	Ni*				Mo*				Al 1475 →1700	Multialloy Si†				Multialloy Ni and Mn†
	1475	1525	1575	1700	1475	1525	1575	1700		1475	1525	1575	1700	
1.15	1.82	2.02	2.04	2.28	1.36	1.36	1.35	2.04	⋯	1.41	1.56	1.56	2.08	2.27
1.20	⋯	2.07	2.09	2.33	1.36	1.37	1.35	2.05	⋯	1.41	1.58	1.57	2.12	2.29
1.25		2.14	2.16	2.40	1.38	1.38	1.36	2.06	⋯	1.41	1.59	1.58	2.15	2.32
1.30		2.21	2.23	2.45	1.39	1.39	1.38	2.07	⋯	1.41	1.59	1.58	2.18	2.35
1.35		2.29	2.33	2.49	1.40	1.40	1.39	2.08	⋯	1.41	1.59	1.59	2.21	2.39
1.40		2.37	2.42	2.55	1.41	1.41	1.40	2.09	⋯	1.42	1.61	1.59	2.23	2.44
1.45		2.50	2.54	2.59	1.42	1.43	1.41	2.10	⋯	1.43	1.51	1.59	2.24	2.47
1.50		2.58	2.62	2.62	1.43	1.45	1.43	2.11	⋯	1.44	1.61	1.59	2.25	2.50
1.55				2.66	1.45	1.46	1.45	⋯	⋯	1.45	1.62	1.60	2.26	2.53
1.60				2.69	1.47	1.47	1.46	⋯	⋯	1.45	1.62	1.61	2.27	2.55
1.65				2.72	1.48	1.49	1.48	⋯	⋯	1.46	1.63	1.62	⋯	2.57
1.70				2.75	1.50	1.51	1.51	⋯	⋯	1.47	1.64	1.64	⋯	2.60
1.75				2.79	1.52	1.53	1.53	⋯	⋯	1.47	1.65	1.65	⋯	⋯
1.80				2.82	1.54	1.55	1.55	⋯	⋯	1.48	1.67	1.67	⋯	⋯
1.85				2.86	1.57	1.58	1.58	⋯	⋯	1.49	1.68	1.69	⋯	⋯
1.90				2.89	1.60	1.60	1.60	⋯	⋯	1.52	1.69	1.70	⋯	⋯
1.95				2.93	1.62	1.62	1.62	⋯	⋯	1.54	1.72	1.72	⋯	⋯
2.00				2.95	1.66	1.66	1.65	⋯	⋯	1.56	1.74	1.74	⋯	⋯

Per-cent	Ni*				Mo*				Al 1475 →1700	Multialloy Si†				Multialloy Ni and Mn†
	1475	1525	1575	1700	1475	1525	1575	1700		1475	1525	1575	1700	
0.05	1.00	1.00	1.00	1.00	1.05	1.05	1.05	1.13	1.02	1.01	1.04	1.04	1.04	See Tables 8, 9, or 10
0.10	1.01	1.01	1.01	1.01	1.10	1.10	1.10	1.27	1.05	1.06	1.06	1.06	1.06	
0.15	1.03	1.03	1.02	1.03	1.15	1.15	1.17	1.42	1.08	1.10	1.10	1.10	1.10	
0.20	1.04	1.04	1.04	1.04	1.20	1.20	1.26	1.56	1.12	1.14	1.14	1.14	1.14	
0.25	1.05	1.05	1.05	1.05	1.24	1.24	1.35	1.73	1.15	1.18	1.18	1.18	1.18	
0.30	1.07	1.07	1.07	1.07	1.29	1.29	1.45	1.90	1.18	1.26	1.26	1.27	1.27	

Continued

429

Table VII.7 Continued

Per-cent	Ni*				Mo*				Al 1475 →1700	Multialloy Si†				Multialloy Ni and Mn†
	1475	1525	1575	1700	1475	1525	1575	1700		1475	1525	1575	1700	
0.35	1.09	1.09	1.09	1.11	1.34	1.34	1.55	2.09	1.22	1.31	1.32	1.33	1.36	
0.40	1.11	1.11	1.11	1.14	1.39	1.39	1.65	2.27	1.27	1.35	1.36	1.36	1.46	
0.45	1.12	1.13	1.12	1.16	1.44	1.44	1.75	2.45	1.31	1.41	1.40	1.40	1.54	
0.50	1.13	1.14	1.13	1.18	1.49	1.49	1.86	2.64	1.35	1.45	1.45	1.45	1.67	
0.55	1.14	1.15	1.14	1.20	1.54	1.54	1.97	2.82	1.40	1.47	1.48	1.47	1.80	
0.60	1.15	1.16	1.15	1.22	1.60	1.60	2.09	3.03	1.45	1.49	1.50	1.49	1.92	
0.65	1.16	1.17	1.16	1.24	1.66	1.66	2.21	3.26	1.48	1.52	1.53	1.52	2.06	
0.70	1.16	1.18	1.17	1.25	1.72	1.72	2.32	3.52	1.53	1.54	1.55	1.54	2.21	
0.75	1.17	1.18	1.18	1.26	1.80	1.80	2.44	3.80	1.57	1.56	1.56	1.56	2.35	
0.80	1.18	1.19	1.19	1.27	1.87	1.87	2.55	4.08	1.61	1.58	1.58	1.58	2.51	
0.85	1.19	1.19	1.20	1.29	1.92	1.92	2.67	4.40	1.65	1.59	1.59	1.59	2.68	
0.90	1.20	1.20	1.21	1.31	2.07	2.07	2.78	4.80	1.70	—	—	—	—	
0.95	1.21	1.21	1.22	1.34	2.18	2.18	2.91	5.20	1.73					
1.00	1.22	1.23	1.23	1.35	2.33	2.33	3.03	5.50	1.77					
1.05	1.22	1.24	1.23	1.36					1.80					
1.10	1.23	1.24	1.24	1.37					1.84					
1.15	1.24	1.25	1.25	1.39					1.87					
1.20	1.25	1.26	1.25	1.41					1.90					
1.25	1.26	1.27	1.26	1.43					1.93					
1.30	1.26	1.28	1.26	1.45					1.95					
1.35	1.27	1.29	1.27	1.48					1.97					
1.40	1.28	1.30	1.28	1.52					1.99					
1.45	1.29	1.31	1.29	1.56					2.00					
1.50	1.31	1.32	1.31	1.58					2.00					

1.55	1.32	1.33	1.32	1.62
1.60	1.33	1.34	1.33	1.66
1.65	1.34	1.35	1.35	...
1.70	1.35	1.36	1.36	...
1.75	1.37	1.37	1.37	...
1.80	1.38	1.39	...	...
1.85	...	1.41	...	...
1.90	...	1.43	...	...
1.95	...	1.45	...	...
2.00	...	1.49	...	...

*These are factors applicable to "single" alloy compositions and to *all* multiple alloy steels when quenching from 1700 F (927 C), however, when heat treating from 1475 to 1575 F (802 to 857 C), use only for those multialloy compositions that do not have a *combined* Ni and Mo content above 1.0 and 0.15%, respectively; for example, a steel with 1.5 Mn, 2.0 Ni, but 0 Mo conforms to this rule.

†When hardening directly from carburizing, use "Carburizer Steels" Cr factor in place of Cr*.

‡When the Ni and Mo contents exceed 1.0 and 0.15% together, use these multialloy factors for Ni, Mn, and Si with Cr*, Mo, and Al.

431

Table VII.8 Nickel + Manganese Multiplying Factors (when Quenched from 1475 F)

Nickel (%)	Manganese (%)					
	0.30	0.40	0.50	0.60	0.70	0.80
0.9	1.50	1.60	1.70	1.80	1.90	2.00
1.0	1.50	1.60	1.70	1.85	2.00	2.15
1.20	1.60	1.70	1.85	2.00	2.25	2.45
1.40	1.70	1.80	2.05	2.30	2.55	2.85
1.60	1.95	2.10	2.30	2.60	3.00	3.55
1.80	2.30	2.50	2.70	3.10	3.70	4.90

Table VII.9 Nickel + Manganese Multiplying Factors (when Quenched from 1525 F)

Nickel (%)	Manganese (%)					
	0.30	0.40	0.50	0.60	0.70	0.80
0.9	1.50	1.70	1.90	2.00	2.20	2.40
1.00	1.55	1.75	1.90	2.05	2.20	2.40
1.20	1.70	1.85	2.00	2.20	2.50	2.75
1.40	1.85	2.00	2.25	2.50	2.70	3.05
1.60	2.05	2.25	2.50	2.75	3.10	3.75
1.80	2.35	2.60	3.00	3.40	4.00	4.20

Table VII.10 Nickel + Manganese Multiplying Factors (when Quenched from 1575 F)

Nickel (%)	Manganese (%)					
	0.30	0.40	0.50	0.60	0.70	0.80
0.9	1.50	1.70	1.90	2.00	2.20	2.40
1.00	1.60	1.80	1.95	2.10	2.30	2.50
1.20	1.70	1.90	2.00	2.20	2.50	2.95
1.40	1.85	2.00	2.20	2.50	2.80	3.20
1.60	2.05	2.25	2.55	2.90	3.25	3.90
1.80	2.30	2.65	2.95	3.50	4.00	5.50

Table VII.11   Multiplying Factors for Boron

	Carbon Content (%)				
	0.60	0.70	0.80	0.90	1.00
Hardened from 1525 F	1.50	1.75	1.30	1.25	1.00
Hardened from 1700 F	1.75	1.60	1.45	1.30	1.20

hardening, and for calculating the hardenability of hypereutectoid steels, which are to be normalized before rehardening from 1700 F, follow the above procedure but use the regular Cr* factor.

4 For calculation of case hardenability of carburizing steels and/or high carbon steels, which are normalized or quenched before rehardening from 1475 to 1575 F, use the factors as follows:

(a) If a composition is pearlitic, use the factors labeled Mn,* Si,* Cr,* Ni,* Mo,* and Al with the appropriate carbon and boron factors.

(b) If bainitic, use the factors Cr,* Mo,* and Al with Si.† Also include the appropriate nickel and manganese factor from Tables VII.8, VII.9, or VII.10 and the appropriate carbon and boron factors.

5 For the calculation of hardenability of high carbon steels hardened from a spheroidize-annealed prior structure use the procedure as set forth in (4) but convert the D_I value obtained to the annealed D_I by using Fig. VII.1.

When employing this procedure, the following facts must also be recognized:

1 The reheat time at the austenitization temperature during rehardening of prior normalized or spheroidize-annealed high carbon steels and carburizing steels should be reasonably close to the 35 to 40 min used by Jatczak.

2 Although grain-size variations need not be taken into consideration with these factors, it is recognized that slight variations exist in hardenability due to variation in grain size, particularly in pearlitic steels.

3 The boron factors shown apply for boron additions protected with titanium, aluminum, and zirconium. A lower effect would probably result if other boron alloys were used.

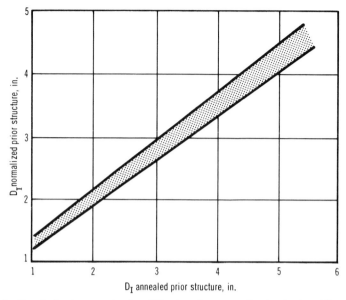

Fig. VII.1   Correlation between hardenability based on normalized and spheroidize-annealed prior structures in alloyed 1.0% carbon steels.

After the multiplying factors have been properly selected and multiplied together, the result is the D_I for 10% transformation. The location of this 10% transformation on the end-quench hardenability specimen can be determined from Fig. VII.2. The location of 1% transformation (60 to 62 Rockwell C) can then be determined from Fig. VII.3; for example, the point at which 10% transformation has taken place for a calculated D_I of 4.0 is 13/16 in. from the quenched end of the Jominy specimen. To determine the Jominy position when only 1% transformation has taken place in a bainitic steel Fig. VII.3 is used. In this example, with a D_I of 4.0, proceed horizontally to an intersection with the 10% bainite line, then vertically downward to the 1% bainite line, then back to the Y axis, at which the D_I for 1% bainite is 3.0. Going to Fig. VII.2, it will be found that 1% bainite occurs at approximately 8.5 sixteenths on the Jominy specimen.

D   The R. A. Grange Method

This method as described in technical reports of the Research Laboratory of the U.S. Steel Corporation, November 1969 and June 1972 involves subjecting a small specimen of steel to various cooling rates by quenching it in brine solu-

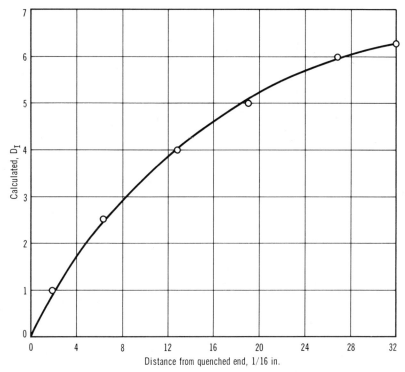

Fig. VII.2 Relationship between Jominy distance and D_I (D. J. Carney, Transactions ASM, Vol. 46, 1954, p. 882).

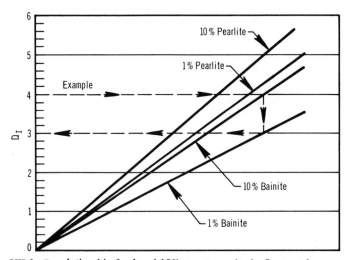

Fig. VII.3 D_I relationship for 1 and 10% structure criteria. See text for example.

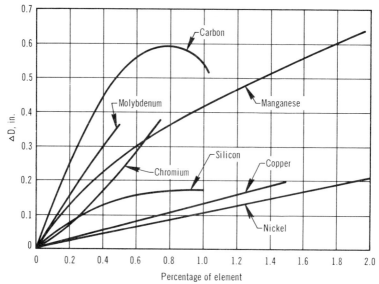

Fig. VII.4   Grange additive factors for hardenable diameter (inch).

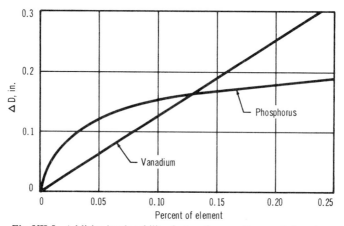

Fig. VII.5   Additive hardenability factors for vanadium and phosphorus.

tions of increasing temperatures rather than end quenching as in the Jominy test. The results are expressed as the diameter of a steel that will harden to a microstructure of 90% martensite at the center when brine-quenched (except for boron steels which were water-quenched). The factors shown are for relatively shallow hardening steels and are additive; for example, HD (hardenable diameter) = D_C (for carbon) + D_{Mn} (for manganese) + D_{Si} (for silicon) (see Fig. VII.4) This result can then be corrected for phorphorus and vanadium contents by using Fig. VII.5 and for boron, by Fig. VII.6.

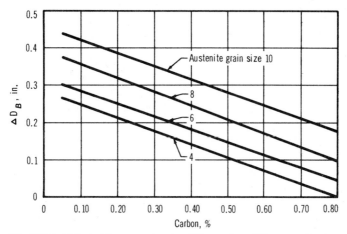

Fig. VII.6 Effect of boron on hardenability at four different grain sizes.

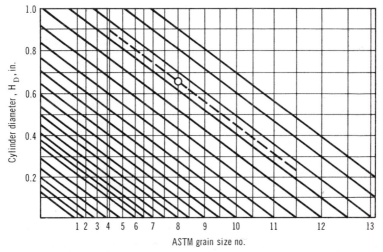

Fig. VII.7 Chart for predicting hardenable diameter (water-quenched steel, 90% marten-site) for any grain size different from that at which hardenability was measured or esti-mated. See text for example.

To predict the hardenable diameter for grain sizes other than that of the steel used in the above calculations, use Fig. VII.7 as follows:

Suppose one has made a hardenability determination that shows a hardenable diameter of 0.65 in. at grain size of 8 (indicated by the small circle). To determine the hardenable diameter at any other grain size draw a line like the broken one shown or lay a straight edge parallel to the angular lines. It can be seen, for example, that the hardenable diameter is 0.9 for a grain size of 4 and only 0.3 for a grain size of 11.

INDEX